CALCULUS

THE ELEMENTS

CALCULUS

THE ELEMENTS

MICHAEL COMENETZ

World Scientific
New Jersey • London • Singapore • Hong Kong

Published by

World Scientific Publishing Co. Pte. Ltd.

5 Toh Tuck Link, Singapore 596224

USA office: Suite 202, 1060 Main Street, River Edge, NJ 07661

UK office: 57 Shelton Street, Covent Garden, London WC2H 9HE

Library of Congress Cataloging-in-Publication Data
Comenetz, Michael.
    Calculus--the elements / Michael Comenetz.
    p. cm.
    Includes index.
    ISBN 9810249039 (alk. paper) -- ISBN 9810249047 (pbk. : alk. paper)
    1. Calculus. I. Title

    QA303.2.C66 2002
    515--dc21   2002069008

British Library Cataloguing-in-Publication Data
A catalogue record for this book is available from the British Library.

Figures by the author and Monotype Composition Co.

Cover art by Charles Jones.

Printed in Singapore by World Scientific Printers

Preface

This book is for anyone acquainted with the rudiments of analytic geometry who wishes to learn the fundamental ideas of calculus and a few of their applications. It aims to present the elements clearly and fully enough that they may be understood. To this end concepts are everywhere emphasized, and physical interpretations of mathematical ideas are given a prominent role; but no prior study of physical science is expected of a reader. A feature of the exposition (which I note here for experts) is that infinitesimals are cautiously introduced and employed in parallel with limits, so that the lucid symbols of calculus can be read as they were meant to be, and as most scientists and engineers have always read them. The plan of the book is sketched at the beginning of the first chapter.

The "questions" (exercises, problems) accompanying the text supplement it and provide the practice in calculation and problem-solving that is needed to test one's understanding. In the interest of active reading, many questions call for arguments and details omitted from the discussion. There are answers in the back, most of them including solutions full enough to be of service to the learner studying alone.

The Reader's Guide which follows this preface explains the reference system and provides other information and advice. For translations of the epigraphs that precede Chapter 1 see Appendix 2.

In both matter and form the book owes much to the classic lectures *Differential and Integral Calculus* of R. Courant (who would have condemned the unregenerate infinitesimals). Its writing was partially supported by St. John's College and the Beneficial-Hodson Trust. I thank Christopher Colby, Thomas Slakey, and Stewart Umphrey for their generous help, and I remember with gratitude George Comenetz, my father, whose reading of the original manuscript led to many improvements. My wife Sandy and sons Joshua and Aaron gave me heart for the work.

MICHAEL COMENETZ

v

Reader's Guide

Reference system. Each chapter is divided into articles, which are divided into sections. Article 5.3 is the third article of Chapter 5, and §5.3.6 is the sixth section of that article. The reference §§5.3.6–8 is short for §§5.3.6–5.3.8.

Some equations, propositions, etc., are labeled with numbers in parentheses at the left margin. Numbering begins at (1) with each article. Reference to such an item is by the number in parentheses, or if necessary by section and number: §1.6.8(2) is the item labeled (2) in §1.6.8.

Some sections are sets of questions, which are numbered Q1, Q2, etc. A reference in the text such as Q4 is to the indicated question in the *next following* set. Other references to questions specify the section: §2.9.5.Q4 is the fourth question in §2.9.5.

Figure 2-5 is the fifth figure in Chapter 2; Fig. A-5 is the fifth figure in the Answers. Similarly for tables.

How to find things. Look in the Contents, the Topical Summary that follows Chapter 8, and the Index, and use the cross-references in the text.

Structure of the book. The first four chapters form the basic sequence; the remaining four are essentially independent of one another. (See the introduction to Chapter 1.) At the end of each of the first four chapters calculus has been portrayed as a whole, with progressively greater fullness and definition.

Alternate routes. A. What can be skipped. Passages in small print, including a few entire sections, are remarks that can be omitted at the first reading of a chapter or until they are cited. Many other sections can be skipped, postponed, or read in part. These are signaled by footnotes, which give further guidance when needed. The most important cases in the early chapters are Art. 1.4, on force, which can be postponed until Chapter 5, and Art. 3.7, on the mean value theorems and the proof of the Fundamental Theorem, which can be reduced to avoid theory and proof. As for the questions, answer as many as you can. In each set the earlier ones are thought to be easier.

B. Shortcuts to certain topics. When I say here that part Z of a sequence XYZ can follow part X, part Y being omitted, I do not mean that Z contains no references to Y, only that such references are inessential or occur in passages that can be omitted. This applies to the *text* in Z, but not necessarily to all the *questions* there, some of which may depend on Y. The following are the chief options.

1. The *chain rule, inverse functions,* and *partial derivatives* (Chapter 4 through §4.5.6) can follow the arithmetic differentiation rules (Art. 3.4); then the Fundamental Theorem (Art. 3.5) will be done next.

2. After the Fundamental Theorem (Art. 3.5) the theory of *maxima and minima* (Art. 3.9) requires only the reduced version of Art. 3.7, together with §3.8.1.

3. *Integration in the plane and in space* (Art. 4.8) can follow §4.5.1.

4. After Chapter 3, the topic of *differential equations of motion* (Chapter 5) requires only the chain rule (Art. 4.1), inverse functions (Art. 4.3), and the limit of $(\sin h)/h$ (§4.4.2), if the value of an integral at the very end of Chapter 5 is accepted without proof.

5. The *differential equation of intrinsic (exponential) growth* (Chapter 6) requires only the chain rule (Art. 4.1) and inverse functions (Art. 4.3), except for the second part of the proof in §6.2.2 (which uses §§4.6.2–3).

6. Articles 7.1–7.4, on *arc length, curvature, local geometry,* and *parametric representation,* can be read after §4.6.3 (omitting Art. 4.5).

7. Chapter 8, on the *Taylor series,* can be read after Art. 4.4.

C. Ill-advised reduction. Overcome the temptation to reduce concepts of calculus to their geometrical interpretations (area, slope, etc.). Calculus is more than geometry.

D. Physical units. Appendix 1 shows how to convert from the International System used here to the older but still important cgs system.

E. Newton's lemmas. Interesting versions of some basic propositions of calculus by an originator of the subject are found at the beginning of Book 1 of his *Principia* (1687). They can be read along with the present book, which has notes relating them to arguments here (see the Index under Newton, *Principia*).

F. The computer. If you can understand what is written in this book you will be able to illustrate it and explore further with the aid of a graphing calculator or the like. But there is no substitute for thinking, calculating, and sketching on one's own.

Abbreviations. Cf. = compare, e.g. = for example, i.e. = that is to say, q.e.d. = which was to be proved.

The Greek alphabet.

A	α	alpha	I	ι	iota	P	ρ	rho
B	β	beta	K	κ	kappa	Σ	σ	sigma
Γ	γ	gamma	Λ	λ	lambda	T	τ	tau
Δ	δ	delta	M	μ	mu	Y	υ	upsilon
E	ε	epsilon	N	ν	nu	Φ	ϕ	phi
Z	ζ	zeta	Ξ	ξ	xi	X	χ	chi
H	η	eta	O	o	omicron	Ψ	ψ	psi
Θ	θ	theta	Π	π	pi	Ω	ω	omega

Contents

(Omitted sections consist of questions. A few questions of special interest are listed.)

CHAPTER 1

THE PROBLEM OF CALCULUS

CHAPTER 2

INTEGRAL AND DERIVATIVE

CHAPTER 3

DIFFERENTIATION AND INTEGRATION: I

CHAPTER 4

DIFFERENTIATION AND INTEGRATION: II

CHAPTER 5

DIFFERENTIAL EQUATIONS OF RECTILINEAR MOTION

CHAPTER 6

THE DIFFERENTIAL EQUATION OF INTRINSIC GROWTH: THE EXPONENTIAL AND LOGARITHMIC FUNCTIONS

CHAPTER 7

LENGTH AND CURVATURE OF PLANE CURVES

CHAPTER 8

REPRESENTATION OF FUNCTIONS BY INFINITE TAYLOR SERIES

οὐκ ἔστι βασιλικὴ ἀτραπὸς ἐπὶ γεωμετρίαν.

Euclid

Allez en avant, la foi vous viendra.

d'Alembert

CHAPTER 1

The Problem of Calculus

INTRODUCTION

As a mathematical theory of wide scope and great power, calculus is remarkable for the unity, simplicity, and intelligibility of its elements. The heart of it can be regarded as a direct answer to a natural question, yet from that answer springs the modern world of exact science. Such ramification from a single stem does not occur without roots that extend throughout our experience of nature, geometry, and number. To expose some of them is the purpose of this first chapter, in which a variety of physical and mathematical perplexities will be shown to issue in a common problem. The solution to this problem will be developed in the next three chapters, and the last four will be devoted to extensions and applications of the fundamental theory.

Metaphor: The points of a line of thought.

1.1 MASS AND LINE DENSITY

1.1.1 Mass of a segment and line density at a point. Imagine a straight wire of unvarying cross-sectional area. I wish to consider two quantities associated with places on the wire, one of them pertaining to segments of the wire and the other to points along it. Let me say at the outset that in this example and the following ones my intention is to propose lines of thought rather than to give adequate accounts of phenomena. Some of the objections that can be raised to the treatment here will be dealt with in subsequent chapters.

"another metaphor." "lines of thought,"

1

The first quantity is **mass**, which belongs to each segment of the wire. It will be measured in kilograms (kg).[1] Mass is that property of a body by virtue of which it has weight and inertia, and in familiar speech we often treat the kilogram as a measure of weight. Since weight, unlike mass, can vary with location—we weigh less on the moon—this use of the kilogram is valid only locally, where weight keeps a fixed proportion to mass.

If the wire is **uniform**, in the sense of being everywhere the same in composition, the mass of a segment is proportional to its length. In general this will not be the case; for example, one segment of a wire of copper and silver may contain a higher proportion of silver than another segment of the same length and consequently outweigh it.

The second quantity is **line density**, which is associated with each point of the wire. (By "point" I mean a location along the wire, not a geometrical point on or within it.) Line density at a point is measured in kilograms per meter (kg/m), the meter being the measure of length, but does not in general refer to whole meters of the wire; rather, it describes the state of affairs at the single point in question.[2] We might say, of a copper and silver wire: at point A, where the wire is mostly copper, the line density is 0.08 kg/m; from that value it rises continuously to 0.09 kg/m at point B, half a meter away, where the wire is mostly silver. This requires an explanation.

To begin with, in the case of a uniform wire every meter of it is just like every other, so the number of kilograms per meter of wire is unambiguously defined. The line density at any point is said to be that number of kilograms per meter. Thus we know the meaning of line density for the uniform wire, and understand that it is the same at all points.

Now let P be a point of an arbitrary wire; what meaning can be attached to "the line density at P"? A cross-section taken at P reveals the composition of the wire at that point—the ratio of copper to silver, say. Imagine another wire, a uniform one, which has *everywhere* the same composition as the given wire has *at* P. As a uniform wire it has a definite line density (kg/m); this line density we take to be the line density at P. You see that the idea is to describe the state of affairs at P by imagining it extended along a whole wire. To summarize,

---

[1] See the note at the end of §1.1.1.

[2] Line density is to be distinguished from "density" without qualification, which ordinarily signifies volume density (kilograms per cubic meter). Silver is denser than copper: it has more mass, volume for volume.

Line density at a slice of the wire.

Line density at the wire? Greater or lesser density.

What about the width of the wire? Wider wire means greater density?

9 July 2010)

the line density of a wire at a point is the line density (kg/m) of
the *uniform* wire which *matches* the given wire at the given point.

Or, in other words: the line density of a wire at a point is the line density
(kg/m) the wire would have if it were everywhere the same in composition as
at the point.

> The notion of line density is of doubtful meaning where composition changes
> abruptly, as at the boundary between a copper and a silver segment. In order
> to avoid this difficulty, and analogous difficulties in the succeeding examples,
> we can assume that all quantities vary smoothly.
> Here and elsewhere I have freely assumed that *exact numerical values can be
> assigned to properties of a physical object*, even though in some cases, at least,
> this is clearly an idealization (as it is for line density at a geometrical cross-
> section through atomic matter). By this assumption the "physical object" is
> rendered mathematical to suit our convenience in the present study.

1.1.2 Relation between mass and line density. Total and specific
quantity. Having defined the two quantities, let us consider the relation
between them. For a uniform wire it is easily stated. If M is the mass of a
segment of length L, and λ is the line density, we have

$$M = \lambda L$$

—the number of kilograms in a segment is the number of kilograms per meter
times the number of meters in the segment. Equivalently,

$$\lambda = \frac{M}{L}$$

—the number of kilograms per meter is the number of kilograms in a segment
divided by the number of meters in the segment. Mass is the *product* of line
density and length, line density the *quotient* of mass by length, or *ratio* of mass
to length. The elementary mathematical concepts of product and quotient
suffice to express either of the quantities in terms of the other, when the wire
is uniform.

What if it is not uniform? Plainly, mass and line density cannot be as simply
related as before. Yet there can be little doubt that they still mutually determine one another. Line density indicates the *specific concentration* of matter
at each *point*, the quantitative aspect of the wire by which mass is accumulated
over distance. Mass represents the *totality* of matter in each *interval*, however
small, and so establishes the distribution of matter along the wire of which

line density is the ultimate measure. If either is known everywhere, the other must be determined; but by what mathematical relations?

We have arrived at a first formulation of the central problem of calculus: to express the relation between what may be called a **total** quantity (mass) and a corresponding **specific** quantity (line density), by means of suitable generalizations of the notions of product and quotient. The importance of this task will become clearer as we proceed with the examination of a series of diverse situations in which the same problem presents itself again and again to the eye that knows how to look. The ability to recognize its essential features in a given set of circumstances is the greater part of understanding calculus, and effort directed to grasping what is common to the different examples will be amply repaid.

nice phrase

1.1.3   QUESTIONS

Q1.   (a) A uniform wire 3 m long has a mass of 9 kg. What is its line density at any point?[3] (b) The line density of a uniform wire 8 m long is 4 kg/m. What is its mass? What is the mass of a 2 m segment?

Q2.   A length of round copper and silver wire, uniform in diameter, is pure copper at its midpoint, but the proportion of copper falls off steadily in both directions until pure silver is reached at the ends. How is the variation in composition reflected by the line density at different points, and by the mass of different segments?

Q3.   Consider the collection of all the possible segments that could be marked off on a given wire. Does the line density at a particular point P depend on the mass of every segment? Explain.

Yes

1.2   VOLUME AND GROWTH RATE

1.2.1   Volume and growth rate.
Suppose that a quantity is growing by continuous addition to its substance—a line being drawn, a shadow spreading, water rising in a tank fed by a pipe. To be definite I will consider the last of these instances, although the discussion will apply to growth generally. The water will be measured by **volume** in cubic meters (m^3), and time in seconds (sec). By a **uniform** flow of water into the tank is meant a steady flow, an unchanging motion of liquid.

---

[3]To keep the numbers simple, here and subsequently we must admit wires which if made of real metal would be very thick. Even platinum (round) wire with this line density would be half an inch in diameter.

Here — time comes in intervals

In each interval of time a certain volume of water is added to the tank. If the flow into the tank is uniform, the volume added in a time interval is proportional to the length of the interval. In general this is not the case, since a stronger flow will add more water in a given time than a weaker one.

At each moment of time the water in the tank has a certain **rate of growth**, or **growth rate**, which we measure in cubic meters per second (m³/sec). If the flow is uniform, the rate of growth is the same at all times, and equals the number of cubic meters that enters the tank each second. For an arbitrary flow, the growth rate at a moment t is understood to be the growth rate (m³/sec) produced by a uniform flow which matches the given flow at the moment t— that is, the growth rate the volume of water would have if the flow were always the same as at t. (Matching flows can be defined directly in terms of matching velocities of their particles, for which an image is suggested in §1.3.1.)

Evidently volume and growth rate form a pair of quantities analogous to mass and line density. Volume is the total quantity, accumulated over a time interval according to the rate of growth, growth rate the specific quantity, indicating the tempo of change in volume at a particular instant. Time plays here the role that was taken in the previous situation by space—the limited one-dimensional "space" along a line, that is.

As before, we can express each quantity in terms of the other when the flow is uniform. Let U be the volume[4] that enters the tank in time interval T, and let γ be the rate of growth; then

$\gamma = gamma$
$$U = \gamma T, \quad Volume = rate\ of\ growth \times Time$$

or equivalently

$$\gamma = \frac{U}{T}.$$

In the uniform case volume is the product of growth rate and time, growth rate the quotient of volume by time. How the two quantities of the pair determine one another when growth is not uniform is the problem of calculus.

1.2.2 QUESTIONS

Q1. Reformulate Q1 of §1.1.3 as an analogous problem about water flowing into a tank, using the same numbers. Solve in the new terms.

---

[4]The letter V is reserved for another use.

The moment of correspomds to the point on the wire.

Q2. The same problem with respect to Q3 of §1.1.3.

Q3. (a) A water tank has two inlets. When only inlet 1 is open, the volume of water in the tank grows uniformly at γ_1 m^3/sec; in a time interval of length T, U_1 m^3 of water is added. When only inlet 2 is open, the corresponding numbers are γ_2 and U_2. Find the growth rate and the volume added in time T when both inlets are open. (b) What can be said if the flows are not assumed uniform?

1.3 MEASURES OF RECTILINEAR MOTION

1.3.1 Distance and velocity. Let a particle move—what is a "**particle**"? A movable point, an imaginary physical object which has location without dimension, and may be assigned mass, electric charge, or other qualities as convenient. Small things, such as gas molecules, are often regarded as particles in order to facilitate the study of their behavior. Since a particle has an exact location and no parts, its motion admits of simpler mathematical description than that required for the motion of an extended body.

Let a particle move along a straight line, in a definite direction—towards the right, say. In any interval of time it traverses a certain **distance**, or length of line; at any moment it is moving at a certain **speed**, or **velocity**, which we measure in meters per second (m/sec).[5] If the motion is **uniform**, or steady, distance covered is proportional to length of time interval and speed is constant, the number of meters covered each second. In variable motion, the velocity at an instant t means the velocity of a uniformly moving particle whose motion matches that of the given one at t.

To assist the imagination with the idea of matching motions intended here it is useful to think not of two particles but of two long parallel trains, of which one advances at a uniform pace while the other accelerates steadily from an initial state of rest so as to pass it. At a certain moment the trains are moving exactly together (to a passenger in one, the other momentarily appears motionless); at that moment their motions match, in the sense we need. The velocity of the accelerating train at that moment is the velocity of the uniformly moving train.

If the particle is regarded as tracing out or drawing the line of its motion, it becomes clear that this situation differs only in terminology from the one last considered. Distance covered is length of a segment, and speed is rate of

---

[5] "Speed" and "velocity" are synonymous here, although a distinction between them, irrelevant at present, will be made later (§2.10.5).

growth of length. Thus we already know that distance and velocity are a total-and-specific-quantity pair. Nevertheless, in view of the importance of motion it is advisable to think the argument through again with the moving particle in mind (Q1).

The relations that hold when motion is uniform between distance L, velocity v, and time T are well known:

$$L = vT, \quad \text{distance} = \text{velocity} * \text{time}$$

$$v = \frac{L}{T}. \quad \text{velocity} = \frac{\text{distance}}{\text{time}}$$

That distance and speed mutually determine one another in non-uniform motion as well is shown by the cooperative behavior of the odometer and speedometer of an automobile. The relation between such instruments is mechanical or electrical; what is the mathematics of it? And that is the problem of the calculus.

1.3.2   QUESTIONS

Q1.   Express in careful language the character of distance as a total quantity, and of velocity as a specific one, without referring to growth.

Q2.   (a) A particle travels at the speed of sound, 330 m/sec. How long does it take it to go 1 km (= 1000 m, about 0.6 mile)? (b) If a particle has traveled 1 km in that time, was it moving at the speed of sound?

1.3.3   Velocity and acceleration.

Suppose now that our particle is not only moving towards the right, but moving more rapidly as it goes. In any interval of time it receives a certain **increase in velocity**, and at any moment its velocity is increasing at a certain rate, which we call the **acceleration** of the particle. Although this new pair of quantities is, formally speaking, just another instance of growth, as a rather difficult one it deserves careful examination.

In a given time interval a certain quantity of velocity, if I may use the phrase, is acquired by the particle, just as the water tank acquired a certain quantity of volume—quantity of water, that is, which was regarded then as a thing with volume, much as the particle's motion is being regarded now as a thing with velocity. Velocity added, or **increment** of velocity, is measured in meters per second: in any time interval a certain number of meters per second is added to the particle's motion. If this augmentation takes place uniformly over time, we have so-called **uniformly accelerated** motion, in which velocity added during a time interval is proportional to the length of the interval. In particular, the

[handwritten top margin: I think this is exactly the consideration on which I hung up in Dr. Walkers physics class at UConn.]

*[handwritten: → and not m/sec/sec ? Normally sec * sec = sec²]*

same amount of velocity is added during each second. The acceleration, or rate of increase of velocity, is then a certain number of *meters per second, per second*, or m/sec^2 as the unit is written.

Let the velocity acquired by a uniformly accelerating particle in time T be V; if a is the acceleration, we have

[handwritten left margin: It seems to me we have 2 kinds of velocity here, uniform & accelerating.]

$$V = aT, \quad \text{Velocity} = \text{acceleration} * \text{Time}$$

$$a = \frac{V}{T}. \quad \text{acceleration} = \frac{\text{Velocity}}{\text{Time}}$$

The meaning of acceleration at an instant for non-uniformly accelerated motion can be explained in the way by now familiar, through an instantaneous match between the given motion and a uniformly accelerated one (Q2).

1.3.4   Velocity as specific and total quantity.

What may be surprising here is that velocity, which in the first discussion of the particle's motion (§1.3.1) was a "specific" quantity, now turns up as a "total" one. Instead of velocity v at an instant we have velocity V added over a time interval—which is why I changed the notation, as I have been using capital letters for quantities belonging to intervals of time or space, small letters for quantities belonging to points and moments. Quantities are not, in fact, intrinsically total or specific, but are one or the other in relation to other quantities. Our objects of study at present are *pairs* of quantities, in which a single quantity can in principle play either of the two roles.

> At the same time there may be mathematical limitations on this possibility (cf. the note at the end of §1.1.1), and in physics it often happens that only one pair involving a given quantity arises naturally. Mass, for example, normally occurs as a total quantity only, as in our first example (Art. 1.1).

In speaking of the two roles of velocity in this way I am being a little careless. Instantaneous velocity is not really the same quantity as added velocity, for the very reason that the one is associated with a moment, and the other with an interval, of time. But they appear to be two aspects of the same thing, and there is a simple expression for their exact relation: the increment of velocity over a time interval is the *difference* of the velocities at its final and initial moments,

$$V = v_2 - v_1.$$

For example, a particle that accelerates from 3 m/sec to 5 m/sec has gained 2 m/sec of velocity.

[handwritten left margin vertical: This leads inevitably to the Twins Paradox in General Relativity.]

1.3.5　Application to a falling body.　When the problem of calculus is solved we will know how to determine velocity from acceleration, and distance from velocity; and likewise in the other direction, velocity from distance and acceleration from velocity. The use of this is illustrated by its most famous application, to free fall under the influence of gravity. A brief description is as follows.

Bodies are observed to fall together, independently of their masses. Upon measurement it is found that distances and times of fall indicate a certain uniform acceleration—the connection between distance and acceleration being made by calculus. From this **acceleration of gravity** g, whose value is about 9.8 m/sec^2, the distance of fall in any given time can then be predicted. Although accurate for terrestrial purposes, this description of gravitation proves to hold good only near the earth's surface; over greater distances astronomy shows a more complex law of acceleration to be at work. Nevertheless, calculus is again able to determine it and account for the motion it prescribes.

In Chapter 5 the arguments from the acceleration laws to the motions will be worked out in detail. We next take up the general subject of motion under the influence of force, of which gravitational motion is only a noteworthy species.

1.3.6　QUESTIONS　　　$1^2 = 1$　　$2^2 = 4$　　$3^2 = 9$

Q1.　Let a heavy particle fall from rest with the acceleration of gravity, 9.8 m/sec^2. (a) What is its velocity after 1 sec? After 2 sec, 3 sec, n sec? After $n + \frac{1}{2}$ sec? (b) Attempt to describe how the distance of fall increases with time.

Q2.　Use the two trains (§1.3.1) to illustrate the idea of uniform and non-uniform accelerations matching at an instant.

1.4　FORCE AND ITS TOTALS[6]

1.4.1　Velocity produced by force over time.　Let a particle move in a straight line as before, but now suppose it has mass m and is under the influence of some force which urges it on in the direction of motion. The particle may be drawn downward by its weight, or pulled along by a thread, or pushed by an expanding spring; whatever the nature of the force, I mean it

---

[6]Article 1.4 is more difficult than the rest of the chapter, and can be omitted at a first reading. It will be needed for Chapter 5. If it is skipped now, one should also ignore a few passages of Chapter 2: §§2.3.4 and 2.8.4, and allusions to force and work in §§2.10.2 and 2.10.7. There is a summary of the article at the end (§1.4.9).

to operate continuously. I further assume that this force is the only one acting on the particle so as to affect its motion. As an ideal example we might think of a small heavy block pulled along a frictionless horizontal table, in vacuum, by a cord in which tension is maintained by a winch. The downward force of gravity on the block, and the upward force exerted on it by the table, are in balance and have no effect on its motion.

As the particle moves along, it picks up speed. How much velocity does it acquire in a given time interval, say from a time t_1 to a later time t_2? That depends not only on the strength of the force, but also on m, for everyone knows by experience that a given force will impart a greater velocity in a given time to a smaller mass than to a larger one.

This does not contradict what is also known from experience, that bodies of different masses subject to the same force of gravity fall together (§1.3.5), because the expression "same force of gravity" is misleading. Bodies of different masses have different weights, that is, feel different downward pulls; in fact gravitation nicely proportions weight to mass (§1.1.1), and that is why the bodies fall together.

I am taking for granted the fundamental physical principle that force *accelerates* mass. If this seems to be denied by the experience of pushing a block along a horizontal surface at *constant* speed by application of a steady force, reflect that two forces, not one, are really at work: the surface acts on the block by friction so as to resist the applied force. Our ideal example excludes friction. *and therefore doesn't exist on the surface of this planet.*

During the interval from t_1 to t_2 the force on the particle may vary with time, or be **constant**, or **uniform**, as when exerted by an unchanging cause. However it varies, the force will impart some definite velocity V. If we let it act in exactly the same way, over an equal time interval, upon a particle of any other mass m', we will find experimentally that the velocity V' acquired by that particle stands to V in the inverse ratio of the masses:

$$\frac{V'}{V} = \frac{m}{m'},$$

or

$$mV = m'V'. \quad m \times V = m' \times V$$

For example, suppose that a winch pulls unevenly on a block of mass m, so that the tension in the cord varies; if it pulls on another block of mass m' with the same irregularity over the same length of time, so that the histories of the actions on the blocks are identical, the added velocities V and V' will bear the stated relation to the masses.

Oooh, a really bad inadvertent pun!

1.4.2 Momentum and impulse. It is clear from the preceding result that the quantity mV may be thought of as generated, or amassed, by the action of force over time. This product, rather than the added velocity V alone, depends only on the force and the time interval over which it operates. If the velocity of the particle is v_1 at time t_1 and v_2 at time t_2, then (§1.3.4) $V = v_2 - v_1$ and consequently

$$mV = mv_2 - mv_1 \,,$$

the increase in the product mv of mass and velocity. We call this latter quantity the **momentum** of the particle, and denote it by p:

$$p = mv.$$ *momentum = m * v*

Like velocity v, momentum is associated with an instant of time. The capital letter P can be used to represent its increase over time,

$$P = mV = p_2 - p_1 \,,$$

He uses capital letters to denote intervals and lower case to denote instants; e.g. P & p.

where $p_1 = mv_1$, $p_2 = mv_2$.

Directing our attention to the force rather than the particle we can also speak of the **impulse** I of the force on the particle over the time interval. Then impulse, which refers to force and time, equals increment of momentum, defined by mass and velocity of a particle: $I = P$.

Momentum is measured in kilogram meters per second: one kg m/sec is the momentum possessed by a mass of 1 kg moving with a velocity of 1 m/sec.

1.4.3 Increment of momentum, and force. The identification of momentum as a thing increased by force over time raises the question whether **increment of momentum** and **force** form another total-and-specific-quantity pair. Increment of momentum is associated with a time interval; let us review how force is associated with an instant. One kind of *constant* force, namely weight, can be measured with a balance, and other constant forces by equating them to weights—e.g., the force of a compressed spring equals the weight that compresses it. The magnitude of a variable force at an instant can then be understood in our usual way in terms of a matching

constant force. Thus the irregular operation of the winch pulling on the block
(§1.4.1) produces a definite force at each moment.

If now we restrict ourselves to constant forces, we find by experiment that:

(1)   for any given force, momentum P added during a time interval is
       proportional to the length T of the interval,[7]

$\propto = $ "is proportional to" $P \propto T$;

(2)   in any given length of time, momentum P added is proportional to
       force f,

$$P \propto f.$$

From (1) and (2) it follows that in general the increment of momentum is
proportional to the product of (constant) force and time,

$$P \propto fT$$

(Q5). Equivalently, $f \propto P/T$, a relation which makes it possible to measure
force as *momentum added per time*. Accordingly we define the **newton**, our
unit of force, to be one kilogram-meter-per-second (of momentum) per second
(of time):

$$1 \text{ newton} = 1 \text{ kg m/sec}^2.$$

Otherwise said,

> a force of one newton is one which, acting steadily for one second,
> imparts one kg m/sec of momentum.

The impulse of such an action can then be called one newton second—one newton acting for one second. Thus 1 newton sec of impulse produces 1 kg m/sec
of momentum.

With the newton as unit our proportionalities become equalities, so that
when force is constant we have

$$P = fT,$$

---

[7]That is, P/T is constant (for $T \neq 0$). The symbol $\propto$ means "is proportional to."

or

Here f = force not function

$$f = \frac{P}{T}.$$

The first equation shows that the impulse of a constant force on a particle, by which momentum P is gained, is the *product* of force and time, $I = fT$; the second equation exhibits force as momentum gained per time. These relations strongly suggest, even if they do not demonstrate, that we do have a new pair of quantities analogous to the pairs already considered. Although variable momentum and force are both too difficult of intuitive access for us to be quite confident that this is so, the fact is asserted as a fundamental postulate of physics, part of Newton's second law of motion. As momentum is accumulated by the instantaneous action of force over time, so force specifies the rate of momentum-acquisition at an instant; and the mathematics, as yet unexplained here, that connects the quantities is the same as that needed to relate the mass of a wire to its line density.

1.4.4   Force the product of mass and acceleration.   On the one hand, we have now identified force as the time-specific quantity corresponding to the increment of momentum,

$$mV \leftrightarrow f.$$

On the other, in §1.3.3 we recognized acceleration as similarly corresponding to the increment of velocity,

$$V \leftrightarrow a.$$

From the latter relation it follows that mass times acceleration corresponds to the increment of momentum,

$$mV \leftrightarrow ma;$$

for multiplication by the constant m cannot affect the relation of total to specific quantity (Q6). Hence

$$f = ma,$$

the better-known form of Newton's law.

1.4.5 QUESTIONS

Q1. Express in proper units the momentum of a 5 kg particle traveling at 10 m/sec.

Q2. Identical time-varying forces act during the same time interval on particles of mass 2 kg and 3 kg. The first is accelerated from rest to 6 m/sec. If the second was initially moving at 4 m/sec, to what speed is it accelerated?

Q3. In the same ten-second time interval force f_1 imparts momentum of 20 kg m/sec to one particle, while force f_2 imparts momentum of 30 kg m/sec to another. What can be said about f_1 and f_2 (a) if they are constant forces? (b) if they are not constant?

Q4. Since weight is a force, it is correctly measured in newtons, not kilograms. How many newtons does a kilogram weigh at the surface of the earth?

Q5 (§1.4.3). In order to establish the proportionality of P and fT, prove the following lemma.

> Lemma. *If Q depends on x and y, which vary independently of one another, in such a way that*
>
> (i) *Q is proportional to x, when y is fixed, and*
> (ii) *Q is proportional to y, when x is fixed,*
>
> *then Q is proportional to xy.*

Q6 (§1.4.4). Explain.

Q7. Prove $f = ma$ when force and acceleration are uniform by considering the velocity and momentum acquired in a time T.

1.4.6 Squared velocity produced by force over distance.
In the preceding discussion we took the force on a particle to be associated with moments of time, and identified momentum as the corresponding quantity accumulated over a time interval. Let us now change our point of view, and regard the force as associated with points of space instead—that is, points of the line on which the particle travels. This may at first seem to be no change at all, since as the particle moves it occupies moments of time and points of space simultaneously; can it matter which the force is associated with in our thought?

The answer is yes, because intervals of time and intervals of space do not maintain a constant relation to one another. When velocity is uniform, time elapsed and space traversed are proportional; but owing to the presence of the force, the velocity of our particle is continuously increasing. If the particle moves a distance L in a time interval of length T, in the next time interval of the same length T it will move not L, but a greater distance L'. This

"Simultaneously": Does it really? —P.

means that momentum cannot be the total quantity accumulated over *distance*. For consider the uniform case, in which the force f is constant. Under its impulsion the particle acquires in each of the two equal time intervals the same momentum, equal to fT. But whatever the total quantity accumulated over distance may be (assuming that there is one), its value for the first interval is fL, and for the second fL', since in the uniform case a total quantity is always a product. These values being different from one another, the total quantity cannot be the same as momentum, or even determined by it.

We must therefore begin afresh in order to determine the total quantity that corresponds to force regarded as specific to points of the line.

Following §1.4.1, we allow a given force, possibly varying with location, to operate over an interval of the line from x_1 to x_2 on particles of mass m and m', and look for an inverse relation between the masses and the changes in velocity. The result is similar to the earlier one, but not the same. Instead of the added velocities it is now the added *squares* of velocity that are inversely as the masses. For the particle of mass m let q denote the square of instantaneous velocity, $q = v^2$, and let Q be the increase in the square of velocity as the particle travels from x_1 to x_2: $Q = q_2 - q_1$, where q_1 is measured at x_1 and q_2 at x_2. Similarly, for the other particle let $Q' = q'_2 - q'_1$. Then experiment shows that

$$\frac{Q'}{Q} = \frac{m}{m'},$$

or

$$mQ = m'Q'.$$

One should not confuse the increase in the square of velocity with the square of the increase in velocity: in general, $Q \neq V^2$. For example, suppose that a force causes a particle of mass m to accelerate from $v_1 = 3$ m/sec to $v_2 = 5$ m/sec. The increase in velocity is $V = 2$, whose square is 4. On the other hand $q_1 = 3^2 = 9$ and $q_2 = 5^2 = 25$, so the increase in the square of velocity is $Q = q_2 - q_1 = 16$.

If another particle of mass m' is accelerated by the same force, over the same interval of the line, from 4 m/sec to 6 m/sec, we will have

$$Q' = q'_2 - q'_1 = 6^2 - 4^2 = 20,$$

and our assertion is that $m/m' = 20/16$.

1.4.7   Kinetic energy and work.   Continuing as in §1.4.2, we think of mQ as a quantity accumulated by the particle by the action of force over distance. It depends only on the force and the interval of the line traversed by the particle. The same is true, of course, of any constant multiple of mQ; and there happens to be advantage in working with $\frac{1}{2}mQ$ rather than mQ itself. We have

How so? {

$$\frac{1}{2}mQ = \frac{1}{2}mq_2 - \frac{1}{2}mq_1 = \frac{1}{2}mv_2^2 - \frac{1}{2}mv_1^2\,;$$

this is the increase in the quantity

$$k = \frac{1}{2}mv^2\,,$$

which is called the **kinetic energy**, or energy of motion, of the particle. Its increment can be denoted by K:

$$K = \frac{1}{2}mQ = k_2 - k_1\,.$$

As we did when defining impulse, we can direct our attention to the force rather than the particle. We then speak of the **work** W done by the force on the particle over the interval of the line. Work, which refers to force and distance, equals increment of kinetic energy, defined by mass and velocity: $W = K$. (This is a restricted meaning of the word "work," by the standard of ordinary English usage (Q4).)

The unit of kinetic energy is the kg m^2/sec^2, twice the kinetic energy possessed by a mass of 1 kg moving with a velocity of 1 m/sec. It is also called the **joule** (rhymes with *cool*).

1.4.8   Increment of kinetic energy, and force.   When force is constant we can show that the increment of kinetic energy is proportional to force times distance,

$$K \propto fL\,,$$

in the same way as we earlier showed the increment of momentum proportional to force times time (§1.4.3). Then also $f \propto K/L$, so force can be measured by *kinetic energy added per distance*. If the newton is retained as the unit of force, the constant of proportionality turns out to be unity: *Eh?*

$$K = fL$$

and

$$f = \frac{K}{L},$$

where K is measured in joules.

The same result in terms of force alone is that the work done by a constant force on a particle (by which kinetic energy K is gained) is the product of force and distance, $W = fL$. A force of 1 newton, working for 1 m, is said to do 1 newton m of work; the effect on the particle is to increase its kinetic energy by 1 joule.

{ We now have reason to believe that **increment of kinetic energy** and **force** form yet another of our pairs. In Chapter 4 (§4.2.8) it will be shown that } this statement is in fact equivalent to the corresponding one about momentum and force. Force at a point specifies the intensity with which the particle is gathering kinetic energy; kinetic energy is accumulated by the action of force on the particle at each point.

1.4.9   Summary.   The conclusions of this article can be summarized as follows.

Let a continuously acting force cause a particle of mass m to accelerate along a straight line. The force can be regarded as a specific quantity in two ways: with respect to time, and with respect to place.

1. During an interval of time the force gives a certain **impulse** I to the particle equal to its **increment of momentum**

$$P = mV = mv_2 - mv_1,$$

momentum being the instantaneous quantity $p = mv$. When the force f is constant, its impulse over time T is fT; hence $P = fT$, or $f = P/T$. In general, P is the total quantity corresponding to the time-specific quantity f—this is Newton's second law (as applied to the particle), also expressible as $f = ma$.

The unit of momentum is the kg m/sec, the momentum of 1 kg moving at 1 m/sec. The unit of force is the newton, which produces 1 kg m/sec of momentum in 1 sec. One newton acting for 1 sec has 1 newton sec of impulse, which equals 1 kg m/sec of momentum.

2. Along an interval of the line the force does a certain amount of **work** W on the particle equal to its **increment of kinetic energy**

He still hasn't told us why the $\frac{1}{2}$ was introduced.

$$K = \frac{1}{2}mQ = \frac{1}{2}mv_2^2 - \frac{1}{2}mv_1^2,$$

kinetic energy being the instantaneous quantity $k = \frac{1}{2}mv^2$. When the force f is constant, the work it does over distance L is fL; hence $K = fL$, or $f = K/L$. In general, K is the total quantity corresponding to the location-specific quantity f—a fact equivalent to $f = ma$.

The unit of kinetic energy is the joule, or kg m^2/sec^2, twice the kinetic energy of 1 kg moving at 1 m/sec. One newton acting over 1 m does 1 newton m of work, which equals 1 joule of kinetic energy.

1.4.10   QUESTIONS

Q1.   A constant force of 1 newton accelerates a particle of mass 1 kg from rest. When the particle has traveled 1 m, how fast is it moving?

Q2.   (a) Near the earth's surface, how far must a body of mass m fall (from rest) to reach a speed of 5 m/sec? (b) Explain why the answer is independent of m.

Q3.   Calculate the increase in kinetic energy of a 1400 kg car when it accelerates from 0 to 10 mph; from 55 to 65 mph (1 mph = 0.447 m/sec).

Q4.   The "work" done by a force, as we have employed the word, involves motion and does not depend on time. Give examples of ordinary English usage in which "work" seems not to involve motion and does depend on time.

1.5   GEOMETRICAL QUANTITIES ASSOCIATED WITH CURVES

1.5.1   Curves and quantities.   From physics we now turn to geometry, where the problem of calculus arises in connection with features of curves in the plane. Because of the capacity of plane curves to represent physical quantities in graphs, this study has more than a purely geometrical interest: whatever can be discovered about curves has application to the quantities whose graphs they are. We will be able to supplement the conceptual structure that has guided us so far, the scheme of total and specific quantities relative to a third quantity (space or time), by a geometrical scheme that represents it—or rather, two schemes, in one of which it is the specific quantity, in the other the total quantity, that appears as a graph.

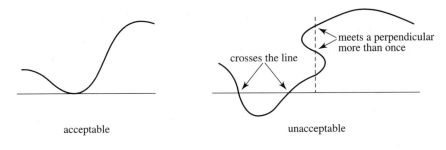

acceptable   unacceptable

Fig. 1-1

1.5.2   Area and height. A. The uniform case.   Let there be a straight line in the plane, and a continuous curve lying on one side of it; the curve may touch the line but not cross or lie along it, and must not have more than one point in common with any perpendicular to the line (Fig. 1-1). It is convenient to draw the line horizontally and call it the x-axis, even though we will not be using Cartesian coordinates as yet. The word "curve" is to be understood as comprehending straight lines and lines with corners, but (in view of the assumed continuity) not broken lines.

With each point of the line let us associate the **height** of the curve above the line at that point; that is, the length of the perpendicular from the point cut off by the curve (Fig. 1-2). With each segment of the line let us associate the **area** under the curve on that segment; that is, the area of the figure bounded by the line, the curve, and the perpendiculars from the endpoints of the segment to the curve (Fig. 1-3). I assert that area and height form a total-specific pair: area is the total quantity relative to height, height the specific quantity relative to area.

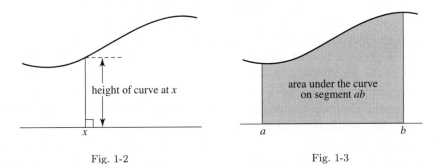

Fig. 1-2   Fig. 1-3

The uniform case is that in which the curve is a straight line parallel to the x-axis (Fig. 1-4). Obviously the height is the same everywhere and the area on a segment is proportional to its length. If the height is h and the area on a segment of length L is A, we have

$$A = hL,$$

or

$$h = \frac{A}{L},$$

the usual relations between total and specific quantity.

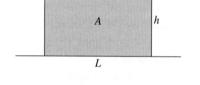

Fig. 1-4

1.5.3 B. The general case.

In the general case, it is evident that height determines area, since the heights at all points of a segment together fix the upper boundary of the figure whose area is to be found, and thereby determine that figure and its area completely. Area is accumulated along a segment according to the heights of the curve above it, much as mass is accumulated along a segment of wire according to the line densities at its points.

The converse, that area determines height, is a little more difficult to see. Why is the height at a point x determined by areas under the curve? The answer is perhaps best given indirectly: if the height were other than it is, certain of the areas would be other than they are. For let h be the height at x. The continuity of the curve implies that its height at points *near* x is *near* h—it cannot *suddenly* rise or fall. If the height at x were $k \neq h$, the height near x would similarly be near k. This would produce a different area over a little segment containing x, as Fig. 1-5 illustrates. Much as the distribution of mass in the vicinity of a point of a wire determines the line density at the point, the presence of area above the vicinity of x determines the height of the curve at x.

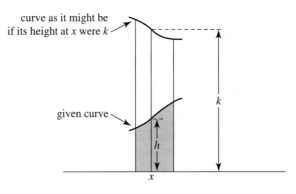

Fig. 1-5

In Fig. 1-5 the segment has been taken so short that above it the two curves (the true one at height h and the imagined one at height k) are wholly separated from one another. This must be possible, although it may require a *very* short segment.

1.5.4   Applications to figures and graphs.

We can say, then, that solving the problem of calculus will enable us to express area and height in terms of one another, and determine one from the other. Of great interest to the geometer is the prospect of a general method of finding areas of curved figures. For example, the curved portion of the boundary of a parabolic figure as in Fig. 1-6 can be defined by specifying its height above all points of the straight portion (that is, giving a rule whereby h is determined by x); therefore we expect calculus to supply the means of calculating the area of the figure, a problem which, in the absence of a general method, once demanded the ingenuity of Archimedes. Areas that are not "under curves" can be calculated as sums and differences of areas that are, so we will not be confined to a special class of problems (Q2).

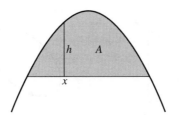

Fig. 1-6

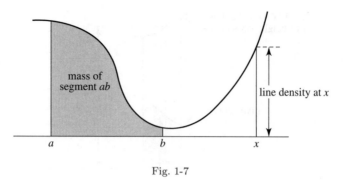

Fig. 1-7

A different application is to graphs of physical quantities. Recognition of area and height as a total-specific pair invites the simultaneous representation of two quantities in a single picture. In Fig. 1-7, the x-axis is identified with a wire of varying line density. The height of the curve at any x represents the line density there, and the area under it on any segment ab represents the mass of the segment. It is as though the mass were piled up on the line like sand. You will agree that this is a great aid to the imagination—and will hasten to produce like representations of the other pairs of quantities we have considered (Q3, Q4). When the quantities are associated with time t rather than place x, the axis may be called the t-axis, and is to be thought of as a geometrical representation of the continuum of time.

1.5.5   QUESTIONS

Q1.   Find the area under the straight line $y = \mu x$, $\mu > 0$, between $x = a > 0$ and $x = b > a$ (Fig. 1-8).

Q2.   (a) If areas under curves can be found, how can the area of a circle be determined? (b) A closed curve is **convex** if it bulges outward everywhere; that is, if every chord, apart from its endpoints, lies entirely inside the curve (Fig. 1-9). Show that the area enclosed by a convex curve is a difference of two areas under curves. (c) What if the closed curve is not convex?

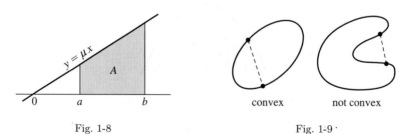

Fig. 1-8   Fig. 1-9 ·

Q3. Draw figures like Fig. 1-7 and label them so that they apply to the other pairs of physical quantities we have identified. Include the appropriate physical units.

Q4. The velocity of a certain particle is always proportional to the time it has been traveling since $t = 0$. How far does it go between $t = a$ and $t = b$?

1.5.6   Rise and slope. A. The uniform case. The second geometrical scheme also begins with the x-axis and a curve which has at most one point in common with any perpendicular to the axis. However, instead of requiring the curve to lie above the axis, we will now assume that it always rises as we move along the axis towards the right (Fig. 1-10).

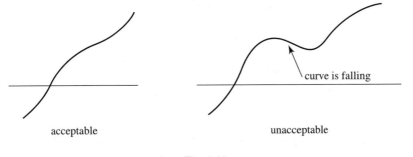

acceptable   unacceptable

Fig. 1-10

With each segment of the x-axis we associate the **rise** of the curve that takes place over that segment: the distance by which the curve rises between the perpendiculars to the axis at the left and right endpoints of the segment (Fig. 1-11). With each point x of the axis we associate the **slope** of the curve at the point P on it determined by the perpendicular at x; that is, the slope of the tangent line at P (Fig. 1-12). I assert that rise and slope form another

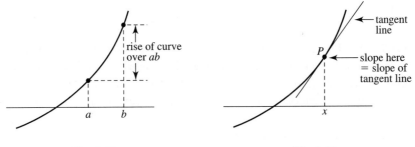

Fig. 1-11   Fig. 1-12

pair: that rise is the total quantity relative to slope, slope the specific quantity relative to rise.

> The slope is represented in Fig. 1-12 only in so far as the visible inclination of the tangent line is able to indicate it. We are assuming that the curve has a definite tangent line and slope at each point P; hence we exclude corners, where there is no well-defined tangent, as well as vertical tangents, which have no slope.

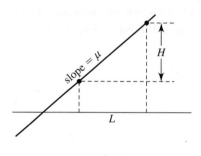

Fig. 1-13

The uniform case is again that in which the curve is a straight line, but this time the line rises (Fig. 1-13). The slope of the line is everywhere the same, and the rise over a segment is proportional to its length. If the slope is[8] μ and the rise over a segment of length L is H, we have

$$H = \mu L \,,$$

or

$$\mu = \frac{H}{L} \,,$$

as expected.

1.5.7   B. The general case.   The definition of slope in the general case, by means of a tangent line, is in keeping with our usual method of describing a non-uniform thing at a point by matching it to a uniform one: the tangent line at P fits the curve better than any other straight line through P. For

---

[8]Avoiding the letter m, already used for mass.

the tangent line, as for any straight line, slope signifies rise per **run**, or distance measured along the x-axis; hence we interpret the slope of a curve at P in the same way, as rise per run "at P"—the slope of the uniform curve, or straight line, that matches the given curve at P. It should be noted that we can speak of the slope "at x" rather than "at P," regarding the point of the curve as specified by the corresponding point on the axis.

This definition of slope shows quite clearly that its relation to rise is the same as that of line density (mass per length) to mass, or velocity (distance per time) to distance. In order to exhibit rise and slope as determining one another, we can proceed as follows (for simplicity confining our description to points where the curve lies above the axis). Suppose first that the rise is known for every segment. The slope at an arbitrary point x will surely be determined if the course of the curve is known in the vicinity of x. Let a be any point to the left of x. Since the rise is known over every interval whose left endpoint is a, the height of the curve is known at every point to the right of a (Fig. 1-14). This means that the course of the curve is determined, and with it the slope at x.

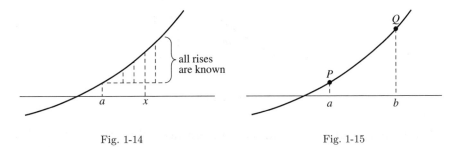

Fig. 1-14   Fig. 1-15

Conversely, suppose that the slope is known at every point. Imagine flying in an airplane above the x-axis along the curve from point P, above a, up to point Q, above b (Fig. 1-15). At each point of the flight the plane is aiming upward at a certain angle to the horizontal. This angle of climb, known at all points, determines the course of the flight from P, since to deviate upward or downward from arc PQ the plane would have to turn so as to ascend more or less steeply. Consequently it determines the total increase in altitude from P to Q, which is the rise over ab. Now slope is not the same as angle, but it determines angle, and hence rise; moreover, as rise *per horizontal distance* it determines rise exactly as a quantity gained relative to distance along the x-axis: at each point the slope directs the airplane as to the rate at which it is to gain altitude relative to its horizontal progress. Thus slope not only determines rise, but determines rise as its corresponding total quantity.

1.5.8   Rise and height.   You will have noticed that the total quantity "rise," denoted H, is closely related to the specific quantity "height," denoted h, which was introduced in §1.5.2. Their relation is analogous to that between velocity V added over a time interval and velocity v at an instant (§1.3.4): for a rising curve above the axis, the rise over a line segment is the difference between the heights at its endpoints, $H = h_2 - h_1$ (Fig. 1-16). If we allow ourselves to regard H and h as aspects of the same quantity, then we find this quantity appearing in one geometrical pair (together with area) as a specific quantity, and in another (together with slope) as a total quantity, in the same way as velocity occurred in two physical pairs.

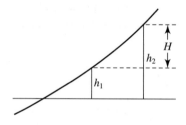

Fig. 1-16

1.5.9   Applications to curves and graphs.   With the solution of the problem of calculus will be solved the geometrical problem of finding the tangent line to a curve at a given point, the curve being defined in relation to some straight line. The parabola shown in Fig. 1-17, for example, can be described by specifying its height h above an axis at each point; from this its

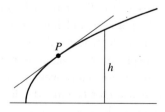

Fig. 1-17

rise over every interval is known, and thereby the slope of the tangent line through P, and the line itself. Like the problem of finding areas of figures, the tangent problem is difficult or impossible when the general method of calculus is unknown.

Rise and slope serve also to represent pairs of physical quantities in a new way. Like Fig. 1-7, Fig. 1-18 has the x-axis identified with a wire. This time, however, the mass of segment ab is represented by the rise between a and b, and the line density at x by the slope at that point. The physical situation portrayed does not differ from the one pictured in Fig. 1-7; what is different is the mode of portrayal.[9] Which mode to prefer in a given case will depend on which curve can be drawn from available data, among other considerations.

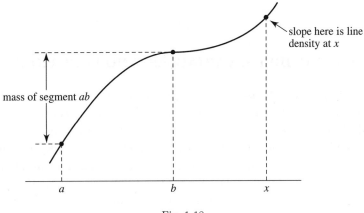

slope here is line density at x

mass of segment ab

Fig. 1-18

1.5.10   QUESTIONS

Q1. What correlation is there between the shapes of the curves in Figs. 1-7 and 1-18?

Q2. Draw figures like Fig. 1-18 for each of the other pairs of physical quantities. Include the appropriate physical units.

Q3. Suppose that the curve in Fig. 1-18 is shifted vertically upward or downward—that is, its height is everywhere increased or decreased by the same quantity. What effect does this have on its representation of mass and line density by rise and slope?

---

[9]The curve in Fig. 1-18 is so drawn that its rise and slope correspond to the area and height in Fig. 1-7, except that the vertical scale is reduced.

Q4. In Fig. 1-19, PQ is a quarter-circle, PR a horizontal tangent. As x moves from P to R, how do the following quantities change? (i) α, the angle of inclination to the horizontal (the sloping line is a tangent); (ii) the slope at x.

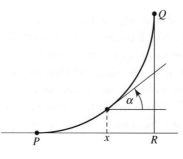

Fig. 1-19

1.6   NUMBERS, VARIABLES, AND FUNCTIONS

1.6.1   Variables. Independent variables.   Each of the pairs of quantities we have looked at has been founded upon a linear continuum—a line in physical space, the continuum of time, or a geometrical line—with whose intervals and points the total and specific quantities have respectively been associated. As you know, it is customary to represent such a continuum by a coordinate axis, usually an x-axis for a line in physical or geometrical space and a t-axis for time. The **variable** x or t assumes as its values the numbers that are assigned to the points of the axis. Coordinate representation requires certain choices to be made; thus in an example involving time, where the values of t are numbers that name moments or points of time, we have to choose a zero-point, a unit, and a positive direction. The positive direction is always taken to be towards the future, and we will always employ the second as unit; commonly $t = 0$ is chosen as the moment when motion begins.

Particular values of a variable that are not specified as numbers can be denoted by other letters or with the aid of subscripts: "the value a of x," or "$x = a$"; "the value x_1 of x"; "time t_2." The plain letters x and t ought in strictness to be reserved for denoting the variables, but I will constantly violate this rule and say, e.g., "the value x," in order to avoid multiplication of symbols. Such deliberate carelessness is often expedient in calculus, provided one remembers that there is no imprecision in the mathematics, only in writing and speech.

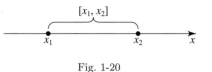

Fig. 1-20

Variables such as x and t take their numerical values at points. As for intervals, these are designated by reference to their endpoints: given $x_1 < x_2$, the interval (or segment) from x_1 to x_2, which includes all x-values such that $x_1 \le x \le x_2$, is denoted $[x_1, x_2]$ (Fig. 1-20). (Read "x_1, x_2"; the simpler notation $x_1 x_2$ would cause confusion with a product.) So, for example, we might say, referring to some given wire:[10] the line density at x_1 is 3 kg/m; the length of $[x_1, x_2]$ is 5 m, and its mass is 6 kg. Points of an interval that are not endpoints are called **interior** points. The word "interval" can also apply to a region such as the non-negative x-axis (all x such that $x \ge 0$), for which we need not introduce a notation.

If the values assumed by x or t can be arbitrarily specified, as when we choose a point x at which to evaluate line density, the variables are called **independent**, a designation which serves to distinguish them from the variables to be considered next.

1.6.2   Dependent variables. Functions.   Quantities that depend on x or t are, or can be made into, numerical variables as well, each denoted by its own letter. They are called **dependent** variables, because we think of their values as determined by values of x or t. Dependent variables which are *total* quantities have values determined by *intervals* of x or t, ones which are *specific* quantities have values determined by *single values* of x or t. Table 1-1 lists the variables we have encountered so far.

Each of the specific quantities is a **function** of its independent variable, which means that to each value of the independent variable is assigned a value of the dependent one. If $x = 3$, λ has a certain value, say 2; if $x = 4$, it has again a certain value, possibly the same one; in general, each value of x determines a value of λ. We write $\lambda(3)$ (read "λ of 3") for the value of λ associated with 3; then $\lambda(3) = 2$. Similarly $\lambda(4)$ is the line density at $x = 4$, and in general $\lambda(x)$ ("λ of x") denotes the value of λ at any value x. We may also write $\lambda(x)$ instead of λ just to remind ourselves that λ is a function

---

[10]Recall the footnote to §1.1.3.Q1(a).

Table 1-1

Where symbols for dependent variables were introduced	Independent variable	Dependent variables	
		Total quantity	Specific quantity
§1.1.2	x (location on wire)	M (mass; kg)	λ (line density; kg/m)
§1.2.1	t (time)	U (volume; m^3)	γ (growth rate; m^3/sec)
§1.3.1	t (time)	L (distance; m)	v (velocity; m/sec)
§1.3.3	t (time)	V (velocity added, or increment of v; m/sec)	a (acceleration; m/sec^2)
§§1.4.2–3	t (time)	P (momentum added, or increment of momentum p; kg m/sec) I (impulse; newton sec)	f (force; newton)
§§1.4.7–8	x (location on line of motion)	K (kinetic energy added, or increment of kinetic energy k; joule) W (work; newton m)	f (force; newton)
§1.5.2	x (abscissa)	A (area)	h (height)
§1.5.6	x (abscissa)	H (rise, or increment of h)	μ (slope)

of the variable x; then we refer to "the function $\lambda(x)$," and similarly "the function $v(t)$."

The relation between a total quantity and its independent variable is a little more complicated. A value of the total quantity is associated not with each value, but with each interval of values of the independent variable. The quantity M, for instance, is a function not of x, but of the intervals of values of x.

To the interval $[3, 5]$, M assigns a certain value, say 4—the mass, in kilograms, of the 2 m segment of wire from $x = 3$ to $x = 5$; to the interval $[5.7, 6.1]$ it likewise assigns a certain value, possibly the same one; so in general, to each interval of x-values corresponds a value of M. We write $M([3, 5])$ ("M of 3, 5") for the value of M associated with $[3, 5]$, so that $M([3, 5]) = 4$. In general, $M([x_1, x_2])$ is the value of M on any interval $[x_1, x_2]$.[11]

1.6.3   Cumulative quantities and functions. The notation last introduced is clumsy, and the truth is that functions of intervals are themselves not as handy as functions of points (that is, of single numbers). Fortunately it is possible to express functions of intervals in terms of functions of points; I will now show how this is done.

Consider first the example of the wire. Let its left end be at $x = 0$, and for any x let $m(x)$ be the mass of the segment of wire from 0 to x; that is,

$$m(x) = M([0, x])$$

(Fig. 1-21). So defined, m is a function of x: to each x it assigns a number $m(x)$. We can call m a **cumulative** quantity or function, since $m(x)$ is the

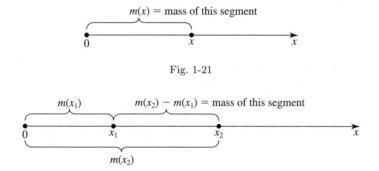

Fig. 1-21

Fig. 1-22

mass accumulated in passing along the wire from 0 up to x. The mass of any segment $[x_1, x_2]$ is evidently given by

$$M([x_1, x_2]) = m(x_2) - m(x_1)$$

---

[11]Here letters with subscripts are being used to denote arbitrary values of the variable.

(Fig. 1-22). Thus M is fully expressed by means of m: the value of M on any interval is the difference of the values of m on the endpoints of the interval.

In the uniform case, since the length of $[0, x]$ is x, its mass is λx, where λ is the constant line density; that is, the cumulative function has the simple form

$$m(x) = \lambda x\,,$$

a so-called **linear** function of x.

> We are at liberty to choose a different point from 0 as the starting-point for accumulation; the resulting cumulative function will differ by a constant from m as defined above (Q3). If we do this, or more generally if accumulation does not begin at the left end of a wire, we have to deal with points to the left, as well as to the right, of the starting-point. This complication will be addressed in Chapter 2 (§2.10.4).

Other total quantities can be handled similarly. In the case of growth, we obtain a cumulative function by defining $u(t)$ to be the volume added to the water tank from the beginning of filling at $t = 0$ until time t, $u(t) = U([0, t])$; then $u(t)$ is the volume at time t, and the volume added in any interval $[t_1, t_2]$ is $u(t_2) - u(t_1)$. In the case of added velocity V, the cumulative function $v(t)$ is the velocity acquired since the beginning of motion at $t = 0$, or simply the velocity at time t, and V added between t_1 and t_2 is $v(t_2) - v(t_1)$, as we noticed before (§1.3.4). Similarly, for a curve which rises from the axis at $x = 0$ its height $h(x)$ serves as cumulative function with respect to rise H (§1.5.8 and Fig. 1-23).

In view of the close connection between total and cumulative functions, it is apparent that our examination of pairs of corresponding total and specific quantities might have been carried out with regard to pairs of cumulative and specific quantities instead. The latter type of pair will play the chief part in the work to come, as most of it will have to do with functions of numbers.

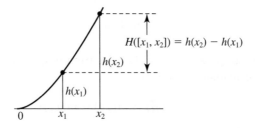

Fig. 1-23

1.6.4　QUESTIONS

Q1.　Referring to Table 1-1 and our examples, express in words the meaning of the following symbols: $\gamma(3.5)$, $L([5,6])$, $a(1)$, $f(t_2)$, $K([a,b])$, $W([a,b])$, $A([0,1])$, $\mu(\pi)$.

Q2.　(a) Find the cumulative function $m(x)$ for a uniform wire of line density 3 kg/m with left end at $x = 0$. (b) For the same wire, find $m_1(x)$, the cumulative mass measured towards the right from $x = 1$.

Q3.　Let x_0 be a point on an arbitrary wire with left end at $x = 0$, and let $m_0(x)$ be the cumulative mass measured towards the right from x_0. Express $m_0(x)$ in terms of the function m.

Q4.　In the pair area and height, what is the geometrical interpretation of the cumulative function?

Q5.　Is momentum $p = mv$ a cumulative function?

1.6.5　The meaning of "function."　When we say of two variables x and y that y is a function of x, we are making mention not of two entities, but three. There are

(i)　the independent variable x,

(ii)　the dependent variable y, and

(iii)　the relation between them, or law of dependence of y on x.

The third of these is what the word "function" primarily refers to.

In the most general meaning of the term, a function is any law by which, to each thing in one class of things, a thing in another class of things (possibly the same class) is assigned. The function is said to be **defined** for the members of the first class and to take **values** in the second class. Each body has a definite mass, in the sense of a number of kilograms; hence the assignment of mass to body specifies a function defined for all bodies and taking values in the class of numbers (of kilograms). That two bodies may have the same mass is immaterial; what matters is that the law specifies one definite mass for each given body. Of interest to us in the realm of purely numerical variables are functions that assign numbers to numbers, or numbers to intervals of numbers. The former of these types is the most common kind of function, and is always intended when the term is used without qualification.

One can think of a function as a black box with a hopper for input and a chute for output (Fig. 1-24). If the function is the one that assigns mass to a body, we imagine putting in any body—a piece of wire, say—and getting out its mass in kilograms, or more precisely a statement of the mass—"0.5 kg," perhaps printed on a slip of paper. An ordinary metric weighing scale is an (imperfect) embodiment of this function.

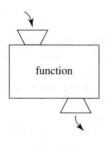

Fig. 1-24

An example of a numerical function is "square," which can be written $(\ )^2$, the blank indicating where the number to be squared, or input, will go (Fig. 1-25). If "y is a function of x" by the law of squaring, the variable x denotes an input *in general*, the variable y denotes an output *in general*, and we write

$$y = x^2$$

to indicate the same thing as is meant by Fig. 1-26. Here the box is labeled with the symbol for the function, and the hopper and chute with the general symbols for the input and the output. Put in any particular value x_0 of x and you get out a certain value y_0 of y, namely the square of x_0. A simpler but less exact picture (recall the remark on symbols, §1.6.1) is Fig. 1-27.

Since I do not mean to suggest by the image of the box that an input vanishes into it and is used up, or that an output once produced is gone from it and no longer available, perhaps it would be better to replace the hopper by a "reader" or "scanner" which examines inputs, and the chute by a "writer" or "producer" which supplies outputs. One might then compare a function to an ordinary bar-code device, which reads numbers written in one form, as bar codes, and

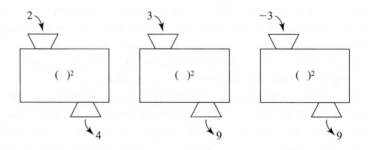

Fig. 1-25

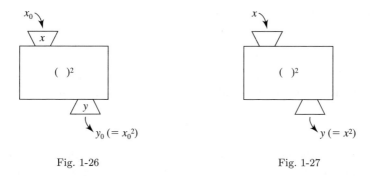

Fig. 1-26　　　　　　　　　　　　　Fig. 1-27

produces corresponding numbers written in another form, as printed prices on a sales slip. Whichever way it is conceived, a function is nothing but a certain correlation of inputs and outputs. The purpose of the images is only to assist the mind in recognizing that such a correlation is an entity in its own right.

1.6.6　Domain and range.　The set of all the allowable inputs to a function, the class of things for which it is defined, is known as its **domain**. The set of all its possible outputs, the collection of values it takes, is called its **range**.

Specifying the domain is part of defining a function. Sometimes the domain is limited by the functional law of dependence, as in the case of $y = 1/x$, for which it cannot include 0. (We may then choose to augment the law in order to extend its domain, as in Example 2, §4.2.9.) A restriction can also be imposed. If we were interested in whole numbers only, say, we might view "square" as applying to them alone. We would then have a different function from "square" as defined for all numbers: its domain would be smaller, and consequently its range too.

The range is determined by the domain and the functional law together. In a given case it may be the law, rather than the domain, that is so chosen as to yield a desired range. An example of this is the definition of "square root" (Q3; see also §4.3.2).

The sign of independent and dependent variables is treated in Art. 2.10.

1.6.7　Notation for functions and representation by graphs.　A function is often assigned its own symbol, usually a letter, and frequently the letter f. If we have assigned f to the squaring function, then

$$y = f(x)$$

means the same thing as

$$y = x^2 \,;$$

thus $4 = f(2)$, etc. In other words, f, or $f(\ )$, is the same as $(\ )^2$. If, in a given context, the function is understood, we can simply write

$$y = y(x)\,,$$

which expresses the fact that y is a function of x—dependent on x by some law—without naming the function. This notation has already been used in writing $m(x)$, $v(t)$, etc.

It is important to realize that a function as such can be thought of apart from the names of any particular variables it may relate. The function $(\ )^2$ is the same whether we write $y = x^2$ or $z = w^2$ (where both x and w can be any number)—the same squaring machine is at work in both cases.

Functions which relate numbers to numbers admit the familiar geometrical representation provided by a Cartesian graph. A graph is a kind of picture of a function, in which the whole law of dependence of one variable on another is exhibited by a pattern in the plane, ordinarily a curve (Fig. 1-28). From the input x the output y is derived by following the indicated path. Since to each x must correspond one determinate y, no vertical line can intersect the curve more than once.

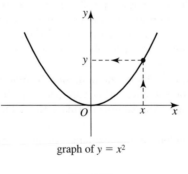

graph of $y = x^2$

Fig. 1-28

1.6.8 Monotone functions. A function whose graph only rises (as in §1.5.6), or only falls, over some interval, is called **monotone** or **monotonic** on the interval; it is **increasing** if the graph rises, **decreasing** if it falls. In the former case an increase in x always results in an increase in $f(x)$,

(1) $$\text{if } x_1 < x_2 \text{ then } f(x_1) < f(x_2);$$

in the latter, an increase in x produces a decrease in $f(x)$,

(2) $$\text{if } x_1 < x_2 \text{ then } f(x_1) > f(x_2).$$

As Fig. 1-28 shows, the function $y = x^2$ is increasing for $x \geq 0$ (that is, (1) holds for any non-negative x_1 and x_2) and decreasing for $x \leq 0$.

A less stringent condition than (1) is:

(3) $$\text{if } x_1 < x_2 \text{ then } f(x_1) \leq f(x_2).$$

It is satisfied by any constant function, for example, whereas (1) is not. A function satisfying (3) can be called increasing *in the weaker sense*. Analogously we define a function decreasing in the weaker sense; both kinds are monotone in the weaker sense.

1.6.9 Functions and calculus.

The laws of dependence called functions will be central to our study. In fact calculus *is* a study of functions. Once number scales have been chosen on which length, time, abscissa, mass, velocity, force, ordinate, etc., are to be measured, all of these general quantities are represented by numerical variables. *Particular* quantities—the line density, or mass distribution, or cumulative mass, of a certain wire; the force on a certain particle, or its acquisition of momentum over time intervals, or its cumulative momentum—all of them are nothing but functions, laws of the dependence of certain of the variables on others. A wire has line density λ; that is, a function $\lambda(x)$ assigns a number λ to each number x. To distinguish this function from others I can give it a name: $\lambda = g(x)$. Then the function g is the mathematical description of the line density. Again, the wire has mass distribution M; that is, a function $M([x_1, x_2])$ assigns a number M to each number-interval $[x_1, x_2]$. And the wire has cumulative mass m; that is, a function $m(x)$ assigns a number m to each number x. It is in virtue of the numerical-valued functions which describe it that the wire becomes a subject of our mathematics.

Consequently the problem of calculus can be put as follows: to determine the nature of that relation between *functions* which expresses the relation common to all pairs of total (or cumulative) and specific quantities, as we have seen it in many examples; to express one function of a related pair in terms of the other; and to find one when the other is known.

1.6.10 QUESTIONS

Q1. (a) As applied to all human beings x, is "y is the mother of x" a function? (b) The same question for "daughter of," as applied to all women.

Q2. What is the difference between "the value $y = 2$" and "the function $y = 2$"?

Q3. Is "square root" a function? Explain.

Q4. A closed curve, such as a circle, is not the graph of a function. How can such a curve be treated by calculus, if calculus studies functions?

Q5. Give an example of a function, defined for all numbers, whose graph is not a continuous curve.

CHAPTER 2

Integral and Derivative

INTRODUCTION

We are now to begin the solution of the problem of calculus, which has been found to arise in a variety of situations united by analogy and admitting a common description in terms of functions and their relations. Although as a purely mathematical theory calculus has properly to do with abstract functions alone, its applications to geometry and physics provide much of its significance, and their study is indispensable for achieving a sound understanding of the subject. In this book, far from presenting calculus as a pure theory of functions, I will rely at many points upon the ordinary intuition and experience that are formalized in physics and geometry, and will frequently interpret mathematical concepts and results with their aid. Nevertheless, applications will receive too little attention. To supplement what is given here, you should try to interpret and understand every new idea as it applies to each of the examples that were set out in the first chapter.

To call geometry a field of applications of calculus is to make a distinction in some respects out of date. Modern geometry has the same kind of foundation as number has, functions are very much a part of it, and calculus enters into its very definitions. But it is still correct to say that the objects of geometry are studied by means of calculus.

The question before us has two symmetrical parts:

1. How does specific quantity determine total or cumulative quantity?
2. How does total or cumulative quantity determine specific quantity?

I will take them up in this order, and treat them in parallel fashion, the first
in Arts. 2.1–2.5 and the second in Arts. 2.6–2.9. A concluding article will
generalize the investigation to take account of signed quantities.

2.1   APPROXIMATE CALCULATION OF MASS

2.1.1   Formulation of the problem.
Let us consider again the wire of
Art. 1.1. In the uniform case, when the line density (specific quantity) is constant, we know how it determines mass (total quantity), namely by formation
of a product:

$$M = \lambda L \,,$$

$$\text{mass} = \text{line density} \times \text{length} \,.$$

Suppose now that we have a non-uniform segment of wire $[a, b]$ along the x-axis,
one whose line density increases from left to right, say (Fig. 2-1). I assume that
we possess complete knowledge of the line density everywhere (as we would if
we knew the composition of the wire at all points); that is, we know the value
of the function $\lambda(x)$ for all points x on the segment. How can we find the mass
of $[a, b]$?

The answer to this question begins with two ideas, of which the first is
approximation of the non-uniform by the uniform.

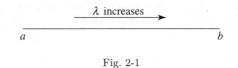

Fig. 2-1

2.1.2   Approximation by successive estimates.
If the variation in
$\lambda(x)$ along $[a, b]$ is slight, a good approximation to the mass can be obtained
by choosing any point ξ_1 in the interval and calculating the mass of a *uniform*
wire of the same length whose line density *matches* that of the given wire at
ξ_1. This uniform wire, which we can think of as also extending from a to b,
has constant line density $\lambda(\xi_1)$ and length $b - a$ (Fig. 2-2); hence its mass is

$$\lambda(\xi_1) \times (b - a) \,.$$

Since $\lambda(x)$ is not very different from $\lambda(\xi_1)$ anywhere on the original wire, the
two wires are quite similar, and the mass so calculated is close to the true mass.
It is not likely to be exactly equal to it, however; for example, if we choose

ξ_1 to be a, the calculated mass will be too small (because $\lambda(x)$ is larger than $\lambda(a)$ everywhere except at a), while if we choose ξ_1 to be b, the mass will come out too large. Other choices of ξ_1 will give values in between. Any uniform wire being only an approximation to the non-uniform one, it is natural that the mass of the former should only approximate that of the latter.

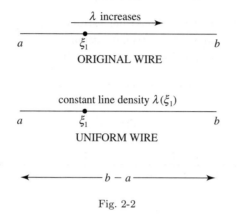

Fig. 2-2

Seeking to improve this estimate of mass, we recognize that if the variation in the original wire's line density were smaller, the wire would be more nearly uniform and the approximation better. But the wire is given, so the variation seems to be out of our power. From this quandary the thought of *division* rescues us: the variation of $\lambda(x)$ along any *part* of the segment of wire is less than that along the *whole* of it; hence there is reason to hope that if the segment is divided into parts, and the mass of each part separately is found by uniform approximation, the sum of the resulting masses will be closer to the true mass than was the original estimate.

To put this idea into effect, let the wire be divided at a point x_1 into two segments, **subintervals** of the x-interval $[a, b]$ (Fig. 2-3). In each a point is chosen, ξ_1 in the first and ξ_2 in the second. (The present ξ_1 need

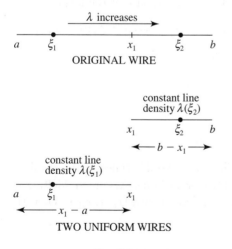

Fig. 2-3

not be the same as the point denoted by that letter a moment ago; with the new division of the line points ξ_1, ξ_2 are newly chosen.) The first segment is approximated by a uniform segment of line density $\lambda(\xi_1)$, the second by a uniform segment of line density $\lambda(\xi_2)$, and the estimate of mass is

$$\lambda(\xi_1)(x_1 - a) + \lambda(\xi_2)(b - x_1).$$

This is, we hope, better than the earlier estimate; in any case, it is not hard to see that it lies between the extreme estimates $\lambda(a)(b - a)$ (smallest) and $\lambda(b)(b - a)$ (largest) obtained without division (Q2).

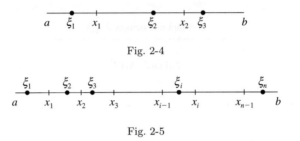

Fig. 2-4

Fig. 2-5

Trying for a still better estimate we can use three subintervals, separated by points of division x_1, x_2 (Fig. 2-4). The resulting estimate of mass is

$$\lambda(\xi_1)(x_1 - a) + \lambda(\xi_2)(x_2 - x_1) + \lambda(\xi_3)(b - x_2).$$

In general, n subintervals (Fig. 2-5) will give an estimate of the form

$$\lambda(\xi_1)(x_1 - a) + \lambda(\xi_2)(x_2 - x_1) + \lambda(\xi_3)(x_3 - x_2) + \cdots + \lambda(\xi_n)(b - x_{n-1}).$$

It is convenient to introduce the notation

$$x_0 = a, \quad x_n = b,$$

and further to write, for any integer i,

$$\Delta x_i = x_i - x_{i-1}$$

(Fig. 2-6). (The letter Δ stands for *difference*, the difference of the x-values: it is an abbreviatory symbol, not a number or variable; Δx_i, read "delta $x\,i$," is not a product, but the length of the ith subinterval $[x_{i-1}, x_i]$.) Then the estimate of mass becomes

$$\lambda(\xi_1)\Delta x_1 + \lambda(\xi_2)\Delta x_2 + \cdots + \lambda(\xi_n)\Delta x_n.$$

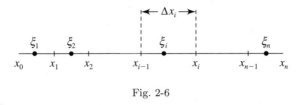

Fig. 2-6

For ease of calculation the n subintervals can be made equal in length, but this is not necessary. Provided that all are short we can expect the variation of $\lambda(x)$ along each to be very small—*how* small, for a given subinterval, depends on the behavior of λ there—and the estimate of mass to be correspondingly accurate, regardless of exactly where in the different subintervals the points ξ_i at which λ is evaluated are located. It may suit us to choose these points to be the left or right endpoints of their subintervals, $\xi_i = x_{i-1}$ or x_i. As mentioned before, since λ increases from a to b, selecting left endpoints will underestimate mass and selecting right endpoints will overestimate it.

An example is provided in Q1, and a slightly more difficult one in §2.1.4.

2.1.3   QUESTIONS

Q1.   Let the line density of a wire (in kg/m) increase uniformly from the value 1 at $x = 0$ to the value 2 at $x = 1$; in other words, for $0 \leq x \leq 1$ let $\lambda(x) = 1 + x$. (a) Calculate estimates of the mass for $n = 1$ (whole interval), $n = 2$ (two subintervals), and $n = 3$ (three subintervals), using equal subintervals and their (i) left endpoints, (ii) midpoints, (iii) right endpoints. What happens as n increases? (b) In view of the symmetry of the wire about the midpoint $x = \frac{1}{2}$, what must its mass be?

Q2 (§2.1.2).   Demonstrate this.

2.1.4   Example.

Consider a wire whose line density (in kg/m) increases from 1 at $x = 0$ to 1.16 at $x = 4$ according to the law

$$\lambda(x) = 1 + 0.01x^2$$

(Fig. 2-7). To avoid writing 0.01 repeatedly, we can replace it by c (for centi-).[1] Then $\lambda(x) = 1 + cx^2$.

---

[1] The 0.01 is in $\lambda(x)$ merely to make the numbers plausible. We could let c be any other positive number (as in Q2).

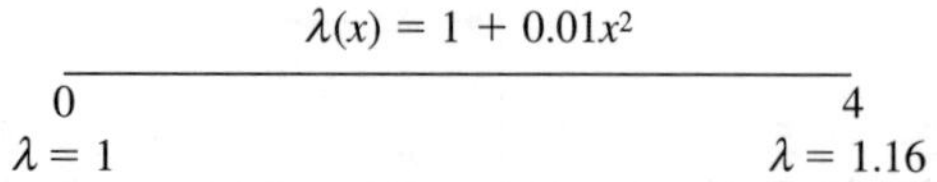

$$\lambda(x) = 1 + 0.01x^2$$

$$
\begin{array}{ll}
0 & 4 \\
\lambda = 1 & \lambda = 1.16
\end{array}
$$

Fig. 2-7

At the left endpoint the line density is $\lambda(0) = 1$; a uniform wire of that line density extending over the same interval has mass

$$1 \text{ kg/m} \times 4 \text{ m} = 4 \text{ kg},$$

so this is a first estimate of the mass of the given wire. If we evaluate λ at the right endpoint instead, the estimate is

$$1.16 \times 4 = 4.64,$$

larger by 16%. Any other point of evaluation will yield an intermediate estimate for the mass; for instance, the midpoint $x = 2$ gives

$$1.04 \times 4 = 4.16.$$

The true value is somewhere between 4 and 4.64.

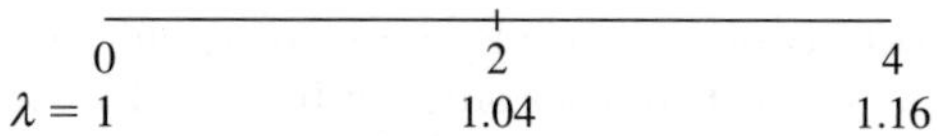

$$
\begin{array}{lll}
0 & 2 & 4 \\
\lambda = 1 & 1.04 & 1.16
\end{array}
$$

Fig. 2-8

Now divide the wire at $x_1 = 2$ (Fig. 2-8). Choosing left endpoints $\xi_1 = 0$ and $\xi_2 = 2$ results in the estimate

$$1 \text{ kg/m} \times 2 \text{ m} + 1.04 \text{ kg/m} \times 2 \text{ m} = 4.08 \text{ kg}$$

based on two uniform wires each of length 2. Use of right endpoints $\xi_1 = 2$, $\xi_2 = 4$ gives

$$1.04 \times 2 + 1.16 \times 2 = 4.40,$$

larger by only about 8%.

Rather than try next a division into three parts, let us subdivide the two subintervals already created. The wire is then divided into four segments at $x_1 = 1, x_2 = 2$, and $x_3 = 3$ (Fig. 2-9). Using left endpoints we get

$$1 \times 1 + 1.01 \times 1 + 1.04 \times 1 + 1.09 \times 1 = 4.14,$$

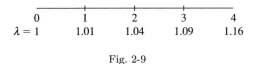

Fig. 2-9

and using right endpoints,

$$1.01 \times 1 + 1.04 \times 1 + 1.09 \times 1 + 1.16 \times 1 = 4.30\,,$$

about 4% higher. The true mass is between these two estimates.

A further subdivision, as far as 64 equal subintervals, yields estimates

$$4.208359375 \quad \text{and} \quad 4.218359375\,,$$

which differ by only 0.01. The larger exceeds the smaller by about 0.2%.

The results are summarized in Table 2-1. As n increases, the underestimates rise, and the overestimates fall, towards the unknown true mass. It appears likely that in this way any degree of accuracy of estimation can be attained by dividing the segment finely enough.

Table 2-1

	$n = 1$	$n = 2$	$n = 4$	$n = 64$
Lower estimate, using left endpoints	4.00	4.08	4.14	4.208359375
Upper estimate, using right endpoints	4.64	4.40	4.30	4.218359375

This is in fact the case, and not only for this example but quite generally, as I will demonstrate presently. Taking it for granted for the moment, we have a method for approximate calculation of mass from line density, but as yet no way of expressing exact mass in terms of that density. To achieve the desired expression we call on a second idea (§2.1.1), which comes in two forms: the concept of *limit*, and the notion of *infinitesimal*.

2.1.5   QUESTIONS

Q1. (a) Calculate the estimate of mass for the λ of the example in the case $n = 2$ when ξ_1 and ξ_2 are taken to be the midpoints of the two equal intervals. (b) The answer to (a) is larger than the estimate of 4.16 that was obtained by using the midpoint when $n = 1$. In §2.1.3.Q1 the two corresponding estimates were the same. Explain the difference.

Q2.   (a) Calculate upper and lower estimates, as in the text, for $\lambda(x) = 1 + x^2$ and $n = 1$, 2, and 4. (b) Can you find ways to shorten such calculations?

2.2   DESCRIPTION OF MASS AS AN INTEGRAL

2.2.1   Mass as the limit of estimates.   The mass M of segment $[a, b]$ is said to be the **limit** of the approximate masses found by our method, because it is the number to which they approach within any given degree of nearness.

In the example just presented, the lower estimates 4.00, 4.08, 4.14, etc., are all smaller than M, but they increase towards it. Since we know (from the estimates for $n = 64$) that M lies between 4.20 and 4.22, the first estimate is within 0.22 of M, the second within 0.14, and the third within 0.08. If we wanted an estimate within, say, 0.01 of M, we could obtain it by dividing the interval more finely; as a matter of fact, 64 equal subintervals would be enough, because the upper and lower estimates for $n = 64$ differ by exactly 0.01. An estimate accurate to 0.001 would require many more subintervals, but could be achieved; and in general, however small a number we choose, we can find a lower estimate of M which differs from M by less than that number. Furthermore, once an estimate satisfactory by some standard has been found, every estimate obtained by further subdivision will also be satisfactory by the same standard, because the estimates keep improving. Thus however small a number we may choose, *all* of the lower estimates from some point on will differ from M by less than that number.

Similarly, the upper estimates 4.64, 4.40, 4.30, etc., descend in such a way that from some point on, all of them are within 0.001 of M, or any other distance that may be named.

More generally, we can obtain a sequence of estimates from *any* series of successively finer partitions of the wire and choices of evaluation points ξ_i. It is not necessary that all subintervals be equal at each stage, or that successive stages be obtained from preceding ones by subdivision; the only condition is that as the stages succeed one another all subintervals eventually become and remain shorter than any given length. (It would not do to keep on choosing one of them longer than 0.01, say, as the approximation would continue to suffer from inaccuracy over such subintervals.) An arbitrary sequence of estimates so calculated may be quite irregular, now underestimating M, now overestimating it, now by chance actually hitting it exactly; the sequence may approach M

only to wander away again—this can happen, despite finer partitioning, if a
new choice of the ξ_i produces inferior local estimates. Nevertheless, for any
such sequence there is *some* term of it such that from that term on, *all* the
terms are within 0.01 of M, and *some* term (probably farther along) such that
from *that* point on all terms are within 0.001 of M, and so generally.

Finally, taking together the whole field of these estimates of M, we can
summarize as follows.

> The mass M of the segment $[a, b]$ of line density $\lambda(x)$ is the limit
> of estimates of the form
>
> $$\lambda(\xi_1)\Delta x_1 + \lambda(\xi_2)\Delta x_2 + \cdots + \lambda(\xi_n)\Delta x_n \,,$$
>
> where in forming successive estimates the subintervals are increased
> in number in such a way that the length of the longest subinterval
> eventually becomes and remains less than any given length.

We say that the estimates **approach** M **as a limit** as the length of the longest
subinterval **approaches zero**.

2.2.2   The concept of limit.   In this language there lurk ever-flowing
sources of confusion. A few cautionary observations will help us toward clarity.

First, by accepting the name of limit M has not lost its character as a
definite number. There is nothing approximate about M, although we have
been approximating it; the segment of wire has an exact mass, which is a
unique number of kilograms.[2]

Secondly, M is the limit *of* the estimates, the estimates *approach* M, but
M itself need not *be* one of the estimates. It is the estimates, not M, that have
the form

$$\lambda(\xi_1)\Delta x_1 + \lambda(\xi_2)\Delta x_2 + \cdots + \lambda(\xi_n)\Delta x_n \,.$$

It may happen that one or more of these estimates equal M, but that is
irrelevant to the definition of limit; what the estimates do is to point at their
limit M, to single it out among all numbers, not to express it as one of them.
Thus we say that 0 is the limit of the sequence $1, \frac{1}{2}, \frac{1}{3}, \frac{1}{4}, \ldots$, but it does not

---

[2] Recall the note at the end of §1.1.1.

occur as a term of that sequence. The same number 0 is the limit of $0, 1, \frac{1}{2}, \frac{1}{3}$, $\frac{1}{4}, \ldots$ and of $0, 0, 0, 0, \ldots$, sequences in which it does occur.

Thirdly, the sequences of which M is the limit are *infinite* sequences, like the sequence of positive integers $1, 2, 3, \ldots$; a sequence does not end, but after each term comes one more. This is necessary, because there is no end to the diminution of subinterval lengths required by the condition that subintervals become *arbitrarily* short. Any one estimate involves a longest subinterval I; but the subintervals used in a sequence of estimates must become and remain shorter than any interval that may be named, shorter therefore than I; hence there must be a next estimate. Even if the estimates repeat, in a sequence like $0, 0, 0, \ldots$, they do not end. In consequence, a limit can never be determined by calculating terms of a sequence one by one. On the other hand, it may be possible to *prove* something about a sequence as a whole that will identify its limit. If, say, we can prove that every term equals zero, we will know that the limit is zero also.

Lastly, despite the use of such words as "approach" and "eventually," *the mathematical definition of limit has nothing to do with time*. Successive terms in a sequence of estimates, like successive terms in the sequence of positive integers $1, 2, 3, \ldots$, succeed one another in the sense of their *ordering*, not in a *temporal* succession. For certain purposes it may be appropriate to apply the ordering temporally, as when we use the integers in counting, but such application belongs to physics or psychology rather than mathematics. The distinction becomes clear in the following description.

Let us call any interval that contains a number u in its *interior* (not as an endpoint) a **neighborhood** of u. Then we can say that M is the limit of the sequence because

> in any neighborhood of M, however small, *all* the terms of the sequence *from some point on* are to be found

(Q5). In other words,

> in any neighborhood of M, however small, *all but a finite number* of the terms are to be found.

We are permitted to say that all the terms are found in a given neighborhood "eventually," if we keep in mind that such temporal language suggests more than is really intended. The sequence $1, \frac{1}{2}, \frac{1}{3}, \ldots$ "approaches zero," its terms are "eventually" smaller than any given number; this does not mean that they move toward zero in time, or that we move along the sequence, but only that any neighborhood of zero, e.g. the interval $-0.001 \leq x \leq 0.001$, contains all but a finite number of the terms, all the terms from some point on—in the example at hand, from the term $\frac{1}{1000}$ on.

What may still be obscure is the reason for our interest in the idea of limit. It may be thought that by introducing the term I have done no more than give a name to the object of approximation. In fact, something more has been accomplished: we now have an exact account of what it means for M to be approximated "to any degree of accuracy," so that we know just how M is singled out among all numbers by the sequences of estimates. In short, we have a new *concept*, by means of which our initial question (as applied to the wire) gets an answer. How is mass determined by line density λ? As the *limit* of the approximating uniform estimates

$$\lambda(\xi_1)\Delta x_1 + \lambda(\xi_2)\Delta x_2 + \cdots + \lambda(\xi_n)\Delta x_n \,.$$

2.2.3   Insufficiency of the concept of limit for determining mass. This answer, such as it is, invites two related objections. The first has to do with the numerical determination of M.

While M is precisely identified as the limit of any sequence of numbers obtained in a certain way, there is no suggestion that its exact value is to be found from such a sequence. The estimates approach the number M, but what is the number they approach? In certain cases the value can in fact be determined by use of an appropriate sequence (Q3), but a general method is lacking. Even if we do not need an exact value, we cannot as a rule hope to calculate very accurate estimates of M, at least by hand. It would seem that in addition to having to settle for approximations we must rely on machines to compute them.

The second objection arises from the generality of our aims. We are looking for a way of determining the mass of an *arbitrary* interval $[a, b]$, and this implies that we want to know the total *function* that assigns values to intervals, or (equivalently) the cumulative function that assigns them to points. To know a function ordinarily means to have a formula for it, an expression which not only directs calculation but also gives the function a place in relation to other known functions. The bare idea that individual values M are limits of estimates does not suggest how to obtain formulas for the functions $M([x_1, x_2])$ and $m(x)$; yet if the function $\lambda(x)$, upon which these estimates are based, has a known formula, one would hope that the other functions would have related formulas somehow derivable from that one. As with particular numerical values of M, formulas can sometimes be found from sequences of estimates (Q4), but not by a widely applicable method.

Thus while identification of M with the limit of sums

$$\lambda(\xi_1)\Delta x_1 + \lambda(\xi_2)\Delta x_2 + \cdots + \lambda(\xi_n)\Delta x_n$$

is a considerable achievement, both in concept and as guidance for calculation by computer, it falls short of the goal of full knowledge of the total quantity as a function. That goal will be reached only after further development of our ideas.

2.2.4　QUESTIONS

Q1.　Suppose that a sequence of estimates of M begins 0.3, 0.33, 0.333, and is known to continue endlessly according to this pattern. (a) What is M? (b) Argue from the meaning of limit that the answer to (a) *is* the limit of the sequence.

Q2.　What is the limit of the sequence $2, \frac{3}{2}, \frac{4}{3}, \frac{5}{4}, \ldots$?

Q3.　Consider the line density $\lambda(x) = x$ on the interval $[0, 1]$. We aim to compute M by using the sequence of upper estimates based on equal subintervals. (a) Show that if n equal subintervals are used, the estimate is

$$\frac{1}{n} \cdot \frac{1}{n} + \frac{2}{n} \cdot \frac{1}{n} + \frac{3}{n} \cdot \frac{1}{n} + \cdots + \frac{n}{n} \cdot \frac{1}{n}$$

(Fig. 2-10). Simplify this to $\frac{1}{n^2}(1 + 2 + 3 + \cdots + n)$. (b) By dividing the $n \times (n + 1)$ array of dots (Fig. 2-11) by a suitable diagonal, show that

$$1 + 2 + 3 + \cdots + n = \frac{1}{2}n(n + 1).$$

(c) Putting together (a) and (b), show that the nth estimate is $\frac{1}{2} \cdot \frac{n + 1}{n}$. (d) Hence the sequence of estimates is $\frac{1}{2} \cdot 2, \frac{1}{2} \cdot \frac{3}{2}, \frac{1}{2} \cdot \frac{4}{3}, \frac{1}{2} \cdot \frac{5}{4}, \ldots$. In light of Q2, what is the value of M?

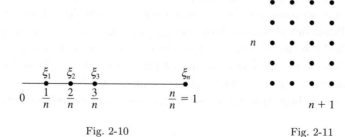

Fig. 2-10　　　　　　　　　　　　　　Fig. 2-11

Q4. Generalize Q3 to the interval $[0, x]$, x arbitrary, and thereby determine the function $m(x)$.

Q5 (§2.2.2). Why must we specify a *neighborhood* of M rather than allow an interval of which M is an endpoint?

2.2.5 Mass as a sum based on infinitesimal subintervals.

Although cumbersome, the preceding representation of the total quantity M as a limit of estimates was quite straightforward, a new application of familiar notions of product, sum, and difference (the last as a measure of nearness). I now turn to a different representation, one which requires a certain suspension of disbelief, and rewards the effort by providing the intuition with a degree of direct access to the thing represented.

We begin again with estimates of M calculated as sums

$$S = \lambda(\xi_1)\Delta x_1 + \lambda(\xi_2)\Delta x_2 + \cdots + \lambda(\xi_n)\Delta x_n \, .$$

As we know, the reason that such an estimate fails in general to *equal* M is that it is based on replacing the true wire along each subinterval $[x_{i-1}, x_i]$ by a segment of uniform wire that only approximates it there. As the subintervals are made shorter the errors so introduced become less, less indeed than any *given* number, but at no stage of approximation do they vanish altogether.

I now ask you to entertain the paradoxical idea that the subintervals can be taken so small that along each of them the wire can be regarded as uniform for the purpose of calculating mass. Any effect of the wire's variation along one of these tiny intervals upon its mass there is to be neglected, ignored, dismissed from the mind: it bears no comparison to the mass, it is beneath the notice of perfectly accurate measurement. With such a choice of subintervals all error is, or is conceived to be, eliminated, so that the corresponding sum S is exactly equal to M.

How small must one of these subintervals be? Its length cannot be a number greater than zero, for along any finite segment, no matter how short, the line density $\lambda(x)$ increases by some finite amount (since it grows steadily from a to b), and its increase makes a definite contribution to the mass of the segment. Nor can it be zero, or there would be no subinterval, but only a point. So the length is not a number at all. If the little subinterval is to be granted existence, its length must be conceived as a new species of quantity, less than every positive number and yet non-zero.

Without subjecting such quantity to critical examination (under which it would seem unlikely to fare well), I will proceed to give it a name, **infinitesimal**, the infinitely small, and justify its introduction by demonstrating that if cautiously employed it illuminates our subject and does not interfere with properly based reasoning. I may as well assure you right now that any mathematical truth that can be established with the aid of infinitesimals can also be established without them—this can actually be proved—so they are not a sign of something rotten in the foundations of calculus.

> Infinitesimals as we will be using them are associated with the name of Leibniz, co-discoverer with Newton of calculus. To everyone's surprise, in the mid-twentieth century infinitesimals were rendered respectable by founding them upon mathematical logic; these were similar to, but not quite the same as, the infinitesimals of Leibniz, which are too good for this best of all possible worlds. The cited proposition concerning truth and proof refers to the new infinitesimals.

2.2.6 QUESTION

Q1. Relate the idea of the infinitely small to the well-known fact that physical limitations on measurement and perception render small differences unobservable. Consider, for example, visual acuity as it affects the appearance of a regular polygon having many short sides.

2.2.7 Notation for the sum. The definite integral.

The symbol Δ, which indicates a difference, has been used so far only in combinations of the form $\Delta x_i = x_i - x_{i-1}$. When one desires to signify a difference in x-values, or change in the value of x, without specially naming either value involved, one writes simply Δx (read "delta x"); thus we might say, "Let the change in x be Δx," or refer to "the interval from x to $x + \Delta x$." If (as is usually the case) Δx is non-zero, it always denotes a *finite* quantity, a *number*. To symbolize an *infinitesimal* change or difference the letter d is used instead. While Δx is a *difference*, dx (read "dx") is called a **differential**: it is the infinitesimal difference between $x + dx$ and x. To both kinds of difference the term **increment** is also applied, Δx being a finite and dx an infinitesimal increment of x.

Of the sum

$$S = \lambda(\xi_1)\Delta x_1 + \lambda(\xi_2)\Delta x_2 + \cdots + \lambda(\xi_n)\Delta x_n$$

a typical term is $\lambda(\xi_i)\Delta x_i$, or, leaving off the subscripts, $\lambda(\xi)\Delta x$, where ξ is a point in the interval of length Δx. We could shorten the expression to

$$S = \text{Sum of the } \lambda(\xi)\Delta x\text{'s}$$

or just

$$S = \text{Sum } \lambda(\xi)\Delta x\,.$$

This is an approximation to the total quantity M.

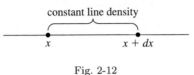

Fig. 2-12

Passing to infinitesimal subintervals, we acknowledge that the multitude of them is infinite, but ignore the difficulty of making sense of an infinite sum of infinitesimals and go ahead to denote a typical term of it by

$$\lambda(x)\,dx\,.$$

This is the infinitesimal mass, or **element** of mass, associated with the infinitesimal interval $[x, x+dx]$ (Fig. 2-12); as ξ we have chosen the left endpoint of that interval—not that the choice matters, since λ is constant along the interval. The sum of these elements is exactly equal to the mass of the whole interval,

$$M([a, b]) = \text{Sum of the } \lambda(x)\,dx\text{'s}\,,$$

or

$$M([a, b]) = \text{Sum } \lambda(x)\,dx\,.$$

Reducing the word "Sum" to an elongated S, we write

$$M([a, b]) = \int \lambda(x)\,dx\,.$$

Finally, the endpoints of the interval are added to the symbol:

$$M([a, b]) = \int_a^b \lambda(x)\,dx$$

(read "the integral from a to b of $\lambda(x)\,dx$," or "... of $\lambda(x)$," or "... of λ").

This is called a **definite integral**: "integral" as expressing a unity composed of parts, "definite" because the interval $[a, b]$ is specified (we will have the "indefinite integral" later). The long S is the **integral sign**: it announces a summation into a whole. The interval $[a, b]$ is the **interval of integration**; a and b are the **limits of integration**, a the **lower limit** and b the **upper limit**.

To find the definite integral is to **integrate** the function λ over the interval; $\lambda(x)$ is the **integrand**, and x the **variable of integration**; integration of λ is done **with respect to** x.

2.2.8 Example. The wire segment of §2.1.4 has $\lambda(x) = 1 + cx^2$ (with $c = 0.01$), $a = 0$, and $b = 4$. Hence the mass of that segment can be written as the integral

$$\int_0^4 (1 + cx^2)\, dx$$

(the parentheses are playing their usual role of marking off one factor in a product). Looking at this symbol I interpret it as follows.

> The interval from $x = 0$ to $x = 4$ is divided into countless immeasurably tiny subintervals, a typical one of which begins at a point x and has infinitesimal length dx. On that typical subinterval the wire is considered uniform, of constant line density $1 + cx^2$; therefore the mass of the tiny segment located there is the product $(1 + cx^2)\, dx$. The whole symbol $\int_0^4 (1 + cx^2)\, dx$ denotes the sum of all such tiny masses, which is the mass of the whole segment of wire between 0 and 4.

2.2.9 The definite integral as limit and sum. We now have two representations of the same mass $M([a, b])$: as a sum of infinitesimal masses $\lambda(x)\, dx$, and as the limit of sums of finite masses $\lambda(\xi)\Delta x$. Let us equate them:

$$\int_a^b \lambda(x)\, dx = \text{limit of } \lambda(\xi_1)\Delta x_1 + \lambda(\xi_2)\Delta x_2 + \cdots + \lambda(\xi_n)\Delta x_n$$

as n increases and the Δx_i approach zero .

On the left is a pregnant symbol, exemplifying an economical and suggestive notation whose great utility will soon be evident; on the right, a description soundly based on arithmetic operations, which incorporates but transcends practical calculation. A conscientious mathematician will simply *define* the left side by the right side; the wise student will bear in mind the meanings intended by both sides, despite the questionable logical antecedents of the expression on the left. Whichever way it is regarded, the quantity referred to is one and the same, the *definite integral of the line density over the interval* $[a, b]$.

To the question, How is mass determined by line density λ? we can now reply concisely: As its definite integral. But if further questions are raised as before—How is the integral calculated? How do integrals over all intervals unite in a single function?—there is as yet nothing to say.

2.2.10　QUESTIONS

Q1.　Generate and reflect upon paradoxes inherent in the idea of total quantity as an infinite sum of infinitesimals.

Q2.　Read the following integrals correctly and interpret them as masses in the manner of §2.2.8.　(a) $\int_1^3 x^3 dx$　(b) $\int_{\sqrt{2}}^2 5\,dx$　(c) $\int_0^{2\pi} (x^2 + x + 1)\,dx$

Q3.　Write integrals for the masses of the following wire segments. (a) A segment from $x = 5$ to $x = 10$ whose line density at each point x equals $\sqrt{x}$. (b) A 2 m segment whose line density increases uniformly from the value 3 at the left end to 7 at the right. (You will have to assign coordinates to the wire.)

Q4.　(a) Show that $\int_a^b dx$ (which means $\int_a^b 1\,dx$) equals $b - a$, arguing both from the point of view of infinitesimals and from that of limits. Interpret in terms of mass. (b) If c is a constant, what is $\int_a^b c\,dx$?

2.3　OTHER INTEGRALS FROM PHYSICS

2.3.1　Generality of the integral.

Although the definite integral has been developed here as an expression of mass in terms of line density, the only essential components of its definition were the function $\lambda(x)$ and the interval $[a, b]$. Consequently it applies equally well to other cases of total and specific quantity. In this article I will treat the other physical examples from Chapter 1, and in the next consider geometry, and functions in general.

2.3.2　Integral of growth rate.

Let water flow into a tank, and let the rate of growth of volume in the tank at any time t be $\gamma(t)$. In a time interval[3] $[c, d]$ the volume $U([c, d])$ is added. I am going to express this total quantity as a definite integral, taking first the viewpoint of limits and then that of infinitesimals.

An approximation to the volume added is obtained by dividing the time interval into subintervals and calculating the volume added during each one by

---

[3]The change of notation from $[a, b]$ is of no importance in itself, but it avoids conflict with acceleration a (§2.3.3) and emphasizes the distinction between time and space in connection with force (§2.3.4). Of course, c here is not the same as in §§2.1.4 and 2.2.8.

a *uniform* flow matching the actual flow at some instant of it. Let the number of subintervals be n. Setting $t_0 = c$ and $t_n = d$ we take points of division $t_1, t_2, \ldots, t_{n-1}$ (in increasing order), and in each subinterval $[t_{i-1}, t_i]$ we choose an instant τ_i. The growth rate at time τ_i is $\gamma(\tau_i)$, while the length of the time interval $[t_{i-1}, t_i]$ is $t_i - t_{i-1} = \Delta t_i$. Hence a uniform flow matching the given one at τ_i will deliver during that time interval a volume of water equal to $\gamma(\tau_i)\Delta t_i$, and this is the quantity by which we estimate the volume actually delivered from t_{i-1} to t_i. Summing these estimates we arrive at the following estimate for the volume added from time c to time d:

$$\gamma(\tau_1)\Delta t_1 + \gamma(\tau_2)\Delta t_2 + \cdots + \gamma(\tau_n)\Delta t_n \,.$$

By dividing the interval into more and smaller subintervals we can improve the approximation, and as the lengths of the subintervals approach zero the estimate approaches as a limit the actual volume added between $t = c$ and $t = d$. In symbols, $U([c, d])$ is the limit of sums

$$\gamma(\tau_1)\Delta t_1 + \gamma(\tau_2)\Delta t_2 + \cdots + \gamma(\tau_n)\Delta t_n$$

as n increases and the Δt_i approach zero, which is to say that

$$U([c, d]) = \int_c^d \gamma(t)\,dt \,.$$

Now for the derivation of the integral by infinitesimals. The time interval $[c, d]$ is made up of infinitesimal subintervals $[t, t + dt]$ on each of which the growth rate $\gamma(t)$ can be regarded as constant. The infinitesimal volume (element of volume) added during time dt is $\gamma(t)\,dt$, so the volume added during the whole time interval from c to d is the sum of these elements, or

$$U([c, d]) = \int_c^d \gamma(t)\,dt \,.$$

The second approach is shorter and simpler, and cannot be said to leave out anything essential. One can (with some mental reservation) accept the argument as it stands, or else regard it as shorthand for the other. In the next section I will use only the briefer mode of derivation.

As an example, suppose that the growth rate from $t = 0$ on is $\gamma(t) = 2t^2$ (a rate which increases with time); between $t = 1$ and $t = 3$ the volume added is

$$\int_1^3 2t^2 \, dt \,.$$

2.3.3   Integrals of velocity and acceleration. A particle that moves along a line at velocity $v(t)$ (that is, whose velocity is given as a function v of time) travels, in an infinitesimal time dt, the infinitesimal distance $v(t)\,dt$. Consequently the distance it goes between time c and time d is

$$L([c,d]) = \int_c^d v(t)\,dt\,.$$

Similarly, if the particle's acceleration is given by $a(t)$, it acquires in time dt the infinitesimal velocity $a(t)\,dt$, and over the time interval $[c,d]$ the total velocity

$$V([c,d]) = \int_c^d a(t)\,dt\,.$$

2.3.4   Integrals of force. In Art. 1.4 the force on a particle was viewed in two different ways: as associated with points of time, and with points of a line. The first way makes the force a function of time, $f_1(t)$; over any time interval $[c,d]$ it imparts to the particle momentum

$$P([c,d]) = \int_c^d f_1(t)\,dt\,,$$

by our now-familiar argument. Here the integral constitutes an expression for the impulse of the force, in which force and time interval, the determinants of impulse, appear explicitly.

The second way has the force a function of location, $f_2(x)$, and as the particle passes from $x = a$ to $x = b$ it acquires kinetic energy

$$K([a,b]) = \int_a^b f_2(x)\,dx\,;$$

the integral is the expression of the work done by the force.

2.3.5   QUESTIONS

Q1. Write integrals to represent the following total quantities. (a) The volume of water that enters a tank during a 6 sec period ending at $t = 11$, in the course of which the flow into the tank is at the rate of $4t + 2$ m^3/sec. (b) The location on the x-axis after 10 sec of a particle that leaves the origin at $t = 0$ and moves at the speed $t/(t^2 + 1)$. (c) (i) The increase in velocity, over a 4 sec interval starting at t_0, of a falling body of mass m (subject to acceleration g, §1.3.5); (ii) the momentum acquired by that body; (iii) the work done upon it by gravity.

Q2. Interpret each of the following integrals as physical quantities in the several ways available. (a) $\int_1^4 (x+1)^2 \, dx$ (b) $\int_0^{t_1} (t^4 + 1) \, dt$ (c) the integral from 0 to 1 of the constant 2.

Q3. (a) Following the model of §2.3.2, derive the integrals of §2.3.3 as limits. (b) Derive the integrals of §2.3.4 both by infinitesimals and as limits.

2.4   INTEGRALS AND GRAPHS OF FUNCTIONS

2.4.1   Area the integral of height. Since our two geometrical instances of total and specific quantity, namely area-and-height and rise-and-slope, have to do with curves which can be regarded as graphs of functions, we can treat functions and geometry together.

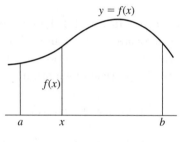

Fig. 2-13

Let the graph of a function $y = f(x)$ be of the kind we admitted when considering area and height (§1.5.2): it is to lie above the x-axis, except where it may touch it (Fig. 2-13). The height of the graph at any x is the ordinate $f(x)$, and height is, as we know, the specific quantity corresponding to the total quantity area-under-the-curve. We therefore expect the area between a and b to be the integral of the height,

$$A([a, b]) = \int_a^b f(x) \, dx \, .$$

That this is actually the case we see by erecting ordinates at any x and $x + dx$ (Fig. 2-14). Since variation in height over the infinitesimal interval can be neglected, the area between these ordinates is the area of a rectangle having sides dx and $f(x)$, or $f(x) \, dx$ (Q3). The whole area is then the sum of these areas, which is the integral $\int_a^b f(x) \, dx$.

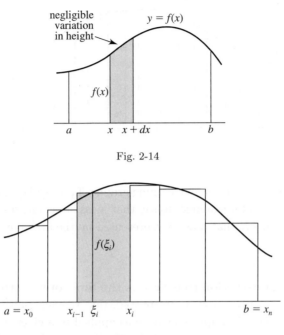

Fig. 2-14

Fig. 2-15

The derivation as a limit goes as follows. For any n, the interval from $a = x_0$ to $b = x_n$ is divided into n subintervals by points of division $x_1, x_2, \ldots, x_{n-1}$, and in each subinterval $[x_{i-1}, x_i]$ a point ξ_i is chosen (Fig. 2-15). The area under the curve between x_{i-1} and x_i is approximated by that of the rectangle of height $f(\xi_i)$ on the same base, which is $f(\xi_i)\Delta x_i$, and consequently the whole area between a and b is approximated by the sum of these areas,

$$S = f(\xi_1)\Delta x_1 + f(\xi_2)\Delta x_2 + \cdots + f(\xi_n)\Delta x_n .$$

This is the area of the polygon bounded by the x-axis, the verticals $x = a$ and $x = b$, and the steps formed by the upper ends of the rectangles. Increasing the number and decreasing the widths of the rectangles will eventually produce a polygon whose area better approximates the desired one, because its upper boundary is closer to the curve (Fig. 2-16); and the approximation can in this way be made as accurate as we may wish. That is to say, the area is the limit of sums S as the widths of the subintervals approach zero, and that limit is $\int_a^b f(x)\, dx$.

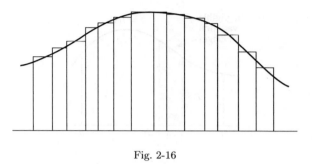

Fig. 2-16

In light of the observations concerning graphs of physical quantities in §1.5.4, it is now clear that, for example, to say that mass is the integral of line density is the same as to say that mass is represented by the area under the graph of line density.

2.4.2   Demonstration that area is the limit of its estimates.   Up to this point I have only asserted, not proved, that the successive estimates produced by our method of approximation as applied to a given specific quantity do approach the corresponding total one to within any degree of accuracy that can be named. In the present case this can easily be demonstrated, provided that the interval from a to b can be divided into a finite number of subintervals along each of which the graph either does not fall or does not rise (Fig. 2-17), i.e. $f(x)$ is monotone in the weaker sense (§1.6.8). Any ordinary function will satisfy this condition. A little thought shows that the different intervals where the function is monotone can be treated separately (Q4); I will accordingly assume $f(x)$ increasing (in the weaker sense), the other case being similar.

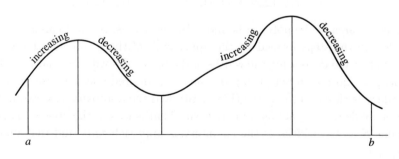

Fig. 2-17

There are graphs of functions which cannot be divided in the way prescribed; examples will be given in §4.2.9. To take account of these a more sophisticated proof is required (see §2.4.3). The present argument follows Newton's *Principia*, Book 1, Lemmas 2 and 3.

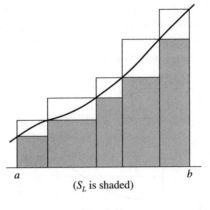

(S_L is shaded)

Fig. 2-18

Let $[a, b]$ be divided into n subintervals as usual. Within each of these we can choose the point ξ_i at which to erect an ordinate. Because the height of the curve increases from left to right, the *shortest* possible rectangle will be formed in each subinterval $[x_{i-1}, x_i]$ if its *left* endpoint x_{i-1} is chosen as ξ_i, and the *tallest* if the *right* endpoint x_i is chosen (Fig. 2-18). Let S_L (L for "lower") be the area estimate obtained by consistently making the former choice, S_U ("upper") that obtained by consistently making the latter:

$$S_L = f(x_0)\Delta x_1 + f(x_1)\Delta x_2 + \cdots + f(x_{n-1})\Delta x_n ,$$

$$S_U = f(x_1)\Delta x_1 + f(x_2)\Delta x_2 + \cdots + f(x_n)\Delta x_n .$$

Then S_L is the least of all possible estimates based on these subintervals, S_U the greatest, so if S is any estimate based on the subintervals, we have

$$S_L \leq S \leq S_U .$$

Furthermore, the area A under the curve also lies between S_L and S_U:

$$S_L \leq A \leq S_U .$$

Now let successive divisions of $[a, b]$ into more and more subintervals be made, in such a way that the length of the longest subinterval approaches zero. At each stage let upper and lower estimates S_U and S_L be formed; also, at each stage let S be any estimate at all, based on an arbitrary choice of the ξ_i. I will show that as the stages succeed one another the difference between S_U and S_L approaches zero. From this it will evidently follow that S approaches A, because S and A are both trapped between the sequence of upper estimates and the sequence of lower estimates, and the terms of these sequences draw closer to one another than any given measure.

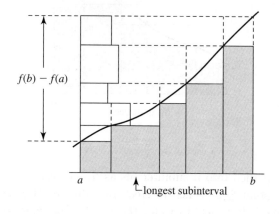

Fig. 2-19

The difference between S_U and S_L is represented in Fig. 2-18 by the area of the unshaded blocks. Imagine sliding them over to the left so that their left sides line up along $x = a$, as shown in Fig. 2-19. The height of the resulting unshaded figure is $f(b) - f(a)$, a number which does not change with successive divisions; its width is nowhere greater than the length of the longest of the subintervals. Hence its area satisfies the inequality

$$S_U - S_L \le [f(b) - f(a)] \times \text{length of longest subinterval}.$$

Since with successive stages the length of the longest subinterval approaches zero, its product by the fixed number $f(b) - f(a)$ does likewise, and so therefore does the smaller quantity $S_U - S_L$. This concludes the proof.

2.4.3 The integral of a function. The foregoing argument can be adapted to show directly that mass is the integral of line density, volume the integral of growth rate, etc. (Q5). It does not in fact depend on geometry, although geometry suggested how it should go. Alternatively, if we simply accept the idea that the nature of the relation between a total and a specific quantity is the same in all cases, and that mass and line density, etc., are such pairs, we can draw the same conclusions on the basis of the geometrical proof.

Suppose now that we have an arbitrary function $f(x)$ (with graph as in §2.4.1), and we aim to define its definite integral $\int_a^b f(x)\,dx$ without reference to geometry or physics, on the grounds that as a law which assigns numbers to numbers f has in itself nothing to do with either, and that intuition is suspect.[4] Clearly we want to say that the integral is the number $\mathscr{S}$ approached as a limit by sums

$$S = f(\xi_1)\Delta x_1 + f(\xi_2)\Delta x_2 + \cdots + f(\xi_n)\Delta x_n$$

when the number of subintervals is increased in the usual way. Here a question arises: *is* there such a number? In the geometrical and physical examples we began not only with a function to be integrated, but also with a total quantity such as A or M; the problem was to show that the sums approached *that* quantity. Now we have to show that all sequences of sums S approach *some* one limit $\mathscr{S}$; then we can define $\int_a^b f(x)\,dx$ to be that number.

Demonstration of the existence of $\mathscr{S}$ from considerations of number alone is possible even without the hypothesis of monotone intervals (§2.4.2), provided that f has a property called *continuity*, which will be defined in Chapter 3 (§3.1.5). It expresses, in the realm of number, something similar to the geometrical idea of a continuous graph. Although the proof is too complex to be given here, it is outlined in §2.4.5.

Its existence established, we call $\int_a^b f(x)\,dx$ the definite integral of the continuous *function* f, without attributing meaning to the function beyond its defining property as a law relating numbers; in this way we define another function $\mathscr{S}$, a function of intervals, which assigns to each interval $[a, b]$ the number

$$\mathscr{S}([a, b]) = \int_a^b f(x)\,dx\,.$$

---

[4]The latter ground presupposes a logical development of number (that is, of the system of real numbers) untainted by geometrical or physical intuition; such a development can be carried out, e.g. by the method of Dedekind (see his *Continuity and Irrational Numbers*, in *Essays on the Theory of Numbers* (1901 and reprints)).

This is a *numerical* "total quantity" function (Q6), whose interpretations in various situations will be the total quantity functions A, M, U, V, etc., with which we are acquainted. Like area and the other quantities, the integral of f can be regarded from the point of view of infinitesimals as a sum of little products $f(x)\, dx$.

The definite integral is wholly determined by the integrand and the interval of integration. The designation of the variable, while useful for placing an integral in context, does not affect its value. Thus for a given f, $\int_a^b f(x)\, dx$ and $\int_a^b f(t)\, dt$ are the same number; if $f = (\ )^2$, this means that

$$\int_a^b x^2\, dx = \int_a^b t^2\, dt\,.$$

The place-holding role of the variable of integration leads to its being called a **dummy** variable.

2.4.4   Additivity of the integral.   A consequence of the definition of the integral, or of the description in terms of infinitesimals, is the following property: for $a < b < c$,

$$\int_a^b f(x)\, dx + \int_b^c f(x)\, dx = \int_a^c f(x)\, dx$$

(Q8; see also §2.10.9.Q3). Its truth is obvious from the interpretation of the integral as area.

2.4.5   Outline of proof of existence of the integral.   There are three steps.

1.   As in §2.4.2, upper and lower sums S_U and S_L are formed for each division of $[a, b]$ into subintervals. Since the function is not assumed monotone, the evaluation points ξ_i used in these sums are not right and left endpoints of subintervals, but instead points at which f has its greatest and least value on each subinterval. Continuity implies that such points exist (§3.7.1).

2.   It is now shown that $S_U - S_L$ approaches zero, by an argument similar to that in §2.4.2, in which however a consequence of continuity called *uniform continuity* (§3.7.12) is used to prove that the quantity corresponding to the height of the pile of blocks in Fig. 2-19 is small enough. (For this step see §3.7.13.Q5.)

3.   Finally, it is shown that any sequence of upper sums S_U approaches a limit $\mathscr{S}$, and that that limit is the same for all sequences. By step 2, the same is then true for lower sums. The proof involves comparing any two sums to a third one formed by using all their points of division together; the existence

of $\mathscr{S}$ follows from a fact about sequences of numbers, that when the terms of a sequence draw near to one another (in a certain precise sense), they are approaching a single number.[5]

2.4.6 QUESTIONS

Q1. Use the area interpretation to find the values of the following integrals. (a) $\int_a^b 3\,dx$ (b) $\int_0^1 x\,dx$ (c) $\int_{-1}^1 \sqrt{1-x^2}\,dx$

Q2. Write an integral for the area of some parabolic figure of the kind shown in Fig. 1-6 (§1.5.4).

Q3 (§2.4.1). Explain carefully why this calculation is analogous to that by which infinitesimal mass was identified as $\lambda(x)\,dx$ in §2.2.7.

Q4 (§2.4.2). Supply an argument.

Q5 (§2.4.3). To show that mass M is the integral of line density $\lambda = f(x)$, proceed as follows. (a) Suppose that two wires have line densities $\lambda_1 = f_1(x)$ and $\lambda_2 = f_2(x)$, and the masses of their corresponding segments $[a,b]$ are respectively M_1 and M_2. Give a physical argument to show that if at every point x in $[a,b]$ $f_1(x) \geq f_2(x)$, then $M_1 \geq M_2$. In other words, of two segments of equal length, if one has greater line density everywhere than the other, its mass is greater. (b) Define S_L, S_U, and S as before, and use (a) to prove $S_L \leq S \leq S_U$ and $S_L \leq M \leq S_U$. (c) The geometrical argument about the blocks can now be applied, because the representation of products (such as $\lambda(x_i)\Delta x_i$) as rectangles is legitimate. Alternatively, calculate $S_U - S_L$ algebraically and show that it is not greater than $[f(b) - f(a)]\Delta x_m$, where Δx_m is the largest of the numbers $\Delta x_1, \Delta x_2, \ldots, \Delta x_n$.

Q6. Show that the numerical total quantity function $\mathscr{S}$ has the property expected of total quantity, that if $f(x)$ is a *constant* function c, then

$$\mathscr{S}([a,b]) = c(b-a).$$

Q7. Explain how the geometrical concept of the area of a curved figure, taken for granted in elementary geometry, could be *defined* in terms of number.

Q8 (§2.4.4). Derive the property.

2.4.7 Rise the integral of slope. Let us now consider the other geometrical representation of total and specific quantity, the pair rise and slope. We begin with a curve which always rises from left to right. At any x it has a certain slope μ, which we will express as a function of x by $\mu = f(x)$. (I am deliberately using the same letter f with a new geometrical meaning, slope not height, since as before it represents the function to be integrated.) From $x = a$

---

[5]The full existence proof can be found in R. Courant, *Differential and Integral Calculus*, Volume I, 2nd edition (1937 and reprints), p. 131, and in books on mathematical analysis.

to $x = b$ the curve rises by $H([a,b])$, the total quantity corresponding to the specific quantity slope, so we expect to have

$$H([a,b]) = \int_a^b f(x)\,dx\,.$$

To verify this we neglect the variation in slope over an infinitesimal interval from x to $x + dx$, so that the curve is regarded as having constant slope $f(x)$ there (Fig. 2-20); then its rise is $f(x)\,dx$, and the integral expression follows.

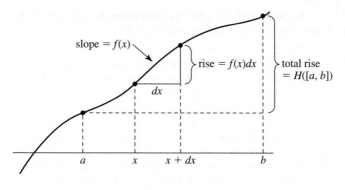

Fig. 2-20

From the other point of view the total rise is obtained as a limit of sums of rises $f(\xi_i)\Delta x_i$ of straight lines matching the curve's slope at points ξ_i (Fig. 2-21).

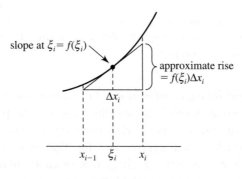

Fig. 2-21

2.4.8   Neglecting infinitesimals.   The preceding argument by infinitesimals is open to objection. Is not the rise, or variation in height, of a curve over an infinitesimal x-interval the very thing we neglected only a moment ago when deriving the area integral (§2.4.1)? If it was negligible then, how can it be the foundation of a total quantity now?

The answer is, briefly, that infinitesimal quantities are negligible not in themselves (which would make them useless), but only in comparison to finite quantities, or, more generally, to quantities to which they bear infinitesimal ratios. In the former situation we were able to neglect the variation in an ordinate because the ordinate entered our area calculation as a finite quantity. For suppose the ordinate to the graph of $y = f(x)$ varies by an infinitesimal amount δ over the interval from x to $x + dx$; we need not distinguish between $f(x)$ and $f(x) + \delta$ in forming the product by dx, because δ enters the calculation only together with $f(x)$, in comparison with which it can be ignored. (This argument has to be modified if $f(x) = 0$.) Another way to see this is to calculate the difference in the product that results from using $f(x) + \delta$ in place of $f(x)$: it is $(f(x) + \delta)\, dx - f(x)\, dx = \delta\, dx$, which bears to the infinitesimal summand $f(x)\, dx$ the infinitesimal ratio $\delta / f(x)$, hence can be neglected in comparison to that summand.

The present case is different: the variation $\mu\, dx = f(x)\, dx$ (remember that $f(x)$ has a different meaning now!) enters the sum as one infinitesimal among equals, so to speak—two such, $f(x_1)\, dx$ and $f(x_2)\, dx$ (taking the infinitesimal x-intervals to be the same) have the finite ratio $f(x_1)/f(x_2)$—hence is no more to be neglected than dx itself, which together with its fellow subintervals constitutes $[a, b]$.

2.4.9   QUESTIONS

Q1.   Write integrals for the rise over the interval $[a, b]$ of (a) a horizontal line; (b) a line of constant slope 2. Evaluate them.

Q2.   Describe and sketch a curve whose rise from $x = 0$ to $x = b > 0$ is given by $\int_0^b x\, dx$.

Q3.   Given a rising curve whose slope $\mu = f(x)$ increases from left to right, draw, for the pair rise and slope, a figure analogous to Fig. 2-18 (which referred to the pair area and height).

2.5   INDEFINITE INTEGRALS

2.5.1   Cumulative quantity expressed as an indefinite integral.
Total quantity is expressed in terms of specific quantity by the definite integral. How is *cumulative* quantity to be expressed? In the case of a wire, we

defined the cumulative mass from 0 to x by $m(x) = M([0, x])$ (§1.6.3), and this can now be written as

$$m(x) = \int_0^x \lambda(x)\,dx\,.$$

Here the upper limit is *variable*. Because the interval of integration $[0, x]$ is not fixed, the expression is called an **indefinite integral** of λ. I say *an* indefinite integral, because the choice of 0 as lower limit is after all quite arbitrary. Any number a will serve (supposing the wire to exist at and beyond a), so we can extend the definition to all integrals of the form $\int_a^x \lambda(x)\,dx$, where a is constant and $x > a$. This is, of course, to extend the notion of cumulative function as well: $\int_a^x \lambda(x)\,dx = M([a, x])$ is the mass of the wire from a to x. Similarly for the other instances of cumulative quantity; for example, $\int_c^t v(t)\,dt$ is the distance covered by a particle from time c to time t.

In the symbol $\int_a^x \lambda(x)\,dx$ the letter x plays two roles. As upper limit it indicates how far along the x-axis integration is to go, while in $\lambda(x)\,dx$ it represents the variable which belongs to that axis. Often a different letter is introduced in one or the other of the two places to avoid confusion (recall the remark at the end of §2.4.3).

2.5.2   The indefinite integrals of a function.   If $f(x)$ is any continuous function (as in §2.4.3) defined for $x \geq x_0$, where x_0 is any number,

$$F(x) = \int_{x_0}^x f(x)\,dx$$

defines a new function of x, an indefinite integral of f.[6] Its value at x can be interpreted as the area under the graph of f from x_0 up to x. Let $x_1 > x_0$; then a different indefinite integral is

$$G(x) = \int_{x_1}^x f(x)\,dx\,.$$

By applying §2.4.4, or simply inspecting Fig. 2-22, we find that for $x > x_1$

$$\int_{x_0}^x f(x)\,dx - \int_{x_1}^x f(x)\,dx = \int_{x_0}^{x_1} f(x)\,dx\,.$$

---

[6]The use of a capital letter for this function of numbers (not intervals), which is an established notation, departs from our practice hitherto (§1.3.4).

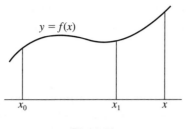

Fig. 2-22

The integral on the right side is a constant; thus

$$F(x) - G(x) = \text{constant:}$$

different indefinite integrals of the same function differ by a constant.

A similar argument provides an expression for a definite integral in terms of an indefinite one: for any a and b with $x_0 < a < b$,

$$\int_a^b f(x)\, dx = F(b) - F(a).$$

The same is true for G (if $x_1 < a < b$); thus *a definite integral is the difference of the values of any indefinite integral at its upper and lower limits of integration* (Q2). This result was obtained for cumulative functions in §1.6.3.

In Chapter 3 (§§3.5.6–7) the definition of indefinite integral and these two propositions will be generalized.

2.5.3   QUESTIONS

Q1. Write a cumulative function for each of the total quantities in Table 1-1 (§1.6.2) as an indefinite integral from x_0 or t_0.

Q2 (§2.5.2). Give the argument, drawing a figure analogous to Fig. 2-22.

Q3. If $x^3 + x + 1$ is an indefinite integral of a function $f(x)$, find $\int_1^2 f(x)\, dx$.

2.6   APPROXIMATE CALCULATION OF LINE DENSITY

2.6.1   Formulation of the problem. The ideas of approximation, limit, and infinitesimal, developed to answer the first part of the question in the introduction to this chapter, prove adequate to the second part as well. While I will take the same course as before, starting with the wire and going on to functions in general, it will be shorter because the tools are at hand.

The line density of a uniform wire at any point x_0 on it is determined by the mass M and length L of any segment, as their quotient:

$$\lambda = \frac{M}{L}.$$

To generalize this determination to an arbitrary wire we assume that the masses of all segments are known—at least, the masses of all small segments containing x_0, which are enough to fix the line density at the single point in question. This means that the values of the total function $M([x_1, x_2])$ are known for such segments $[x_1, x_2]$; or, what is equivalent (§1.6.3), that the cumulative mass $m(x)$ measured from some fixed point a (located to the left of x_0) to any x is known. It will be better to work with the function m than with M; therefore, for the mass of $[x_1, x_2]$ we will write $m(x_2) - m(x_1)$ rather than $M([x_1, x_2])$. From knowledge of such differences of m-values we are to determine $\lambda(x_0)$ for any arbitrary x_0; in other words, to determine the function $\lambda(x)$.

2.6.2 Approximation by successive estimates.

We begin, as we did before (§2.1.2), by approximating the non-uniform by the uniform. Recall that our very conception of line density is based on uniformity: we think of $\lambda(x_0)$ as the line density of the uniform wire that matches the given one at x_0. But this matching wire is unknown; we must proceed from what *is* known, namely the distribution of mass, to find a uniform wire whose line density *approximates* $\lambda(x_0)$.

We have assumed (§1.1.1) that changes in the composition of the wire occur smoothly, not suddenly, along its length. This implies that as we move away from x_0 the wire deviates in character only gradually from the uniform wire that matches it there. Near x_0 the distribution of mass will be quite similar to that of the matching uniform wire, a fact which suggests the following way of estimating $\lambda(x_0)$.

Let $[x_1, x_2]$ be a small interval containing x_0 (possibly as one of its endpoints). On that interval there is a certain mass $m(x_2) - m(x_1)$, which may not be uniformly distributed along it. Imagine taking the same quantity of mass and spreading it uniformly along the interval—in effect, smoothing out the irregularities in the given wire—to form a *uniform* segment of wire $[x_1, x_2]$ with the same mass $m(x_2) - m(x_1)$. Although this uniform wire is probably not the matching wire at x_0, it is likely to resemble it; hence its line density provides an approximation to $\lambda(x_0)$. That line density is

$$\frac{m(x_2) - m(x_1)}{x_2 - x_1},$$

called the **average** or **mean** line density of the original wire over the interval $[x_1, x_2]$.[7] An expression of this kind is called a **difference quotient**, since it is the quotient of a difference of values of a function by the difference of the corresponding values of the independent variable.

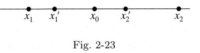

Fig. 2-23

We may hope to improve our estimate of $\lambda(x_0)$ by choosing a shorter interval $[x_1', x_2']$ (Fig. 2-23), because the nearer we are to x_0, the more the wire ought to resemble the matching uniform wire at x_0. The new estimate is the new difference quotient

$$\frac{m(x_2') - m(x_1')}{x_2' - x_1'},$$

the average line density over $[x_1', x_2']$. Seeking a still better estimate we pass to an even shorter interval $[x_1'', x_2'']$, and so on.

2.6.3   Example. Let $a = 0$ and let the cumulative mass function be $m(x) = 0.01x^3 = cx^3$—the number of kilograms of wire from 0 to x. (Here $c = 0.01$ as in §2.1.4.) To estimate the line density at $x_0 = 2$ we calculate the average density over a series of decreasing intervals containing 2. Let us first use intervals centered at 2. For the interval $[0, 4]$ the difference quotient is

$$\frac{m(4) - m(0)}{4 - 0} = \frac{c \cdot 4^3 - c \cdot 0^3}{4 - 0} = \frac{0.64}{4} = 0.16.$$

Similar calculations for smaller intervals produce the results shown in Table 2-2. The numbers are approaching 0.12, whose accuracy as a value for $\lambda(2)$ can be borne out by using other intervals to find upper and lower sequences of estimates, somewhat as we did in §2.1.4 when approximating mass (Q1). It will

---

[7]More will be said about this concept in §§3.7.5 and 3.7.16.

Table 2-2

Interval	Average line density
$[0, 4]$	0.16
$[0.5, 3.5]$	0.1425
$[1, 3]$	0.13
$[1.5, 2.5]$	0.1225
$[1.9, 2.1]$	0.1201
$[1.99, 2.01]$	0.120001

be shown presently that in this way any $\lambda(x_0)$ can be estimated as accurately as we please. The situation is similar to that of §2.1.4, except that the calculation of these estimates is easier.

2.6.4   QUESTIONS

Q1.   Future work will show that in the preceding example $\lambda(x)$ is an increasing function. (a) Assuming this, argue that using intervals of which 2 is the *right* endpoint (such as $[0, 2]$) *underestimates* $\lambda(2)$, while using intervals of which 2 is the *left* endpoint (such as $[2, 4]$) *overestimates* it. (b) Calculate the estimates for the following sequences of intervals, and then apply (a) to specify an interval within which $\lambda(2)$ lies. (i) $[0, 2]$, $[1, 2]$, $[1.9, 2]$, $[1.99, 2]$; (ii) $[2, 4]$, $[2, 3]$, $[2, 2.1]$, $[2, 2.01]$.

Q2.   Can there be a wire, and a point x_0 on it, such that the average line density over *every* small interval containing x_0 overestimates $\lambda(x_0)$?

Q3.   (a) If $m(x) = 1 - (1/x)$ for $x \geq 1$, show that the difference quotient for $[x_1, x_2]$ equals $1/x_1 x_2$. (b) What then is the value of $\lambda(x_0)$ at any x_0?

2.7   DESCRIPTION OF LINE DENSITY AS A DERIVATIVE

2.7.1   Line density as the limit of estimates.   I now assert that $\lambda(x_0)$ is the *limit* of the estimating difference quotients. It does not matter how the successive intervals $[x_1, x_2]$ containing x_0 are chosen, as long as their lengths approach zero. That is,

$$\lambda(x_0) \text{ is the limit of } \frac{m(x_2) - m(x_1)}{x_2 - x_1}, \text{ where } x_1 \leq x_0 \leq x_2, \text{ as } x_2 - x_1$$
approaches zero.

In other words, the line density at x_0 is the limit approached by any sequence of average line densities calculated over intervals containing x_0, provided that the lengths of the intervals diminish so as eventually to become and remain less than any given length. An argument that supports this claim will be made when we examine its analogue in geometry (§§2.9.4 and 2.9.6).

To allow both endpoints of the interval $[x_1, x_2]$ to vary as the interval shrinks results in some complication. For this reason we will employ only intervals of which x_0 is an endpoint—a harmless restriction, since $\lambda(x_0)$ can be determined from *any* sequence of intervals containing x_0 whose lengths approach zero. The other endpoint is denoted $x_0 + \Delta x$, Δx being positive if this endpoint is on the right, negative if it is on the left (Fig. 2-24). The length of the interval is then the absolute value $|\Delta x|$, and we are interested in a limit as this quantity approaches zero. Since it is clearly the same thing for a sequence of numbers which may be positive or negative to approach zero as for the sequence of their absolute values to do so, we can speak more simply of the limit as Δx approaches zero.

Fig. 2-24

With this notation the difference quotient $\dfrac{m(x_2) - m(x_1)}{x_2 - x_1}$ is

$$\frac{m(x_0 + \Delta x) - m(x_0)}{(x_0 + \Delta x) - x_0} \quad \text{if } \Delta x > 0,$$

or

$$\frac{m(x_0) - m(x_0 + \Delta x)}{x_0 - (x_0 + \Delta x)} \quad \text{if } \Delta x < 0.$$

Both of these expressions reduce to

$$\frac{m(x_0 + \Delta x) - m(x_0)}{\Delta x},$$

the ratio of the change in m to the change in x that produces it.

The change in m, as well as the change in x, can be denoted by using Δ:

$$\Delta m = m(x_0 + \Delta x) - m(x_0)$$

—the mass of the interval, if $\Delta x > 0$, or its negative if $\Delta x < 0$. Then the difference quotient will be written

$$\frac{\Delta m}{\Delta x},$$

a concise form in which the value x_0 is not indicated. Introducing an abbreviation for "the limit as Δx approaches zero of ...," we can express the line density at x_0 as follows:

$$\lambda(x_0) = \lim_{\Delta x \to 0} \frac{m(x_0 + \Delta x) - m(x_0)}{\Delta x},$$

or just

$$\lambda(x_0) = \lim_{\Delta x \to 0} \frac{\Delta m}{\Delta x}.$$

The arrow is used with any quantity to mean "approaches (as a limit)," so our statement can also be written this way:

$$\text{as } \Delta x \to 0, \quad \frac{\Delta m}{\Delta x} \to \lambda(x_0).$$

We have now represented the specific quantity $\lambda(x_0)$ as a limit of an expression involving the cumulative function m. In Art. 2.2 we did the opposite, in effect obtaining for $m(b) = M([a, b])$ the formula

$$m(b) = \lim_{\text{all } \Delta x_i \to 0} \lambda(\xi_1)\Delta x_1 + \lambda(\xi_2)\Delta x_2 + \cdots + \lambda(\xi_n)\Delta x_n$$

(§2.2.1). The remarks made in §§2.2.2 and 2.2.3 about the nature and extent of this achievement, and the relation of a limit to the expression whose limit is taken, apply equally here (with the difference mentioned at the end of §2.6.3, that calculation is less difficult).

2.7.2 Line density as a quotient of infinitesimal differentials. In the difference quotient

$$\frac{\Delta m}{\Delta x} = \frac{m(x_0 + \Delta x) - m(x_0)}{\Delta x}$$

let us venture to replace the finite difference Δx by an infinitesimal differential dx. Since the numerator of the resulting fraction is the infinitesimal difference of m-values corresponding to the infinitesimal difference dx of x-values, we call it a differential too, a differential of m, and write[8]

$$dm = m(x_0 + dx) - m(x_0).$$

The difference quotient then becomes a **differential quotient**

$$\frac{dm}{dx} = \frac{m(x_0 + dx) - m(x_0)}{dx}.$$

(The symbol $\dfrac{dm}{dx}$ is read "$dm\,dx$," not "dm over dx," unless it has to be distinguished from the product $dm\,dx$.)

I assume, just as I did when giving an account of the integral by means of infinitesimals, that dx is so small that any departure of the wire from uniformity over the interval between x_0 and $x_0 + dx$ can be neglected. In the former situation this meant that an infinitesimal summand of the *total* quantity, mass, could be expressed as a *product* $\lambda(x_0)\,dx$. At present our interest is in the *specific* quantity, line density, which the uniformity allows us to represent as a *quotient*,

$$\lambda(x_0) = \frac{dm}{dx}.$$

This is the expression for line density in the language of infinitesimals. If for $dx > 0$ we rewrite the equation as $dm = \lambda(x_0)\,dx$ it expresses the relation used before, because dm is the element of mass located on the interval of length dx.

To indicate the value x_0 we can augment the quotient symbol:

$$\left.\frac{dm}{dx}\right|_{x_0} \quad (\text{"}dm\,dx \text{ at } x_0\text{"}) \text{ is the differential quotient at } x_0.[9]$$

We can also replace m by an explicit formula, as in the example which follows.

2.7.3   Example. If $m(x) = cx^3$ as in §2.6.3 we write

$$\frac{d(cx^3)}{dx} \quad \text{or} \quad \frac{d}{dx}\left(cx^3\right).$$

---

[8]With regard to this expression see Q6. In §3.8.8 there is a notion of differential that does not involve infinitesimals.

[9]The notation $\dfrac{dm}{dx}(x_0)$ is also used.

(Either form can be read "$d\,dx$ of cx^3.") The symbol $\dfrac{d}{dx}(cx^3)\Big|_{x_0}$ then has the following interpretation. (I assume $dx > 0$; in the opposite case minor changes are required.)

> An immeasurably tiny interval of infinitesimal length dx begins at the point x_0. On that interval the wire is considered uniform, so that mass there is proportional to length of segment, their ratio being line density. The mass of the interval is the increment dm it contributes to the cumulative mass m, whose expression in terms of x is $m(x) = cx^3$; hence the line density is the ratio $\dfrac{dm}{dx} = \dfrac{d(cx^3)}{dx}$,
>
> or adding a specification of the point x_0, $\dfrac{d}{dx}(cx^3)\Big|_{x_0}$.

2.7.4   Line density as limit and quotient. The derivative.   We now have two representations of the same line density $\lambda(x_0)$: as a quotient of an infinitesimal mass-differential dm by an infinitesimal x-differential dx, and as the limit of quotients of finite mass-differences Δm by finite x-differences Δx. Setting them equal, we have

$$\frac{dm}{dx}\Big|_{x_0} = \lim_{\Delta x \to 0} \frac{m(x_0 + \Delta x) - m(x_0)}{\Delta x},$$

or just

$$\frac{dm}{dx} = \lim_{\Delta x \to 0} \frac{\Delta m}{\Delta x}.$$

Whether defined as differential quotient or limit of difference quotients, this quantity is called the **derivative at x_0 of the function** $m(x)$. The *function* whose value at any x_0 is the derivative at x_0 is the **derivative of the function** $m(x)$ ("derivative" because derived from the given function). Our conclusion can be stated: the function $\lambda(x)$ is the derivative of the function $m(x)$. Forming the derivative of a given function is called **differentiation**; thus $\lambda(x)$ is obtained from $m(x)$ by differentiation, we **differentiate** m to get λ. The independent variable can be specified by saying: differentiate **with respect to** x.

As with the integral, the representation of the derivative as a limit is logically unexceptionable, and can be taken as the definition of the differential symbol

dm/dx. If this is done, dm/dx does not signify a quotient at all—for the *limit* of quotients need not *be* one—but is a single inseparable symbol, standing for the limit of $\Delta m/\Delta x$. Still, one should not forget its other interpretation.

2.7.5   Example of differentiation.   While the twofold account of the derivative is formally analogous to that of the integral, the former is evidently simpler. Instead of the limit of a sum of an ever-greater multitude of products, or an infinite sum of infinitesimal products, there is the limit of one ratio, or a single ratio of infinitesimals. Such simplicity of definition makes direct calculation feasible in many cases of interest.

Consider, for example, the cumulative mass function $m(x) = cx^3$ (with $c = 0.01$) introduced in §2.6.3. To find its derivative at any point x_0 we first form the difference quotient

$$\frac{\Delta m}{\Delta x} = \frac{m(x_0 + \Delta x) - m(x_0)}{\Delta x} = \frac{c(x_0 + \Delta x)^3 - cx_0^3}{\Delta x}$$

and then simplify it, continuing as follows:

$$c \cdot \frac{x_0^3 + 3x_0^2 \Delta x + 3x_0 (\Delta x)^2 + (\Delta x)^3 - x_0^3}{\Delta x} = c \cdot \left(3x_0^2 + 3x_0 \Delta x + (\Delta x)^2\right).$$

Now the derivative is the limit of this as Δx approaches zero,

$$\frac{dm}{dx} = \lim_{\Delta x \to 0} c \cdot \left(3x_0^2 + 3x_0 \Delta x + (\Delta x)^2\right)$$

$$= \lim_{\Delta x \to 0} \left[c \cdot (3x_0^2) + c \cdot (3x_0 + \Delta x)\Delta x\right].$$

Since x_0 is a fixed number, it is clear that as Δx approaches zero so does $c \cdot (3x_0 + \Delta x)\Delta x$ (Q7), while the other term remains unchanged; hence the limit is

$$\frac{dm}{dx} = c \cdot 3x_0^2 = 0.03x_0^2.$$

Remember that the meaning of this statement is only that

$$c \cdot \left(3x_0^2 + 3x_0 \Delta x + (\Delta x)^2\right)$$

approaches $c \cdot 3x_0^2$ as Δx *approaches* zero. The quantity also *equals* $c \cdot 3x_0^2$ if Δx *equals* zero, but that is irrelevant to the notion of limit. In Chapter 3 (§§3.1.5, 3.1.10) I will explain why, in a case like this, it happens that a limiting value can be found by simple substitution. In general it cannot; e.g., in the original difference quotient $\Delta m/\Delta x$ we could not substitute $\Delta x = 0$, because division by zero is forbidden. It was only after dividing through by $\Delta x \neq 0$ that we had the quotient in a form in which $\Delta x = 0$ could be substituted.

If we take $x_0 = 2$, the formula yields the value 0.12:

$$\lambda(2) = \left.\frac{dm}{dx}\right|_2 = 0.12\,,$$

which confirms the apparent limit of the sequence of approximations obtained in §2.6.3. The present value is not approximate, but exact—the limit itself, not an estimate of it or an educated guess at its value. And this is the case not only when $x_0 = 2$, but for any value of x_0. In other words, the whole function $\lambda(x)$, the derivative of the function cx^3, has been determined:

$$\lambda(x) = c \cdot 3x^2\,.$$

With very little effort we have completely solved the problem of finding the specific quantity corresponding to the total quantity $m(x) = cx^3$. The tasks of exact calculation and determination of a function, which appeared so vexing when we wished to obtain total quantity from specific (§2.2.3), have here been easily dispatched by application of the definition of derivative. Far from being an isolated success, this example promises extensive conquest in the realm of derivatives, and—surprisingly—in that of integrals as well.

2.7.6 QUESTIONS

Q1. Read the following derivatives correctly and interpret them as line densities in the manner of §2.7.3. (a) $\left.\dfrac{dx}{dx}\right|_1$ (b) $\dfrac{d}{dx}\left(\dfrac{x}{x+1}\right)$ (c) $\left.\dfrac{d}{dx}[(x+2)^2]\right|_{\sqrt{2}}$

Q2. Write derivatives for the following line densities. (a) The line density at $x = 3$ of a wire whose cumulative mass, measured from $x = 0$, is given by $8x^2$. (b) The line density function of a wire whose cumulative mass increases uniformly from the value 3 at the left end of a 2 m segment to the value 7 at the right end (cf. §2.2.10.Q3(b)).

Q3. Does dm/dx depend on the choice of a (§2.6.1)? Explain.

Q4. Find a point on the wire of §2.7.5 where the line density is 1 kg/m.

Q5. Calculate $\left.\dfrac{d}{dx}(cx^3)\right|_{x_0}$ by means of infinitesimals, first simplifying dm/dx and then neglecting infinitesimal quantities.

Q6. Resolve the following contradiction. According to our calculation of the derivative of $m(x) = cx^3$, $dm = c \cdot 3x^2\, dx$. But we defined dm to be $m(x+dx) - m(x)$, which by direct calculation equals $c \cdot (3x^2\, dx + 3x(dx)^2 + (dx)^3)$.

Q7 (§2.7.5). Make a careful argument.

Q8.   Carry out the limit argument as in §2.7.5 to find the derivatives of the following functions. (a) $m(x) = x$. Interpret the result. (b) $m(x) = x^2$

Q9.   Do Q8(b) in a different way, by finding the limit of the difference quotient for intervals $[x_1, x_2]$ with x_0 the midpoint of each interval; i.e., for $\varepsilon > 0$ let $x_1 = x_0 - \varepsilon$, $x_2 = x_0 + \varepsilon$, and $\varepsilon \to 0$.

2.8   OTHER DERIVATIVES FROM PHYSICS

2.8.1   Generality of the derivative.

Like the integral (§2.3.1), the derivative is applicable to other total-specific pairs than mass and line density, since they can be described by use of functions analogous to $m(x)$. I am now going to review once again the other physical examples from Chapter 1, this time in order to express specific quantities as derivatives of cumulative functions.

2.8.2   Derivative of growing volume with respect to time.

Let water flow into a tank, so that the volume accumulated from a given time c to any time t is $u(t)$. (If the tank is empty at c, i.e. $u(c) = 0$, then $u(t)$ is also the volume of water present in the tank at t.) From time t_0 to time $t_0 + \Delta t$ the volume $u(t_0 + \Delta t) - u(t_0)$ is added. A *uniform* flow would add that volume in the same time interval at a rate given by the difference quotient

$$\frac{\Delta u}{\Delta t} = \frac{u(t_0 + \Delta t) - u(t_0)}{\Delta t}.$$

In the general situation this is only an approximation to the instantaneous rate of flow at any time during the interval, and at t_0 in particular. But the approximation can be improved by shortening the interval, and as its length approaches zero the difference quotient approaches as a limit the actual rate at which the volume is growing at time t_0:

$$\gamma(t_0) = \lim_{\Delta t \to 0} \frac{\Delta u}{\Delta t} = \lim_{\Delta t \to 0} \frac{u(t_0 + \Delta t) - u(t_0)}{\Delta t}.$$

This limit is the derivative of $u(t)$ at t_0, written $\left. \dfrac{du}{dt} \right|_{t_0}$. Dropping the reference to the particular time t_0, we have $\gamma(t) = \dfrac{du}{dt}$, the expression for the function $\gamma(t)$ as the derivative of the function $u(t)$.

From the point of view of infinitesimals, the derivative du/dt is obtained by supposing that over a very brief time interval $[t_0, t_0 + dt]$ the flow can be regarded as uniform. The volume added during that interval is $u(t_0 + dt) - u(t_0) = du$, hence the rate of growth of volume is du/dt. An argument of this kind will suffice for the next physical examples. Note that the subscript 0 is superfluous: we can simply write $du = u(t + dt) - u(t)$.

2.8.3 Derivatives of location and velocity.

Let $l(t)$ be the distance a particle moves along a line towards the right from time c to any time t. This is the location of the particle relative to its location at time c; if $x(t)$ is its coordinate at any time t, $l(t) = x(t) - x(c)$. Hence if $x(c) = 0$, $l(t)$ is the same as $x(t)$.

In any infinitesimal time dt the particle moves a corresponding infinitesimal distance $dl = l(t + dt) - l(t)$; consequently its velocity at time t is

$$v(t) = \frac{dl}{dt}.$$

Thus instantaneous velocity is the derivative of distance covered, or relative location.

In the same way, a particle which acquires velocity $v(t)$ between time c and time t—or, assuming $v(c) = 0$, which *has* velocity $v(t)$ at time t—increases its velocity by dv between t and $t + dt$, hence has acceleration

$$a(t) = \frac{dv}{dt}$$

at any time t.

2.8.4 Force as a derivative.

To derive the force on a particle from its momentum we differentiate with respect to time. If the momentum at time t is $p(t)$, the particle acquires momentum dp in time dt, and the force on it at t is therefore

$$f_1(t) = \frac{dp}{dt}.$$

Since $p = mv$, this implies Newton's law $f_1 = ma$ (Q4).

We can also differentiate the kinetic energy $k(x)$ of the particle with respect to location x to obtain the force as a function $f_2(x)$. Between x and $x + dx$ the particle adds kinetic energy dk, therefore

$$f_2(x) = \frac{dk}{dx}.$$

2.8.5   QUESTIONS

Q1.   Write derivatives to represent the following specific quantities. (a) The rate at which the volume of water in a tank grows if the volume present at any time t is $(t+1)^4$. (b) (i) The acceleration at $t=1$ of a particle whose velocity function is $t^3 + 5t$; (ii) the velocity one minute after $t=0$ of a particle whose x-coordinate is proportional to time. (c) (i) The force which has cumulative impulse $2t$ on a particle; (ii) the force acting on a particle of mass 5 kg whose velocity at any x is $2x$.

Q2.   Interpret each of the following derivatives as physical quantities in the several ways available. (a) $\dfrac{dx^2}{dx}$ (b) $\dfrac{dt}{dt}\Big|_3$

Q3.   (a) Following the model of §2.8.2, represent the derivatives in §2.8.3 as limits. (b) Represent the derivatives in §2.8.4 both as quotients of infinitesimals and as limits.

Q4.   Argue by infinitesimals that $f = ma$ follows from $p = mv$.

2.9   DERIVATIVES AND GRAPHS OF FUNCTIONS

2.9.1   The derivative of a function.

The derivative of an arbitrary function $y = f(x)$ is defined in the same way as the derivative of $m(x)$ and the other cumulative functions already considered, namely as the limit of the difference quotient

$$\frac{f(x + \Delta x) - f(x)}{\Delta x}$$

as $\Delta x \to 0$. (As in §§2.8.3–4 I have omitted the subscript 0; x represents *any arbitrary* x-value, held fixed as $\Delta x \to 0$.) The numerator of the quotient is the change in y corresponding to the change Δx in x, hence can be denoted Δy as well as Δf. The differential notations for the limit are dy/dx and df/dx:

$$\frac{dy}{dx} = \lim_{\Delta x \to 0} \frac{\Delta y}{\Delta x}\,.$$

As the limit of a ratio of changes $\Delta y/\Delta x$, or as a ratio of infinitesimal changes dy/dx, the derivative at any x measures **the rate of change of y with respect to x** at that point. In this descriptive phrase "rate" is in general to be taken not in the temporal sense, but in the sense of comparative quantity: the meaning is y *per* x. Thus the equation $\lambda(x_0) = \dfrac{dm}{dx}\Big|_{x_0}$ asserts that line density at x_0 is the rate of change of cumulative mass m with respect to location x at the point $x = x_0$—the mass added (dm) per locational change

(dx) at that place on the wire. Of course, if the independent variable is time, the derivative does measure temporal rate of change.

The derivative of $f(x)$ can also be denoted by $f'(x)$ ("f-prime of x"), a notation which emphasizes that it too is a function. The result of differentiating $m(x) = cx^3$, for example (§2.7.5), can be written $m'(x) = c \cdot 3x^2$. Sometimes y' is written instead of f' for the derivative of $y = f(x)$.

2.9.2　Derivative of a constant and of $f(x)+$ constant. Two closely related fundamental facts about differentiation are these:

(1)　the derivative of a constant is zero;
(2)　adding a constant to a function does not change its derivative.

The proof of (1) is requested in Q3; the proof of (2) goes as follows. Given a function $f(x)$ and a constant c, we are to show that $f(x) + c$ has the same derivative as $f(x)$. The two functions do have the same difference quotient, for

$$\frac{[f(x + \Delta x) + c] - [f(x) + c]}{\Delta x} = \frac{f(x + \Delta x) - f(x)}{\Delta x} ;$$

consequently their derivatives agree.

2.9.3　QUESTIONS

Q1.　Suppose $y = f(x)$ and $f'(x) = x^2 - 1$. (a) What is $\left.\dfrac{dy}{dx}\right|_1$? (b) If also $z = f(w)$, what is z'?

Q2.　(a) What is the derivative of $cx^3 + 1$? (b) Can the derivative of $c(x + 1)^3$ be found in the same way?

Q3.　(a) Prove (1) above. (b) Can (2) be deduced from (1)?

2.9.4　Interpretation of the derivative as slope. Slope the derivative of cumulative rise. The slope at x of the graph of $y = f(x)$ is given by its derivative there, $f'(x)$. In order to see this, first suppose f increasing, and consider Fig. 2-25. We assume that from x to $x + dx$ the graph rises uniformly, i.e. can be treated as a straight line over the infinitesimal interval (cf. §2.4.7); its slope there, and at x in particular, is then "rise over run," dy/dx, as claimed.

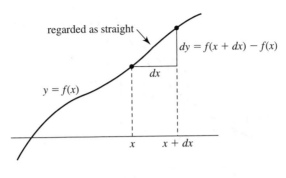

Fig. 2-25

A consequence is that slope is the derivative of cumulative rise, the companion proposition to the assertion of §2.4.7 that rise is the integral of slope. For the rise of the graph from a given x-value a to any $x > a$ is $h(x) = f(x) - f(a)$ (Fig. 2-26). Then rise $h(x)$ is obtained from ordinate $f(x)$ by adding a constant (namely $-f(a)$), and hence (§2.9.2(2)) $h'(x) = f'(x)$, which is the slope at x.

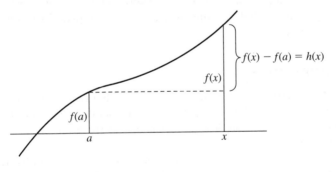

Fig. 2-26

From the point of view of limits, to say that slope is the derivative is to say that it is $\lim_{\Delta x \to 0} \Delta y / \Delta x$. As Fig. 2-27 shows, $\Delta y / \Delta x$ is the slope of the secant line[10] through points P and Q. As Δx approaches zero, Q approaches P along the curve, and secant PQ turns about P so as to approach the tangent line at P (see §2.9.6 for details). The slope of the secant therefore approaches that of

---

[10] A **secant** is a straight line through two points of a curve. The name secant ("cutting") is intended to distinguish it from the tangent line at P, which even if it crosses the curve does not do so at an angle; however, a secant can also be a tangent.

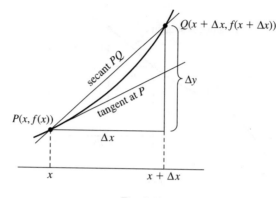

Fig. 2-27

the tangent, which is the slope of the curve at P. Thus the slope of the curve is the limit of the difference quotient.

It is not hard to see that the differential and limit arguments just given are valid whether the change in x (dx or Δx) is positive or negative, and whether the graph is rising, falling, or level at x (Q6). The last two cases mean that the initial hypothesis that f is increasing can be dropped. I will return to this point in §2.10.8.

2.9.5 QUESTIONS

Q1. By §2.7.6.Q8(b) the derivative of x^2 is $2x$. Find the equation of the tangent line to the parabola $y = x^2$ at $x = 3$, and draw the parabola and tangent.

Q2. What is the equation of the tangent line to the graph of $y = f(x)$ at the point (x_0, y_0) on it?

Q3. A particle leaves $x = 0$ at $t = 0$ and travels in the positive direction at velocity $v(t) = t^2$. Graph the velocity on tv-axes, and interpret (a) the slope of the graph at t, (b) the area under it from 0 to t.

Q4. "Slope is the rate of change of ordinate"—is the meaning of this quite clear?

Q5. Interpret (1) and (2) of §2.9.2 geometrically.

Q6 (§2.9.4). Draw figures like Figs. 2-25 and 2-27 to illustrate various cases. Show that the sign of dx does not affect the sign of dy/dx.

Q7. Can every increasing function $f(x)$ be interpreted as the cumulative rise of a curve from some point $x = a$?

2.9.6 Demonstration that the secant approaches the tangent. The meaning of the statement that secant PQ in Fig. 2-27 "approaches" the tangent at P as Q approaches P is that the angle between the lines approaches

zero as the distance of Q from P does the same. This is an assertion about curves in general rather than graphs of functions, and can be established without reference to slope or coordinates. The concept of tangency I take to be understood; similarly, in §2.4.2, to which this section corresponds, the meaning of "area" was assumed known. In the next section there will be a comment on these assumptions.

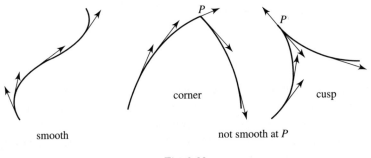

Fig. 2-28

Two further hypotheses are appropriate, as in §2.4.2. As the curve there was continuous, here it is to be **smooth**, which means that it has a continuously turning tangent: the *heading*, or direction of motion, of a traveler along it, as indicated by a tangent arrow, varies without sudden jumps (Fig. 2-28; Q2). ("Continuously turning" includes non-turning—a straight line is smooth.) And as the graph there could be divided into ascending and descending segments, here we assume that any finite portion of the curve can be divided into finitely many sections each of which either is straight or bends only one way (to the right or to the left, from the traveler's point of view; Fig. 2-29). Ordinary continuous curves have these properties (except where smoothness fails at corners and cusps, Fig. 2-28); certain extraordinary ones do not (§4.2.9).

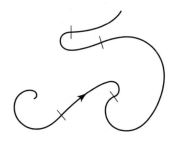

Fig. 2-29

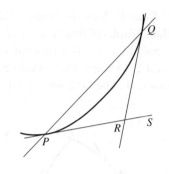

Fig. 2-30

By taking Q close enough to P we can ensure that the stretch of curve from P to Q belongs wholly to a section as described. If the section is straight, secant and tangent actually coincide as Q approaches P within the section. If the section is curved, then for Q near P the situation is as illustrated in Fig. 2-30. Because the curve is smooth, as Q moves to P tangent QR turns continuously into the position of tangent PS. As Q *approaches* P, then, $\angle QRS$, the angle between the tangents, must approach zero. Then the smaller $\angle QPS$, which is the angle between secant PQ and the tangent at P, approaches zero as well.

2.9.7 Intuition and definition. The argument of §2.4.2 to prove that the area of a curved figure is the limit of areas of polygons, and the argument here to prove that the slope of a curve is the limit of slopes of secants, have in common that they introduce artificial simplifying hypotheses and rest upon unexamined intuitive notions, in the one case of area and in the other of tangency. In the interest of rigorous mathematical reasoning it is better first to *define* area and slope as limits, and then to investigate the conditions under which the limits exist (see §2.9.9). This approach is fully justified only when the intuition has been shown preliminary evidence that the thing defined (*definitum*) will be the thing that is to be defined (*definiendum*). Although our arguments may be considered to have accomplished this preparatory work, I will not pursue the matter of rigorous definition in geometry; but the issue will come up again in Chapter 4 (§4.4.4) and in Chapter 7 (§7.1.1).

2.9.8 QUESTIONS

Q1. Is the preceding argument valid if P is a boundary point between two of the sections of the curve?

Q2. Suppose that the cusp in Fig. 2-28 is formed by arcs of circles tangent at P. Is there a tangent line at P? Discuss.

2.9.9   Differentiability.   The derivative brings to our attention a fact hardly suggested by the integral: that a limit may fail to exist. Even a function whose graph is a smooth curve (§2.9.6) can give rise to a difference quotient which fails to approach any number as $\Delta x \to 0$. Take, for example, the function $y = x^{1/3}$. At $x = 0$ the difference quotient is

$$\frac{\Delta y}{\Delta x} = \frac{(0 + \Delta x)^{1/3} - 0^{1/3}}{\Delta x} = \frac{(\Delta x)^{1/3}}{\Delta x} = \frac{1}{(\Delta x)^{2/3}},$$

which becomes larger and larger as $\Delta x \to 0$. A look at the graph (Fig. 2-31) explains this behavior: at the origin P the tangent is a vertical line, to which no slope can be assigned; and as $Q \to P$ the slope of secant PQ, which is $\Delta y/\Delta x$, grows without any limit.[11]

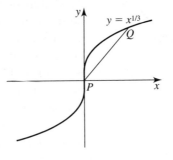

Fig. 2-31

Where a function *has* a derivative, where the difference quotient does approach some number as a limit, the function is said to be **differentiable**. A function differentiable everywhere is a **differentiable function**. The graph of such a function has a non-vertical tangent at every point; if in addition it is a smooth curve, the function too can be called smooth.[12]

All the functions we have to do with will be differentiable, and even smooth, except possibly at isolated points. This will in fact often be taken as a fundamental *hypothesis*, of special significance for physical applications: thus when

---

[11] In §3.1.2 the meaning of the term "limit" will be extended so as to include cases like this one. The extension will not affect the definition I am about to give, which refers only to limits which are numbers.

[12] An analytic definition of smoothness appears in §3.1.7, and an example to show that smoothness does not follow from differentiability in §4.2.9.

I say that line density is the derivative of (cumulative) mass, or acceleration of velocity, I am tacitly assuming that the wire or motion is of such a character that a certain limit exists at every point. As an approximation to reality this assumption is remarkably effective (Q2).

In terms of infinitesimals, differentiability of $y = f(x)$ means that the quotient dy/dx is a *number*; that is, the ratio of dy to dx has a definite numerical value. In the example of $y = x^{1/3}$ at $x = 0$, $dy = (dx)^{1/3}$ and therefore $dy/dx = 1/(dx)^{2/3}$. This quotient of a number by an infinitesimal cannot be a number; if it makes sense at all it must be *larger* than any number (infinite), because its reciprocal is the infinitesimal $(dx)^{2/3}$. Thus there is no derivative at 0. (See also Q1).

2.9.10   One-sided derivatives.

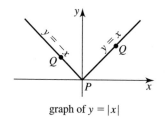

graph of $y = |x|$

Fig. 2-32

The failure of differentiability at a point need not be total, as the example $y = |x|$ illustrates. Although its graph has no tangent at the origin P, where there is a corner (Fig. 2-32), secants PQ approach (in fact, coincide with) the line $y = x$ as $Q \to P$ from the right, and similarly approach $y = -x$ as $Q \to P$ from the left; hence we can say that at P there is a "right-hand tangent," namely $y = x$, and a "left-hand tangent," $y = -x$. This phenomenon is reflected in the difference quotient

$$\frac{\Delta y}{\Delta x} = \frac{|0 + \Delta x| - |0|}{\Delta x} = \frac{|\Delta x|}{\Delta x},$$

which equals 1 if $\Delta x > 0$ and -1 if $\Delta x < 0$. If Δx is confined to positive values as it approaches 0, $\Delta y/\Delta x$ approaches 1; if Δx is confined to negative values, $\Delta y/\Delta x$ approaches -1; there are a "right-hand limit" (limit from the right side) and a "left-hand limit," although there is no limit simply. While the function is not differentiable at 0, we can nevertheless say that at that point it has a **right-hand derivative** equal to 1 and a **left-hand derivative** equal to -1.

2.9.11   QUESTIONS

Q1.   Make the argument about $y = |x|$ using infinitesimals.

Q2.   Why is it reasonable to expect that a physical quantity such as line density can be well represented by a smooth differentiable function?

2.9.12   Height the derivative of cumulative area.   To conclude the introduction of the derivative it remains to establish its role in our other geometrical pair, area and height, as a complement to §2.4.1. Given a curve above the x-axis, let the area under it from a to x be $f(x)$, and its height at x, $h(x)$ (Fig. 2-33). (Note that the curve is the graph of h, not of f.) The function $f(x)$ being the cumulative quantity here, and $h(x)$ the specific, we expect the latter to be the derivative of the former.

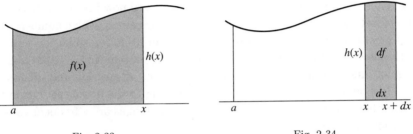

Fig. 2-33   Fig. 2-34

The demonstration by infinitesimals is as follows. We neglect the variation in height between x and $x + dx$ (cf. §2.4.1); then the differential of area df is the area of a rectangle of infinitesimal width dx and height $h(x)$ (Fig. 2-34), so that $df = h(x)dx$, or

$$h(x) = \frac{df}{dx}.$$

For the limit account we can assume as in §2.4.2 that the height is increasing over an interval from x to $x + \Delta x$. From Fig. 2-35 we have

$$h(x)\Delta x \leq \Delta f \leq h(x + \Delta x)\Delta x$$

and therefore

$$h(x) \leq \frac{\Delta f}{\Delta x} \leq h(x + \Delta x).$$

As $\Delta x \to 0$, $Q \to P$ and $h(x + \Delta x) \to h(x)$; therefore also $\Delta f/\Delta x \to h(x)$, i.e. $h(x) = f'(x)$.

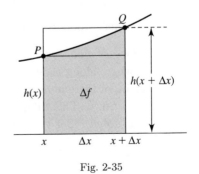

Fig. 2-35

2.9.13   QUESTION

Q1.   Reformulate Q4 of §2.9.5 to fit the present situation, and answer the revised question.

2.10   SIGNED QUANTITIES AND THE INTEGRAL AND DERIVATIVE IN GENERAL

2.10.1   Negative numbers admitted.   Until now I have arranged matters so that most quantities could be described by using positive numbers. The purpose of this artificial limitation was to permit attention to be concentrated on the new concepts of calculus. It is now to be removed, so that the full range of functions may come within the scope of integration and differentiation. Although little intrinsic interest can belong to the details set out in this article, they deserve attention for the reason that signs are a frequent cause of confusion and error.

2.10.2   Sign of the independent variable and its increments.   An independent variable, whether x or t or some other, is in general free to assume any numerical value, positive, negative, or zero. A function such as $y = x^2$ is defined for all values of x; a wire may extend to any location on the x-axis; growth or motion may occur at any time t, as well before the clock reads 0 as after it. Of course, circumstances often limit the domain within which a given variable can take values—the function $y = \sqrt{x}$ requires $x \geq 0$, a physical wire has finite length, growth or motion may begin and end. And it is sometimes convenient, especially when the variable is time, to use non-negative values only. But nothing in the mere idea of a numerical variable demands this.

The only effect upon the integral $\int_a^b f(x)\, dx$ and the derivative $\left. \dfrac{dy}{dx} \right|_{x_0}$ of allowing arbitrary values of x is that the interval $[a, b]$ and the point x_0 can be located anywhere along the axis.

A *change* in, or increment of, the independent variable also may be positive, negative, or zero.[13] This affects the derivative only in its definition, as has already been noted (§2.9.4): when we write $\Delta y/\Delta x$ or dy/dx we must allow

---

[13] "Increment" is regularly used to mean any change, despite its primary sense of "increase."

Δx or dx to be negative as well as positive, if we are not to adopt a one-sided view of the state of affairs at a point (§2.9.10).

With the integral something new is found. Given an interval $[a, b]$, we defined the integral of $f(x)$ *from a to b*, $\int_a^b f(x)\,dx$, as a sum of terms $f(x)\,dx$, where $dx > 0$, or a limit of sums of terms $f(\xi_i)\Delta x_i$, where $\Delta x_i > 0$. In so doing we were thinking of x as increasing from a to b, marking off subintervals as it grew. Suppose instead we similarly form the integral *from b to a*, regarding x as decreasing over the interval; then dx and Δx_i are negative (the latter because now $x_{i-1} > x_i$). Nothing else about the sums changes, so if we write $\int_b^a f(x)dx$ for the new integral we evidently have

$$\int_b^a f(x)\,dx = -\int_a^b f(x)\,dx\,.$$

It is further natural to define

$$\int_a^a f(x)\,dx = 0$$

for any a. These two steps extend the definition of the definite integral to allow any values of its upper and lower limits independently—their designation as "upper" and "lower" now referring to the integral symbol only.

As a result the *indefinite* integral $\int_a^x f(x)\,dx$ is now defined for $x \leq a$ as well as for $x > a$; and we can if we wish let the lower rather than the upper limit vary:

$$\int_x^a f(x)\,dx = -\int_a^x f(x)\,dx\,.$$

As one would expect, intervals of *time* are rarely considered to be traversed in the negative direction. Intervals of *space*, however, are often thought of that way, e.g. when the motion of a particle is described. In particular, if a positively directed force $f(x)$, represented as usual by a positive function, acts on a particle while it moves from x_1 to x_2, the work done by the force is defined to be $\int_{x_1}^{x_2} f(x)\,dx$ regardless of the relative location of x_1 and x_2. This will be negative if $x_1 > x_2$, which can happen even if no other force affects the particle. If, for example, the block of §1.4.1 loses its cord and slides freely across the table so as to compress a fixed spring from x_1 to x_2 (Fig. 2-36), the work done on the block by the expansive force of the spring during the compression is accounted negative. (Its effect on kinetic energy will be considered in §2.10.7.)

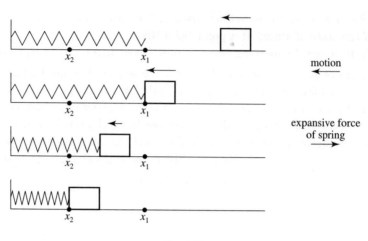

Fig. 2-36

2.10.3   The integral as a function of directed intervals. In §2.4.3 we wrote $\mathscr{S}([a,b]) = \int_a^b f(x)\,dx$, assuming $a < b$. Now we can write, say, $\mathscr{S}_{12} = \int_{x_1}^{x_2} f(x)\,dx$, whether $x_1 < x_2$ or not; then $\mathscr{S}_{21} = -\mathscr{S}_{12}$, and if $F(x) = \int_a^x f(x)\,dx$, $\mathscr{S}_{12} = F(x_2) - F(x_1)$ (Q3). The new $\mathscr{S}_{12}$ is a function of the variable **directed interval** or **displacement** *from* x_1 *to* x_2, rather than of the simple interval between the two points; it represents the change in the indefinite integral F from x_1 to x_2. This idea will be applied to certain total quantities in Arts. 5.3 and 7.1.

2.10.4   Sign of the dependent variable and its increments. A. Negative cumulative quantity. Many functions, including ones which represent total, cumulative, and specific quantities, take on negative values at some values of the independent variable. No real difficulty is caused by this complication, but one must be alert to interpret signed quantities correctly.

The matter of sign arises most simply in a case such as that of the wire. Positive numbers represent both mass and line density, but cumulative mass $m(x)$ is defined as a positive number only for x to the right of the given starting-point a. In order to extend the definition of $m(x)$ to all values of x we can rely on the formula

$$m(x) = \int_a^x \lambda(x)\,dx$$

Fig. 2-37

(§2.5.1), understanding it in the light of §2.10.2. For $x < a$ we then have

$$m(x) = -\int_x^a \lambda(x)\,dx = -M([x, a]):$$

$m(x)$ is the *negative* of the mass of the interval from x to a (Fig. 2-37). If the name "cumulative mass" is retained for the extended function m, it is to be interpreted in this way. Since the fundamental formula

$$M([x_1, x_2]) = m(x_2) - m(x_1)$$

still holds for all x_1 and x_2 with $x_1 \leq x_2$, the relation

$$\lambda(x) = \frac{dm}{dx}$$

is valid everywhere (Q5).

Similarly, let a particle traveling towards the right have location function $x(t)$, and suppose that at time t_0 it is at the origin, $x(t_0) = 0$; *before* t_0 its coordinate $x(t)$ is negative, but the distance covered from t_1 to t_2 is always

$$L([t_1, t_2]) = x(t_2) - x(t_1),$$

and at all times we have

$$v(t) = \frac{dx}{dt}.$$

The notion of x-coordinate extends to all t the idea of cumulative distance to the right of the origin.

2.10.5   B. Negative total and specific quantity. Electric charge.

A more complex situation occurs when total and specific quantities take negative values. As an example, let us consider a continuous linear distribution of **electric charge**, which accumulates like mass but comes in two species, positive and negative. They are symmetrical in their properties, and are measured in the same unit, the **coulomb**, with a sign attached; equal charges of opposite sign in the same location cancel one another's effects. When calculating total charge, therefore, we add "algebraically," allowing for cancellation, as Fig. 2-38 illustrates. (Each sign there represents 1 coulomb, but the distribution is supposed continuous.) *Total* quantity has become *net* quantity. As for specific quantity, the **line density of charge**, it can be positive, negative, or zero, depending on the nature of the charge deposit at a given location. Thus in Fig. 2-38 the line density might be $+10$ coulomb/m near a and -10 coulomb/m near b, and at certain intermediate points it would be 0 coulomb/m.

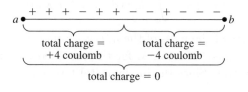

Fig. 2-38

There are corresponding phenomena in the case of the moving particle. Let it be permitted to move towards the left as well as the right; we count a leftward shift as negative distance covered,[14] and attribute negative velocity to motion towards the left. Here we must distinguish *velocity* from *speed*: velocity is a signed quantity, speed its absolute value. Cancellation occurs when the particle moves first one way and then the other during an interval of time. The analogue of the spatial distribution of charge depicted in Fig. 2-38, for example, is a motion to the right 3 m, then left 1 m, then right 2 m, left 2 m, right 1 m, and left 3 m; the farthest point reached is 4 m to the right of the starting-point, and the particle ends up where it began, having achieved zero net change of position. Its velocity has switched sign five times.

2.10.6   Functions and graphs. A. Negative integrand. Net area.

The definition of the integral requires no modification when the integrand is allowed to be negative, but its meaning of total quantity has to be understood algebraically as net signed quantity. Assuming that $a < b$, in $\int_a^b f(x)\,dx$ the infinitesimal summand $f(x)\,dx$ has the same sign as $f(x)$; hence positive terms are found where $f(x) > 0$, negative where $f(x) < 0$, and if $f(x)$ changes sign within the interval $[a, b]$ there will be cancellation within the sum. The same

---

[14]That is, negative distance-covered (distance covered in the negative direction), not negative-distance covered: "distance," like "mass," is intrinsically positive.

remark applies to the sums

$$f(\xi_1)\Delta x_1 + f(\xi_2)\Delta x_2 + \cdots + f(\xi_n)\Delta x_n$$

used in the limit definition. If $a > b$, $dx < 0$ everywhere and all signs are reversed (§2.10.2).

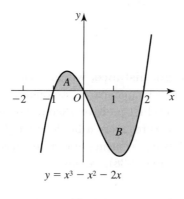

$$y = x^3 - x^2 - 2x$$

Fig. 2-39

Where $y = f(x)$ assumes negative values its graph lies below the x-axis. Figure 2-39 illustrates this in the case of

$$y = x^3 - x^2 - 2x = (x+1)x(x-2),$$

a function which is negative for $x < -1$ and for $0 < x < 2$. In these regions the notion of area under the curve is meaningless, but the more general idea of area *between the curve and the x-axis* can be applied. What the integral does is to calculate the net quantity of such area over a given interval, counting area above the axis ("under the curve") as positive and area below the axis as negative. Area A in Fig. 2-39 is $\frac{5}{12}$ and area B, $\frac{8}{3}$ (as will follow from results of Chapter 3); hence

$$\int_{-1}^{0} (x^3 - x^2 - 2x)\,dx = \frac{5}{12}, \qquad \int_{0}^{2} (x^3 - x^2 - 2x)\,dx = -\frac{8}{3},$$

and $$\int_{-1}^{2} (x^3 - x^2 - 2x)\,dx = -\frac{9}{4}.$$

In each integral $dx > 0$ everywhere, while the sign of the ordinate $x^3 - x^2 - 2x$ depends on x; where it is negative, the product $(x^3 - x^2 - 2x)\,dx$ represents the *negative* of the area of a thin rectangle extending downward from the x-axis to the curve.

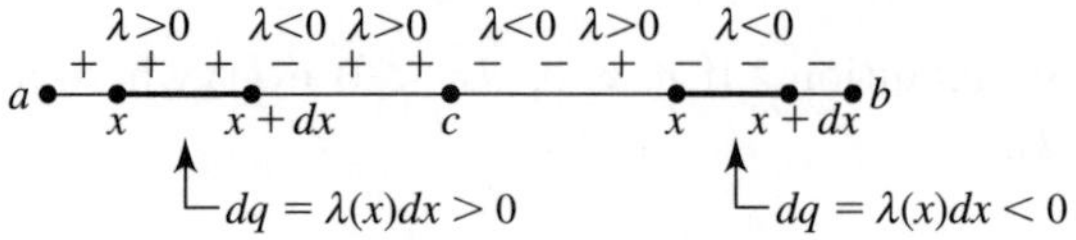

$$\lambda > 0 \quad \lambda < 0 \; \lambda > 0 \quad \lambda < 0 \; \lambda > 0 \quad \lambda < 0$$

$$+ \; + \; + \; - \; + \; + \; - \; - \; + \; - \; - \; -$$

$$a \bullet\!\!-\!\!\bullet\!-\!\!\bullet\!\!-\!\!\bullet\!-\!\!\bullet\!\!-\!\!\bullet\!\!\bullet\, b$$

$$\quad x \quad x+dx \quad c \qquad x \quad x+dx$$

$$\llcorner dq = \lambda(x)dx > 0 \qquad \llcorner dq = \lambda(x)dx < 0$$

Fig. 2-40

2.10.7 B. Net charge, distance, work. The example of charge distribution from §2.10.5 can now be understood as follows. On the interval $[a, b]$ the line density of charge $\lambda(x)$ is first positive, then negative, and so on (Fig. 2-40). Let cumulative charge measured from a be denoted $q(x)$; total charge is then the sum of its differentials dq, elements of charge whose sign depends on that of $\lambda(x)$. In particular (cf. Fig. 2-38), we have

$$q(c) = \int_a^c \lambda(x)\,dx = 4\,, \quad q(b) - q(c) = \int_c^b \lambda(x)\,dx = -4\,,$$

$$\text{and} \quad q(b) = \int_a^b \lambda(x)\,dx = 0\,.$$

A graph of λ displays total charge as signed area determined by the line density curve (Fig. 2-41).

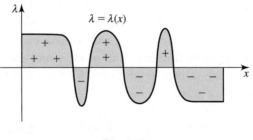

Fig. 2-41

The analogous motion of a particle (§2.10.5) is described in exactly the same way, x, $\lambda(x)$, and $q(x)$ being respectively replaced by t, $v(t)$, and $x(t)$. The sign of $dx = v(t)\,dt$ is that of $v(t)$; dx is an infinitesimal shift rightward if velocity is positive, leftward if negative, and net change of place from t_1 to

t_2 is

$$\Delta x = x(t_2) - x(t_1) = \int_{t_1}^{t_2} v(t)\, dt\,.$$

In the graph of $v(t)$ (Fig. 2-42) area above the axis represents distance covered in the positive direction; area below, distance covered in the negative direction.

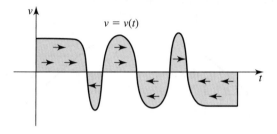

Fig. 2-42

As a final example, consider the work $W = \int_{x_1}^{x_2} f(x)\, dx$ done by a force $f(x)$ on a particle. So far we have always assumed the force positively directed, and we have seen that even so the work will be negative if $x_1 > x_2$ (§2.10.2). Let us now admit force that acts in the negative direction, and represent it by negative numbers; then the sign of $f(x)\, dx$ will be positive or negative according as $f(x)$ and dx are alike or different in sign. In other words, positive work results when force and motion agree in direction, negative work when they are opposed. If in the course of the particle's motion both situations occur, the integral expression for W represents the net quantity of work done.

In order to relate work in this generalized sense to kinetic energy, we first repeat the experiments of §§1.4.6 and 1.4.8, this time allowing force to oppose motion, and then reason as in §2.3.4. The conclusion is as before, that (net) work equals the (net) increment of kinetic energy, $W = K = \Delta k = \Delta \frac{1}{2}mv^2$. In particular, negative work W done on a particle reduces its kinetic energy by $|W|$.

2.10.8   C. Negative derivative and slope. Net rise. We have now removed the restriction to *positive* functions, functions whose graphs lie above the x-axis, that was originally introduced to simplify integration. Its counterpart for differentiation was the restriction to *increasing* functions, the graphs of which rise as x increases. As I have already mentioned (§2.9.4), nothing hinders us from abandoning this condition too. The interpretation of the derivative as slope then makes it clear that a function is increasing on an interval where its derivative is positive, decreasing where the derivative is negative; for if we take $dx > 0$ the sign of dy is the same as that of dy/dx, so y increases or decreases with the increase in x according as the derivative is positive or negative (Fig. 2-43). At a point where the derivative is zero there is a horizontal tangent, but the function can be increasing, decreasing, or neither (Q11); this fact will be investigated in Chapter 3 (Art. 3.9).

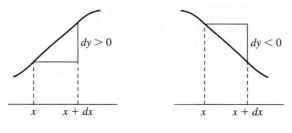

Fig. 2-43

From the preceding discussion of the integral it is apparent that with the admission of negative slope comes the reinterpretation of rise, the total quantity corresponding to slope, as *net* rise.

It should now be sufficiently clear how to interpret and extend the other ideas and results of these first two chapters so that they apply to all kinds of functions. Familiarity with the integral and derivative in general will be improved by the next chapter, in which a variety of functions will be examined and our central arguments recast.

2.10.9   QUESTIONS

Q1.   A wire has charge density $\lambda(x) = x$ coulomb/m. What is the total charge (a) on $[-1, 1]$? (b) on $[-2, 1]$?

Q2.   A 2 kg particle traveling towards the left on the x-axis at a speed of 5 m/sec passes the origin at $t = 0$. What is its momentum at $t = 3$?

Q3.   Show that the additive property of the integral stated in §2.4.4 is valid regardless of the values of a, b, and c (that is, even if we do not have $a < b < c$).

Q4.   (a) When the block in Fig. 2-36 rebounds from the spring, what is the net amount of work that has been done on it? (b) At what speed does it leave the spring?

Q5 (§2.10.4).   Establish the two formulas.

Q6.   As an analogue to electric charge, but closer to the idea of mass, consider the weight under water of a rod composed partly of metal and partly of wood. Develop the analogy.

Q7. The acceleration of a particle on the x-axis, like its velocity and the force upon it, is called positive when directed towards the right and negative when directed towards the left. (a) Give the history of velocity and acceleration for the block in Fig. 2-36 from the time it strikes the spring until it leaves it. (b) Do the two sides of the equation $f = ma$ always agree in sign? (c) Do velocity and acceleration always agree in sign?

Q8. A particle at rest at $x = 0$ sets out on a smooth journey along the non-negative x-axis and eventually returns to rest at the origin. With no further information, what can you say about the behavior of the functions $v(t)$ and $a(t)$?

Q9. Extend the accounts of growth in §§2.3.2 and 2.8.2 to allow for negative quantity.

Q10. Suppose that, as in Fig. 2-39, the graph of a function $y = f(x)$ encloses area both above and below the x-axis. Express as an integral the *total* quantity of such area, as opposed to the *net* quantity of signed area.

Q11 (§2.10.8). Give examples of the three possibilities.

Q12. Newton's third law of motion tells us that if particle A exerts a force on particle B, particle B exerts a force equal in magnitude and opposite in direction on particle A. Let A be the block in Fig. 2-36 and let B be the tip of the spring (a little inflexible part that comes into contact with A). (a) If the force exerted by B on A at any x is $f(x)$ (towards the right, or positive), what is the force exerted by A on B at x? (b) In view of (a), why doesn't B keep accelerating towards the left after being set in motion? (c) Show that the work A does on B during the compression of the spring equals the initial kinetic energy of A.

CHAPTER 3

Differentiation and Integration: I

INTRODUCTION

Having defined the integral and the derivative and demonstrated their capacity to represent total and specific quantities in terms of one another, we have dealt with the problem of calculus in its conceptual and descriptive aspect. We have also shown that approximate numerical values of the integral and derivative can be calculated by successive estimation. The remaining question, how to perform the operations of differentiation and integration upon given functions, has only been touched on. The solution to this part of the problem, so far as it can be solved in general, goes broadly as follows.

First, the derivatives of certain simple functions are found by direct application of the definition, as has already been done in one instance (§2.7.5).

Secondly, the derivatives of complex functions which can be constructed out of the simpler ones are determined, by establishing how differentiation behaves with respect to the operations employed in construction. These operations, which include "composition" and "inversion" as well as addition and other arithmetic means of combination, give rise to a wide class of functions, all of which are brought within the reach of differentiation. The achievement is in fact still greater, since relating differentiation to the other operations is important for theoretical purposes.

Thirdly, it is observed that the derivatives of the members of this class of functions form a great part of the same class again; that is, very many functions of interest appear as derivatives of members of the class. Now the operations of differentiation and integration are essentially *inverse* to one another, as we have seen—together they provide the correspondence back and forth between cumulative and specific functions. Therefore the integral of a function can be found by looking up the function whose derivative it is. By this device we are spared the labor of direct calculation of the complicated limit that defines an integral.

The principal aim of this chapter is to fill in the preceding outline in a preliminary way, deferring to Chapters 4 and 6 parts of each of the three steps. The first two are treated in Arts. 3.1–3.4, and the third in Arts. 3.5–3.7; in the latter articles will be found the Fundamental Theorem of Calculus, in which the inverse relation of integral and derivative is formulated, and the closely related mean value theorem. Article 3.8 considers repeated differentiation, and Art. 3.9 presents a leading application of the derivative, the theory of maximum and minimum values of functions.

3.1 LIMITS AND CONTINUITY

3.1.1 Evaluation of limits. To apply the definition of the derivative we need to know how to evaluate limits. Besides direct examination, there are two basic ways to determine a limit: as the value of a so-called *continuous* function at the point where its limit is taken, and as a combination of known limits. After reviewing and generalizing the meaning of limit I will present both of these.

3.1.2 Limit of a sequence and limit of a function. Suppose that a sequence of x-values $x_1, x_2, x_3, \ldots$ approaches a number a as a limit, a situation which we symbolize by writing $x_n \to a$. For example, the sequence of partial decimal representations of $\sqrt{2}$, namely 1.4, 1.41, 1.414, 1.4142, ..., approaches $\sqrt{2}$, because the first term of the sequence differs from $\sqrt{2}$ by not more than 0.1, the second by not more than 0.01, the third by not more than 0.001, and so on (Q1).

Given a function $y = f(x)$, we can consider the corresponding sequence of y-values $f(x_1), f(x_2), f(x_3), \ldots$. If this sequence approaches a limit l, we say that l is its limit *as* $x_1, x_2, x_3, \ldots$ approaches a: as $x_n \to a$, $f(x_n) \to l$. With the x-sequence as in our example, let the function be $y = x^2$; then the y-sequence is $(1.4)^2$, $(1.41)^2$, $(1.414)^2$, $(1.4142)^2, \ldots$, or 1.96, 1.9881, 1.999396, 1.99996164, \ldots, which appears to be approaching the limit 2. In fact it can be deduced from what we have said about the x-sequence that the nth term of the y-sequence differs from 2 by less than $3/10^n$ (Q9).

Even if we are not given any particular sequence of x-values approaching a, we can speak of the limit of the *function* $f(x)$ as $x \to a$. To say that this limit has the value l means that $f(x)$ will be as near as we please to l, provided only that x is near enough to a; in other words, all values $f(x)$ (not just certain values $f(x_n)$) corresponding to x-values that lie near enough to a differ arbitrarily little from l. This is abbreviated

$$\lim_{x \to a} f(x) = l$$

(the limit symbol is also written $\lim_{x \to a}$). In speaking of x *near* a we do *not* include the value $x = a$ itself: we do not require that $f(a)$ equal l, or even be defined. The reason for this convention will become clear shortly (§3.1.7).

Returning to the function $y = x^2$, we have

$$\lim_{x \to \sqrt{2}} x^2 = 2\,,$$

because x^2 will be as near as we please to 2 if x is near enough to $\sqrt{2}$; the proof is quite easy (Q6).

Another use of the words "limit" and "approach" occurs in describing what happens when a quantity grows "beyond all bounds"—so as to exceed, in the positive or the negative direction, any number that may be specified. The sequence $x_1 = 10$, $x_2 = 100$, $x_3 = 1000$, $x_4 = 10000, \ldots$ has no limit, in the sense in which we have used the word up to now; but we say that it approaches **infinity**, or has limit infinity, meaning by these phrases *only* that the terms of the sequence eventually become and remain larger than any given number. This assertion is symbolized $x_n \to \infty$; here the sign for "infinity" has no independent meaning—in particular, it certainly does not stand for a number—but only serves to remind us of the use of the word in such phrases as have just been employed.[1] Similarly, the sequence $x_1 = -10$, $x_2 = -100$, $x_3 = -1000, \ldots$ is said to approach minus infinity,[2] $x_n \to -\infty$. A limit which is not ∞ or $-\infty$ is called finite.

This usage can be applied in the obvious ways to the behavior of the independent or dependent variable in a function. For example,

---

[1] The symbol is also used to specify an interval, as in the expression $0 \le x < \infty$, which means that x is non-negative.

[2] Or, sometimes, "negative infinity"; but -10 is better read "minus ten" than "negative ten."

(i) $\lim\limits_{x \to \infty} 1 + \dfrac{1}{x} = 1$,   (ii) $\lim\limits_{x \to 0} \dfrac{1}{x^2} = \infty$,   (iii) $\lim\limits_{x \to -\infty} x^3 = -\infty$

—explanations are called for in Q3. As a special case, we can express the limit of a sequence $x_1, x_2, x_3, \ldots$ as

$$\lim_{n \to \infty} x_n.$$

Here the sequence is being regarded as a function s defined for positive integers $n = 1, 2, 3, \ldots$ by $s(1) = x_1$, $s(2) = x_2$, $s(3) = x_3$, and in general $s(n) = x_n$. As the variable n (which can assume only positive integral values) grows beyond all bounds, $s(n) = x_n$ approaches the indicated limit. Thus $x_n \to a$ and $\lim_{n \to \infty} x_n = a$ mean the same thing.

3.1.3   The epsilon-delta definition. The meaning of $\lim_{x \to a} f(x) = l$ can be put formally as follows. Given any positive number ε (however small it may be), there is some positive number δ (which may have to be very small, depending on the size of ε) such that *if* x is within distance δ of a, *then* $f(x)$ will be within distance ε of l. More briefly: for any $\varepsilon > 0$ there exists $\delta > 0$ such that if $|x - a| < \delta$ then $|f(x) - l| < \varepsilon$. This definition, especially in the latter version, frees the concept of limit (§2.2.2) from the vagueness of such language as "approach," "as near as we please," and "near enough." For examples of its application see §2.7.6.Q7 and Q2 & Q6 below. There are similar definitions for the cases when a or l or both are replaced by ∞ or $-\infty$: see Q3 & Q8.

3.1.4   QUESTIONS

Q1 (§3.1.2).   Show that this follows from the meaning of the decimal representation.

Q2.   In each of the following cases, what do you think the limit is? What do you mean by that? How might you prove that it is as you say (cf. Q6)?   (a) $\lim\limits_{x \to 2} (x + 2)^2$
(b) $\lim\limits_{x \to 0} \dfrac{1}{1 - x}$

Q3 (§3.1.2).   (a) (i) How large must x be for $1 + (1/x)$ to be within 0.001 of 1? (ii) How close to 0 must x be for $1/x^2$ to be greater than 1000? (iii) How large in the negative direction must x be for x^3 to be less than -1000?   (b) Explain the meaning of each of the three limit statements.

Q4.   Is it true that $\lim\limits_{x \to 0} \dfrac{1}{x} = \infty$?

Q5.   Explain the difference in meaning between $\lim\limits_{x \to \infty} \dfrac{1}{x}$ and $\lim\limits_{n \to \infty} \dfrac{1}{n}$.

Q6.   Let ε be any small positive number. (a) Explain why the following statements are equivalent: (i) x^2 differs from 2 by less than ε; (ii) $2 - \varepsilon < x^2 < 2 + \varepsilon$. (b) Find a neighborhood of $\sqrt{2}$ (§2.2.2) such that if x is in its interior, the statements in (a) hold. (c) Find a positive number δ so small that if x differs from $\sqrt{2}$ by less than δ, the statements in (a) hold.

Q7.   Restate in terms of neighborhoods the definition of $\lim_{x \to a} f(x) = l$ given in the third paragraph of §3.1.2.

Q8 (§3.1.3).   Give the formal definitions for the cases involving infinite limits.

Q9 (§3.1.2).   Prove this.

3.1.5   Continuity.   In §3.1.2 the value of $\lim_{x \to \sqrt{2}} x^2$, if we had not known it, could have been found by substitution:

$$\lim_{x \to \sqrt{2}} x^2 = (\sqrt{2})^2 = 2 \,.$$

How general is this method? It succeeds whenever

$$\lim_{x \to a} f(x) = f(a) \,,$$

in which case[3] the function $f(x)$ is said to be **continuous** at a; a **continuous function** is a function continuous everywhere. A function whose graph is a single continuous line having no breaks or gaps is continuous. If you imagine x approaching a (from either side), and picture how this motion is transmitted to the y-axis via the graph, you will see why this is so (Fig. 3-1). An argument independent of motion can also be made (Q3).

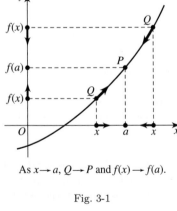

As $x \to a$, $Q \to P$ and $f(x) \to f(a)$.

Fig. 3-1

If we are considering a function f only on an interval $[a, b]$ of values of x, we regard it as continuous there if (i) it is continuous at all interior points of the interval, and (ii) as x approaches each endpoint *from within the interval*, $f(x)$ approaches its value at that endpoint.

---

[3]Here a and $f(a)$ are numbers; the expression $f(a)$ would be meaningless if either a or the limit were ∞ or $-\infty$, since these cannot be values of either variable.

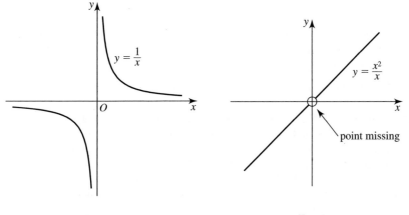

Fig. 3-2   Fig. 3-3

Of the several ways in which a function can fail to be continuous, two should be noticed here. First, it may happen that f is not defined at a, and $\lim_{x \to a} f(x)$ does not exist or is infinite. This is the case, for example, when $f(x) = 1/x$ and $a = 0$: $f(0)$ is not defined, and clearly $1/x$ does not approach any one finite value as $x \to 0$ (Fig. 3-2). Secondly, $\lim_{x \to a} f(x)$ may exist and be finite although $f(a)$ does not exist. Let $f(x) = x^2/x$ and $a = 0$: $f(0)$ is not defined ($0/0$ is meaningless), but at every non-zero x we have $f(x) = x$ (Fig. 3-3). Since the existence of $f(x)$ at $x = 0$ is irrelevant to the limit of the function as $x \to 0$ (§3.1.2), as far as that limit is concerned $f(x)$ is indistinguishable from the continuous function x, and therefore

$$\lim_{x \to 0} \frac{x^2}{x} = \lim_{x \to 0} x = 0 \,.$$

By §3.1.3, continuity of $y = f(x)$ at a means that for any $\varepsilon > 0$ there exists $\delta > 0$ such that if $|x - a| < \delta$ then $|f(x) - f(a)| < \varepsilon$; that is, if $|\Delta x| < \delta$ then $|\Delta y| < \varepsilon$, where $\Delta x = x - a$ is the increment by which x differs from a, and Δy is the corresponding increment by which y differs from $f(a)$. In other words, by making the change in x small you can make the change in y small. (See also Q6.) The analogous statement for infinitesimals is simply that an infinitesimal change in x produces an infinitesimal change in y: to dx corresponds dy. This was assumed in §2.7.2 and afterwards.

As I remarked in §2.4.3, continuity is the property of a function upon which the analytical proof of the existence of the definite integral is based.

3.1.6   QUESTIONS

Q1.   Give a geometrical reason why $y = \sqrt{4 - x^2}$ is continuous for $-2 \leq x \leq 2$.

Q2.   Discuss the continuity of $1/(x - 1)^2$.

Q3 (§3.1.5).   Supply the argument.

Q4.   Show that if f is continuous at a so is $|f|$ (the function whose value at x is $|f(x)|$).

Q5.   Let $f(x)$ and $g(x)$ be continuous functions. If $f(r) = g(r)$ for all rational numbers r (numbers that can be expressed as fractions), what follows?

Q6.   (a) Let $f(x)$ be defined in a neighborhood of a. Show in detail from §3.1.3 that continuity of f at a is equivalent to the following condition: for any neighborhood V of $f(a)$ there is a neighborhood U of a such that if x is in U, $f(x)$ is in V. (b) Extend (a) to the case in which $f(x)$ is defined only on an interval of which a is an endpoint.

3.1.7   Continuity and differentiability.

The significance of the two kinds of discontinuity shows itself as soon as we try to calculate the limit of a difference quotient

$$\frac{f(x_0 + \Delta x) - f(x_0)}{\Delta x}$$

as $\Delta x \to 0$. Here Δx plays the role of the independent variable: the difference quotient is a function of Δx, and the derivative $f'(x_0)$ (if it exists) is the sought-for limit of this function. It was pointed out in §2.7.5 that the limit cannot be found simply by substituting $\Delta x = 0$, for at that value of Δx the difference quotient is undefined. If the discontinuity of the difference quotient is of the first kind, there is no finite limit, i.e. $f(x)$ is not differentiable at x_0; if it is of the second kind, the limit exists and equals $f'(x_0)$. In the latter case we may be able to find the derivative in the same way as we calculated $\lim\limits_{x \to 0} \dfrac{x^2}{x}$.

If a function is differentiable at a point, it is continuous there. For let $y = f(x)$ be differentiable at x_0, so that

$$\lim_{\Delta x \to 0} \frac{f(x_0 + \Delta x) - f(x_0)}{\Delta x}$$

exists and is finite. This can be rewritten

$$\lim_{x \to x_0} \frac{f(x) - f(x_0)}{x - x_0}.$$

Since the denominator of the fraction approaches 0 as $x \to x_0$, the numerator must do the same (otherwise the fraction would be arbitrarily large for values of x near x_0); i.e. $f(x) \to f(x_0)$ as $x \to x_0$, which means that f is continuous at x_0.

Differentiability of a function on an interval $[a, b]$ is defined analogously to continuity on an interval (§3.1.5): the function is to be differentiable at all interior points, and at each endpoint the one-sided derivative (§2.9.10) taken from within the interval is to exist.

A *smooth* function, defined in §2.9.9 as a differentiable function having a smooth graph, can now be characterized as a **continuously differentiable** function—a differentiable function whose derivative is continuous. For continuity of the derivative means that slope varies continuously with respect to x, and this evidently means that direction varies continuously. (A *proof* of the latter equivalence will have to wait until the connection between slope and direction is fully explained (§4.4.9.Q9).) By the preceding result, $f(x)$ will surely be continuously differentiable if its derivative $f'(x)$ is itself differentiable; this latter condition is then sufficient to ensure smoothness of a graph.

3.1.8   QUESTION

Q1. If a function is continuous at a point, is it differentiable there?

3.1.9   Operations with limits.

Suppose that two functions of x, $f(x)$ and $g(x)$, approach finite limits l and m respectively as $x \to a$:

$$\lim_{x \to a} f(x) = l, \quad \lim_{x \to a} g(x) = m.$$

By applying arithmetic operations to f and g we can form new functions:

$$f(x) + g(x), \quad f(x) - g(x), \quad f(x)g(x), \quad \text{and} \quad \frac{f(x)}{g(x)} \text{ (where } g(x) \neq 0).$$

What are the limits of these new functions as $x \to a$? The answer is as simple as can be: they are, respectively,

$$l + m, \quad l - m, \quad lm, \quad \text{and} \quad \frac{l}{m} \text{ (if } m \neq 0)$$

—the limits of the sum, difference, product, and quotient of two functions are the sum, difference, product, and quotient of the limits of the functions. Special cases that often occur are obtained by taking one of the given functions to be constant; in particular, the limit of $-g(x)$ is $-m$, and of $1/g(x)$, $1/m$.

The truth of these assertions is fairly obvious. The first, for example, can be explained as follows: if by taking x close enough to a we can make $f(x)$ as close as we please to l, and $g(x)$ as close as we please to m, surely we can make $f(x) + g(x)$ as close as we please to $l + m$ (for details, see Q5). The others are similarly justified.

Another plausible relation is this (Q6):

(1)   if $f(x) \le g(x)$ for all x near enough to a, then $l \le m$.

All of these rules apply if a is replaced by ∞ or $-\infty$ (and "near enough to a" in (1) by the corresponding condition); in particular, they apply to limits of sequences. If l or m is infinite they do not apply (or even make sense) as stated, but may indicate correct results if suitably interpreted (Q7).

3.1.10   Construction of continuous functions.   From the rules for operations with limits it follows that *the sum, difference, product, and quotient of continuous functions are continuous* (the last where the denominator is non-zero). For assuming f and g continuous, we conclude from $f(x) \to f(a)$ and $g(x) \to g(a)$ that $f(x) + g(x) \to f(a) + g(a)$, and likewise for the other operations. This fact enables us to identify many functions as continuous, and hence to recognize many limits as suitable for evaluation by substitution. In particular, since the function $y = x$ is obviously continuous, so is $x^2 = x \cdot x$, and $x^3 = x \cdot x^2$, and in general any power x^n; and since any constant function is also continuous, so are monomials such as $5x^n = 5 \cdot x^n$, and polynomials such as $3x^5 - 4x^2 + 2$; and also quotients of polynomials, where the denominator does not vanish.[4] Hence, e.g.,

$$\lim_{x \to 1} \frac{3x^5 - 4x^2 + 2}{11x^6 + 4x^3 - 6} = \frac{3(1)^5 - 4(1)^2 + 2}{11(1)^6 + 4(1)^3 - 6} = \frac{1}{9}.$$

3.1.11   QUESTIONS

Q1.   Evaluate the following limits.   (a) $\lim\limits_{x \to -3} \dfrac{x}{x - 3}$ (b) $\lim\limits_{t \to 0} \dfrac{t^2 + 1}{t^2 - 1}$ (c) $\lim\limits_{x \to a} \dfrac{1}{x^2 - 1}$

(d) $\lim\limits_{x \to 5} (x - x)$ (in two ways)   (e) $\lim\limits_{n \to \infty} \dfrac{n}{n + 1}$ (first divide numerator and denominator by n)

---

[4]To **vanish**, in mathematical jargon, means to become or take the value zero.

Q2. If f, g, and h are continuous, is $f - \dfrac{g}{h}$ continuous?

Q3. Name a familiar function, known to be continuous on geometrical grounds, whose continuity cannot be deduced from §3.1.10.

Q4. Suppose that as $x \to a$, $f(x) \to l$ and $g(x) \to m$. (a) Show that if $\dfrac{f(x)}{g(x)} \to 1$, then $l = m$. (In Newtonian language: quantities ultimately in the ratio of equality are ultimately equal.) (b) Show that the converse is false.

Q5 (§3.1.9). Show that $f(x) + g(x)$ will be within ε of $l + m$ if $f(x)$ is within $\varepsilon/2$ of l and $g(x)$ is within $\varepsilon/2$ of m.

Q6 (§3.1.9). Prove (1).

Q7 (§3.1.9). Discuss some cases.

Q8. *L'Hospital's rule.* Suppose that $f(x)$ and $g(x)$ are differentiable at $x = a$, $f(a) = g(a) = 0$, and $g'(a) \neq 0$. Prove that

$$\lim_{x \to a} \frac{f(x)}{g(x)} = \frac{f'(a)}{g'(a)} \,.$$

(For applications see §3.4.4.Q7; the rule is generalized in §8.2.10.Q18.)

3.2 DERIVATIVE OF A POLYNOMIAL

3.2.1 Derivative of a constant and of x^n. Any constant function $f(x) = c$ has derivative zero, as was noted in §2.9.2(1).

The derivative at any x of the function $f(x) = x$, if it exists, is the limit as $\Delta x \to 0$ of the difference quotient

$$\frac{f(x + \Delta x) - f(x)}{\Delta x} = \frac{x + \Delta x - x}{\Delta x} = \frac{\Delta x}{\Delta x} \,.$$

For $\Delta x \neq 0$, $\Delta x / \Delta x = 1$; therefore, by the argument near the end of §3.1.5,

$$\lim_{\Delta x \to 0} \frac{\Delta x}{\Delta x} = \lim_{\Delta x \to 0} 1 = 1 \,.$$

Consequently the function is differentiable and its derivative is the constant function 1, $dx/dx = 1$.

In the future I will suppress reference to the argument just cited from §3.1.5; it is to be understood that $\Delta x \neq 0$ in the difference quotient, so that we can simply say, for example, $\Delta x / \Delta x = 1$. I will also frequently omit to mention that the *existence* of a derivative is at issue, since the question of differentiability is settled when a derivative is calculated.

The derivative of x^2 is the limit of

$$\frac{(x + \Delta x)^2 - x^2}{\Delta x} = \frac{x^2 + 2x\Delta x + (\Delta x)^2 - x^2}{\Delta x}$$

$$= \frac{2x\Delta x + (\Delta x)^2}{\Delta x}$$

$$= 2x + \Delta x.$$

Here x is arbitrary, but fixed as $\Delta x \to 0$; then $2x$ is constant as Δx varies, and $2x + \Delta x$ is a first-degree polynomial in the variable quantity Δx (e.g., if $x = 2$, $2x + \Delta x = \Delta x + 4$).[5] Therefore it is continuous as a function of Δx (§3.1.10), and its limit can be evaluated by substitution of the limiting value of Δx:

$$\lim_{\Delta x \to 0} 2x + \Delta x = 2x + 0 = 2x.$$

Thus

$$\frac{d}{dx}(x^2) = 2x.$$

Similarly, the derivative of x^3 is the limit of

$$\frac{(x + \Delta x)^3 - x^3}{\Delta x} = \frac{x^3 + 3x^2\Delta x + 3x(\Delta x)^2 + (\Delta x)^3 - x^3}{\Delta x}$$

$$= 3x^2 + 3x\Delta x + (\Delta x)^2,$$

a second-degree polynomial in the variable Δx, and this limit is

$$3x^2 + 3x(0) + (0)^2 = 3x^2.$$

Thus

$$\frac{d}{dx}(x^3) = 3x^2.$$

Our results for the first few powers of x can be summarized as follows (with the usual convention that[6] $x^0 = 1$).

---

[5] A polynomial **in** a variable u is a sum of powers of u with coefficients (including a "constant" term) which do not depend on u. In the present case u is Δx, and one of the coefficients (the "constant" term $2x$) depends on x. When considering the variation of Δx, we hold x fixed.

[6] If $x = 0$, x^0 is undefined, but the derivative of x^1 is defined and equals 1.

Function	Derivative
x^1	$1 \cdot x^0$
x^2	$2 \cdot x^1$
x^3	$3 \cdot x^2$

The natural conjecture is that

$$\frac{d}{dx}(x^n) = nx^{n-1}$$

for every positive integer n. This is correct, and can be proved in the same way as the special cases, once the difference quotient

$$\frac{(x + \Delta x)^n - x^n}{\Delta x}$$

has been reduced to manageable form.

3.2.2   Limit of the difference quotient for x^n.   One way to accomplish the reduction is to examine how the nth power $(n > 1)$ of any binomial $a + b$, $(a + b)^n$, is expanded by multiplication. We write out the n factors,

$$(a + b)(a + b)(a + b) \cdots (a + b);$$

form products by choosing one term, either a or b, from each factor, and multiplying the chosen terms together; and then add all the products. That is how we calculate $(a + b)^3$, for example:

$$(a + b)^3 = (a + b)(a + b)(a + b)$$

$$= aaa + baa + aba + aab + bba + bab + abb + bbb$$

$$= a^3 + 3a^2b + 3ab^2 + b^3\,.$$

Observe that:

(i)   When we choose a from *every factor* the resulting product is a^n.

(ii)   When we choose a from *every factor but one* the products are $baa \cdots a$, $aba \cdots a$, etc.; there are n such products, each of them equal to $a^{n-1}b$. So together they contribute $na^{n-1}b$ to the expansion.

(iii)   When we choose the a's and b's in any other way, every product includes at least two b's. Therefore b^2 can be factored out of the part of the expansion remaining after (i) and (ii), which remainder then takes the form

$$b^2 \cdot (\text{a sum of terms of the form } a^p b^q, \ p \text{ and } q \geq 0).$$

Putting together (i), (ii), and (iii), we have

$$(a + b)^n = a^n + na^{n-1}b + b^2 \cdot (\text{sum of terms } a^p b^q).$$

Applied to the difference quotient, this gives[7]

$$\frac{(x + \Delta x)^n - x^n}{\Delta x} = \frac{x^n + nx^{n-1}\Delta x + (\Delta x)^2 \cdot (\text{sum of terms } x^p(\Delta x)^q) - x^n}{\Delta x}$$

$$= nx^{n-1} + \Delta x \cdot (\text{polynomial in } \Delta x).$$

As $\Delta x \to 0$ the polynomial in parentheses approaches some number l, so Δx times it approaches $0 \cdot l = 0$. Hence the difference quotient approaches nx^{n-1}.

3.2.3   QUESTIONS

Q1.   Find the derivatives of the following functions. (a) 0 (b) 1 (c) t (d) x^4 (e) t^{11} (f) $x^3 + 4$

Q2.   Find the derivatives. (a) $\left. \dfrac{dx^5}{dx} \right|_2$ (b) $\left. \dfrac{d}{dt}(t^2 - 1) \right|_{10}$

Q3.   Find the slope of the parabola $y = x^2$ at $x = 3$.

Q4.   (a) Show that the function

$$y = \begin{cases} 0, & x \leq 0 \\ x^2, & x \geq 0 \end{cases}$$

is continuously differentiable (§3.1.7), and find its derivative. (b) Is that derivative a differentiable function?

Q5.   Prove that if n is an integer > 1 and $x > 0$ then $(1+x)^n > 1+nx$. (See also §3.9.9.Q8.)

Q6.   *Binomial theorem.* (a) Show that the coefficient of $a^{n-i}b^i$ in the expansion of $(a + b)^n$ (after terms have been collected) equals the number of different sets of i things each that can be selected from among n things. For example, the coefficient of ab^2 in $(a + b)^3$ is 3, which is the number of different sets of two things each that

---

[7]In the second expression, if $p = 0$, x^p means 1, even if $x = 0$.

can be selected from three things. (b) Show that for $i > 0$ that number is

$$\frac{n(n-1)(n-2)\cdots(n-i+1)}{1\cdot 2\cdot 3\cdots i}.$$

Then

$$(a+b)^n = a^n + \frac{n}{1}\,a^{n-1}b + \frac{n(n-1)}{1\cdot 2}\,a^{n-2}b^2$$

$$+ \frac{n(n-1)(n-2)}{1\cdot 2\cdot 3}\,a^{n-3}b^3 + \cdots + b^n \quad (n+1 \text{ terms in all})\,.$$

3.2.4 Derivative of a product of a function by a constant, of a sum of functions, and of a polynomial.

A polynomial

$$a_m x^m + a_{m-1}x^{m-1} + \cdots + a_1 x + a_0$$

is constructed out of constants $a_0, a_1, \ldots, a_m$ and powers of x. Using what has just been proved we can differentiate it in two steps.

First, we show that *the derivative of a constant times a function is the constant times the derivative of the function,*

$$\frac{d}{dx}\,cf(x) = c\frac{d}{dx}\,f(x)\,.$$

Here is the proof:

$$\frac{d}{dx}\,cf(x) = \lim_{\Delta x \to 0} \frac{cf(x+\Delta x) - cf(x)}{\Delta x} = \lim_{\Delta x \to 0} c\left[\frac{f(x+\Delta x) - f(x)}{\Delta x}\right]$$

$$= c\lim_{\Delta x \to 0} \frac{f(x+\Delta x) - f(x)}{\Delta x} \quad (\text{by a rule for limits (§3.1.9)})$$

$$= c\frac{d}{dx}\,f(x)\,.$$

This enables us to differentiate any monomial ax^n:

$$\frac{d}{dx}\,(ax^n) = a\frac{d}{dx}\,x^n = anx^{n-1} \quad (\text{or } nax^{n-1})\,.$$

Next, we show that *the derivative of a sum of two functions is the sum of their derivatives,*

$$\frac{d}{dx}\,(f(x) + g(x)) = \frac{d}{dx}\,f(x) + \frac{d}{dx}\,g(x)\,.$$

The proof is as follows.

$$\frac{d}{dx}(f(x) + g(x)) = \lim_{\Delta x \to 0} \frac{(f(x + \Delta x) + g(x + \Delta x)) - (f(x) + g(x))}{\Delta x}$$

$$= \lim_{\Delta x \to 0} \left[\frac{f(x + \Delta x) - f(x)}{\Delta x} + \frac{g(x + \Delta x) - g(x)}{\Delta x} \right]$$

$$= \lim_{\Delta x \to 0} \frac{f(x + \Delta x) - f(x)}{\Delta x} + \lim_{\Delta x \to 0} \frac{g(x + \Delta x) - g(x)}{\Delta x}$$

(by another rule for limits (§3.1.9))

$$= \frac{d}{dx} f(x) + \frac{d}{dx} g(x).$$

This proposition extends to a sum of more than two functions; e.g., for three we have

$$\frac{d}{dx}(f(x) + g(x) + h(x)) = \frac{d}{dx}(f(x) + g(x)) + \frac{d}{dx} h(x)$$

(by the original proposition applied to the functions $f + g$ and h)

$$= \frac{d}{dx} f(x) + \frac{d}{dx} g(x) + \frac{d}{dx} h(x).$$

Since the polynomial is a sum of monomials, we can now differentiate it:

$$\frac{d}{dx}(a_m x^m + a_{m-1} x^{m-1} + \cdots + a_1 x + a_0)$$

$$= \frac{d}{dx}(a_m x^m) + \frac{d}{dx}(a_{m-1} x^{m-1}) + \cdots + \frac{d}{dx}(a_1 x) + \frac{d}{dx}(a_0)$$

$$= m a_m x^{m-1} + (m-1) a_{m-1} x^{m-2} + \cdots + 2 a_2 x + a_1,$$

a polynomial of degree one less than the degree of the original polynomial. For example, the derivative of

$$2x^5 + 4x^3 - 7x^2 + 3x - 8 = 2x^5 + 4x^3 + (-7)x^2 + 3x + (-8)$$

is

$$5 \cdot 2x^4 + 3 \cdot 4x^2 + 2 \cdot (-7)x + 3 = 10x^4 + 12x^2 - 14x + 3.$$

3.2.5   QUESTIONS

Q1.   Differentiate. (a) $x^2 + 2x + 1$ (b) $4t^3 - \sqrt{2}\,t^2 + \pi$ (c) $8x - 8x^4 + \frac{1}{3}x^3$

Q2.   Are the following formulas correct? (a) $\dfrac{d}{dx}(x+1)^3 = 3(x+1)^2$

(b) $\dfrac{d}{dx}(x+x)^3 = 3(x+x)^2$

Q3.   If $f'(x) = h(x)$ and $g'(x) = k(x)$ what is $\dfrac{d}{dx}(2f(x) + g(x))$?

Q4.   Show that the statement $f = dp/dt$ of §2.8.4 is equivalent to $f = ma$.

Q5.   As will be shown in Chapter 5 (§5.1.1), the location of a particle falling in the negative direction along the x-axis is given by an equation such as $x = -\frac{1}{2}gt^2 + 3t + 4$, where g is the acceleration of gravity. At $t = 0$, (a) where is the particle? (b) how fast is it moving? (c) what is its acceleration?

Q6.   Prove that $\dfrac{d}{dx}(-f(x)) = -\dfrac{d}{dx}f(x)$.

3.3   ELEMENTARY FUNCTIONS AND THEIR DERIVATIVES

3.3.1   More differentiable functions.   The method by which we have obtained the derivative of a polynomial is now to be generalized in two ways: by enlarging the class of **elementary functions** with known derivatives, such as c and x^n, out of which others can be built; and by establishing more rules for differentiating functions that are constructed by arithmetic operations out of simpler functions. The present article is devoted to the first of these tasks, and the second will occupy the next article.

In the sections that follow I am going to introduce, with only partial explanations, several differentiable functions considered "elementary"; and tell you what their derivatives are, without giving any proofs. My purposes in pursuing this course are to promote early familiarity with important formulas, and to provide material on which to practice differentiation. A fuller account of each function will be supplied in time.

3.3.2   The power functions x^α and their derivatives.   The first elementary functions are the **powers** x^α, one for each value of α. Let us review the definition of these functions as it is given in elementary algebra.

   (i) $x^0 = 1$, if $x \neq 0$ (0^0 is undefined).
   (ii) $x^n = xx \cdots x$ (n factors), if n is a positive integer.

(iii) $x^{1/q} = \sqrt[q]{x}$, if q is an integer > 1 and, in case q is even, $x \geq 0$; i.e., $x^{1/q}$ is the number such that $(x^{1/q})^q = x$, the non-negative such number being chosen if q is even (e.g., $4^{1/2} = 2$, not -2).[8]

(iv) $x^r = (x^{1/q})^p$, if $r = p/q$, p and q are integers > 1 with no common factor (i.e. p/q is in lowest terms), and, in case q is even, $x \geq 0$.

(v) $x^{-r} = 1/x^r$, if r is any rational number (see below), x^r is defined, and $x \neq 0$.

It is a rather complicated definition, with various restrictions as to allowable x-values. Nevertheless, in each case the power x^α is defined at least for $x > 0$. But α is confined to *rational* values—numbers (including the integers) which can be expressed as fractions.

I now *assume* that:

(vi) the definition of x^α can be extended so as to cover all values of α, with the restriction $x > 0$ for α irrational;

(vii) under the extended definition the usual laws of exponents are valid— $x^{\alpha+\beta} = x^\alpha x^\beta$, etc.; moreover,

(viii) for $x > 0$, x^α is a positive function, increasing if $\alpha > 0$; and

(ix) x^α is differentiable, and its derivative follows the pattern for the special case x^n:

$$\frac{d}{dx} x^\alpha = \alpha x^{\alpha-1},$$

whenever x^α and $x^{\alpha-1}$ are both defined; in particular, for $x > 0$. (If $\alpha = 1$ the derivative is 1 everywhere, even at $x = 0$, as we have already shown.)

Thus, e.g., $x^{4/3}$ is defined for any x, and its derivative is $\frac{4}{3}x^{1/3}$; $x^{\sqrt{2}}$ is defined for $x > 0$, and its derivative is $\sqrt{2}\,x^{\sqrt{2}-1}$; x^{-3} is defined for $x \neq 0$, and its derivative is $-3x^{-4}$. The extension to arbitrary α will be justified in Chapter 6 (§6.2.6).

Note that since $(x^{1/\alpha})^\alpha = x$ ($\alpha \neq 0$), by a law of exponents, we have generalized the qth root ((iii) above) as well as the nth power.

A definition of x^α for α irrational can be given in a way I will illustrate for $\alpha = \sqrt{2}$.

---

[8]This part of the definition is justified in §4.3.3.Q1.

As we know (§3.1.2), $\sqrt{2}$ is the limit of the sequence 1.4, 1.41, 1.414, ... Since each term of the sequence is rational, the sequence of powers $x^{1.4}$, $x^{1.41}$, $x^{1.414}$, ... is defined for $x > 0$. In general, any sequence $r_1, r_2, r_3, \ldots$ of rational numbers approaching $\sqrt{2}$ gives rise to a sequence $x^{r_1}, x^{r_2}, x^{r_3}, \ldots$. It can be shown that all such sequences have a common limit, and we define $x^{\sqrt{2}}$ to be that limit.

Although it is easy to accept the idea that $x^{\sqrt{2}}$ should equal the limit of the x^{r_n}, this definition involves certain difficulties, including that of establishing the common limit. For this reason the one given later will be different. (See §6.2.10.Q14 for the limit cited here.)

I have taken the irrationality of $\sqrt{2}$ for granted; for a demonstration of it, see Q7.

3.3.3   QUESTIONS

Q1. Simplify each expression by applying a law of exponents. (a) $x^2 x^{-3}$ (b) $(t^{1/3})^{-3}$ (c) xx^α (d) $1/t^{-p}$

Q2. Differentiate. (a) x^{23} (b) $x^2 + x^{-3}$ (apply §3.2.4) (c) $1/x$ (d) $3\sqrt{t}$ (e) $t^{2\alpha}$ (f) $5x^{-n/2}$

Q3. By considering the signs of their derivatives, determine whether (a) $y = x^{1/5}$ and (b) $y = x^{-5}$ are increasing or decreasing for $x \neq 0$ (§2.10.8).

Q4. For which values of α is x^α the derivative of a monomial (as, e.g., x^2 is the derivative of $\frac{1}{3}x^3$)?

Q5. (a) Show that $\lim_{x \to \infty} x^\alpha = \infty$ if $\alpha > 0$. (b) Deduce that $\lim_{x \to 0} x^\alpha = 0$.

Q6. For any fixed value of x, consider the sequence $x, x^2, x^3, \ldots$. Show that:
(a) if $x > 1$, $x^n \to \infty$ (use §3.2.3.Q5);
(b) if $x = 1$, $x^n \to 1$;
(c) if $-1 < x < 1$, $x^n \to 0$ (for $x \neq 0$ apply (a) to $1/|x|$);
(d) if $x \leq -1$, the sequence has no limit.

Q7. Prove $\sqrt{2}$ irrational as follows. (a) If it is not, we can write $\sqrt{2} = p/q$, where p and q are positive integers with no common factor. (b) Then $p^2 = 2q^2$. (c) Hence p is even, $p = 2r$ for some positive integer r. (d) Then $2r^2 = q^2$. (e) Hence q is even, a contradiction.

3.3.4   The trigonometric functions $\sin x$ and $\cos x$.   A. Radians and directed angles.   As they are used in calculus, trigonometric functions are functions not of *angles* but of *numbers*. The variable x in $\sin x$ and $\cos x$ is the same as in the function x^2. In order that we may be quite clear about how this is so I will review angle-measurement and the definition of the functions.

In the first place, angles are measured in **radians** rather than degrees. (The choice of unit will be justified in §4.4.3.Q1 and §8.2.10.Q4.) An angle is placed at the center of a circle, and its measure θ is then the ratio of the length s of the arc it cuts out to the radius r,

$$\theta = \frac{s}{r}$$

(Fig. 3-4). This ratio is independent of the size of the circle.[9] An angle of 1 radian cuts out an arc equal to the radius ($s = r$). If $r = 1$, $\theta = s$; that is, in a unit circle, an angle of θ radians cuts out an arc of length θ. Since the circumference of a unit circle is 2π, there are 2π radians in a whole circular angle, or $\pi/2$ in a right angle. It follows that

$$1 \text{ radian} = \frac{180}{\pi} \text{ degrees},$$

or about $57.3°$.

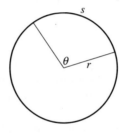

Fig. 3-4

That s/r is independent of r can be deduced from the well-known formula $C = 2\pi r$ for the circumference of a circle (Q6), which I have also used to calculate the circumference of a unit circle. This formula, together with the area formula $A = \pi r^2$ (needed in Q7), will be discussed in §§4.4.4–5.

Next, instead of the angles of elementary geometry we usually employ **directed** angles, angles with a sense of direction counterclockwise or clockwise, counting the former positive and the latter negative (Fig. 3-5). Furthermore, we allow these directed angles to be of any magnitude, so that an "angle" may exceed a whole circular angle. Angles are added algebraically, e.g. $(\pi/3) + (-\pi/2) = -\pi/6$.

It is helpful to think of an angle as specifying a *rotation* about a center. Addition of angles then corresponds to successive performance of rotations

---

[9]See the remark following.

Fig. 3-5

(the order does not matter). In this interpretation $(\pi/3) + (-\pi/2) = -\pi/6$ signifies that a rotation of $\pi/3$ radians counterclockwise, followed (or preceded) by a rotation of $\pi/2$ radians clockwise, results in a net rotation of $\pi/6$ radians clockwise. The fact that angles can be of any size means that there is no limit to the number of radians that can be turned through in a rotation.

3.3.5   B. Definitions of $\sin x$ and $\cos x$.   Now in order to define $\sin x$ and $\cos x$ for any number x, we interpret x as the radian measure of an angle and proceed as follows. In a coordinate plane let a unit circle be drawn about the origin O as center, and let A be the point $(1,0)$ where the circle cuts the positive horizontal axis (Fig. 3-6).[10] Measure the angle x about O from radius OA to some radius OP, possibly circling the origin several times to reach OP; in other words, rotate a radius from position OA through x radians, and let OP be its final position. Then the numbers $\cos x$ and $\sin x$ are defined to be the coordinates of P: P is the point $(\cos x, \sin x)$.

When $0 < x < \pi/2$ this definition agrees with the elementary one based on right triangles, in the sense that the cosine or sine of the *number* x, as defined here, equals the cosine or sine of the acute *angle AOP* of x radians, as defined in plane geometry (Q3). A definition independent of geometry will be given in §4.7.4.

3.3.6   QUESTIONS

Q1.   How long an arc does an angle of $\sqrt{2}$ radians cut out on a circle of radius $\sqrt{2}$?

Q2.   Determine and memorize the radian measures of the following common angles: $0°$, $30°$, $45°$, $60°$, $90°$, $180°$, $270°$, $360°$.

Q3 (§3.3.5).   Verify this.

---

[10]The coordinate variable of this axis, unnamed here, is not the same as the variable x of which the sine and cosine are to be functions.

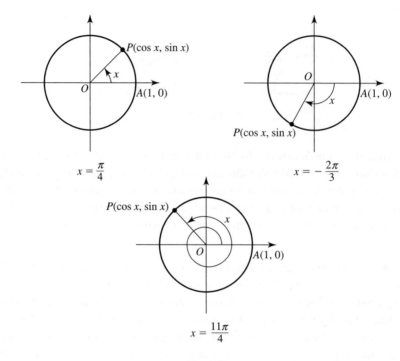

Fig. 3-6

Q4. Determine and memorize the sines and cosines of the numbers found in Q2.

Q5. In Fig. 3-6, if the circle is not necessarily of unit radius, what are the coordinates of P?

Q6 (§3.3.4). Perform the deduction.

Q7. If A is the area of the sector of a circle of radius r with central angle θ (radians), show that $\theta = 2A/r^2$.

3.3.7 Properties and derivatives of trigonometric functions.

The definitions of $\sin x$ and $\cos x$ imply that their values repeat after any x-interval of length 2π; that is, we have the following **identities**, or equations true for every value of the independent variable:

$$\sin(x + 2\pi) = \sin x, \quad \cos(x + 2\pi) = \cos x.$$

We say that the functions are **periodic** with **period** 2π. This periodicity distinguishes their graphs from those of polynomials, which have only finitely many hills and valleys (Fig. 3-7; Q15).

Other identities easily derivable from the definitions are[11]

$$\sin(-x) = -\sin x, \quad \cos(-x) = \cos x,$$
$$\sin^2 x + \cos^2 x = 1$$

(Q4). Any function $f(x)$ which satisfies the identity $f(-x) = -f(x)$ is called an **odd function**, because in this respect it behaves like an odd power of x; thus $\sin x$ is an odd function. Similarly, $\cos x$ satisfies $f(-x) = f(x)$, hence is an **even function**.[12] The graph of an odd function is symmetrical about the origin, that of an even function about the y-axis (Q9(a)); these symmetries are evident in Fig. 3-7.

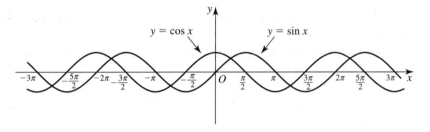

Fig. 3-7

Such identities as

$$\sin\left(\frac{\pi}{2} - x\right) = \cos x, \qquad \cos\left(\frac{\pi}{2} - x\right) = \sin x,$$
$$\sin\left(x + \frac{\pi}{2}\right) = \cos x, \qquad \cos\left(x + \frac{\pi}{2}\right) = -\sin x,$$
$$\sin(x + \pi) = -\sin x, \qquad \cos(x + \pi) = -\cos x,$$

can also be demonstrated directly; but they as well as the others I have given and many more are consequences of the fundamental **addition formulas**

$$\sin(u + v) = \sin u \cos v + \cos u \sin v,$$
$$\sin(u - v) = \sin u \cos v - \cos u \sin v,$$

---

[11] Recall that $\sin^2 x$ means $(\sin x)^2$, etc.

[12] The full significance of the association of these functions with odd and even powers will come out in §8.1.5.

$$\cos(u+v) = \cos u \cos v - \sin u \sin v,$$

$$\cos(u-v) = \cos u \cos v + \sin u \sin v,$$

together with the values of the functions at 0 and $\pi/2$. In particular, there are the double-angle formulas

$$\sin 2x = 2\sin x \cos x, \quad \cos 2x = \cos^2 x - \sin^2 x,$$

and the half-angle formulas

$$\sin^2 \frac{x}{2} = \frac{1}{2}(1 - \cos x), \quad \cos^2 \frac{x}{2} = \frac{1}{2}(1 + \cos x).$$

A proof of the addition formulas, and derivations based on them, are called for in Q14.

The remaining trigonometric functions are the tangent, cotangent, secant, and cosecant, defined by the formulas

$$\tan x = \frac{\sin x}{\cos x}, \quad \cot x = \frac{\cos x}{\sin x},$$

$$\sec x = \frac{1}{\cos x}, \quad \csc x = \frac{1}{\sin x},$$

wherever the denominators are not zero. Among other identities we have

$$\tan\left(\frac{\pi}{2} - x\right) = \cot x$$

and

$$\sec^2 x = 1 + \tan^2 x,$$

the latter obtained by division from $\sin^2 x + \cos^2 x = 1$.

The tangent is represented geometrically in Q10. The definitions of the sine and cosine imply that the tangent of the **angle of inclination** of a non-vertical straight line—the angle α, $-\pi/2 < \alpha < \pi/2$, *from* the horizontal *to* the line—is the slope of the line. In consequence we have a useful relation: if the tangent line to the graph of a differentiable function $f(x)$ has angle of inclination α at x,

(1) $$f'(x) = \tan \alpha.$$

I now make the *assumption* that the derivatives of $\sin x$ and $\cos x$ are as follows:

$$\frac{d}{dx} \sin x = \cos x \,, \qquad \frac{d}{dx} \cos x = -\sin x \,.$$

A proof will be given in Chapter 4 (§§4.4.1–2). As you will see in §3.4.4.Q2, the derivatives of the other four functions are deducible from these; the results are

$$\frac{d}{dx} \tan x = \sec^2 x \,, \qquad \frac{d}{dx} \cot x = -\csc^2 x \,,$$

$$\frac{d}{dx} \sec x = \sec x \tan x \,, \qquad \frac{d}{dx} \csc x = -\csc x \cot x \,.$$

3.3.8   Polar coordinates. Trigonometric functions relate our customary xy-coordinates to another coordinate system in the plane, the **polar coordinates** r and θ, which are defined as follows. Given a point P, let r be its distance from the origin O and θ an angle from the positive x-axis to radius OP, as in Fig. 3-8; in this diagram O is called the **pole**, and OP the **radius vector**. The two numbers r and θ clearly suffice to locate P, so P can be regarded as having coordinates (r, θ) in this scheme of representation, just as it has coordinates (x, y) in the Cartesian scheme. The new coordinates have, however, certain peculiarities: (i) $r \geq 0$ everywhere; (ii) θ is not defined at the pole O, which is identified only by $r = 0$; (iii) the θ-coordinate of a given point P is not unique, because there are many angles from the positive x-axis to OP, differing one from another by multiples of 2π. For example, the point P in Fig. 3-9 has polar coordinates $(5, \pi/3)$, and also $(5, -5\pi/3)$, $(5, 7\pi/3)$, etc.

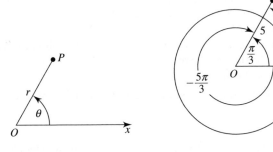

Fig. 3-8   Fig. 3-9

The definitions of the sine and cosine imply that rectangular coordinates are expressed in terms of polar coordinates by the equations

$$x = r \cos \theta,$$

$$y = r \sin \theta$$

(if $r = 0$ we understand the right-hand sides to be 0). In the other direction, an identity gives

$$x^2 + y^2 = r^2 (\cos^2 \theta + \sin^2 \theta) = r^2,$$

so that

$$r = \sqrt{x^2 + y^2},$$

as we would have expected. We also have $y/x = (r \sin \theta)/(r \cos \theta) = \sin \theta / \cos \theta$, so that

$$\tan \theta = \frac{y}{x}$$

where $x \neq 0$, and similarly

$$\cot \theta = \frac{x}{y}$$

where $y \neq 0$.[13]

Sometimes it is advantageous to write the equation of a curve in polar coordinates; e.g., the circle of radius a centered at the origin has the very simple polar equation $r = a$. A generalization of this equation will be given in §3.3.12.

3.3.9   QUESTIONS

Q1.   Derive Fig. 3-7 by following P around the circle in Fig. 3-6.

Q2.   Use identities to determine the sines and cosines of the following numbers from the values found in §3.3.6.Q4. (a) $5\pi/2$ (b) $-5\pi/2$ (c) $2\pi/3$ (d) $-5\pi/6$ (e) $-\pi/4$ (f) $\pi/12$ (g) $x + (\pi/2)$

Q3.   State an identity which asserts that the graph of $\cos x$ is obtained from that of $\sin x$ by shifting it $\pi/2$ to the left. Explain.

Q4 (§3.3.7).   Verify these identities.

Q5.   Differentiate. (a) $x^2 + \sin x$ (b) $3 \tan t + 4 \cos t$ (c) $\sin^2(x/2)$ (first apply an identity)

---

[13]For an analytic proof that x and y determine θ up to an additive multiple of 2π see §3.7.17.Q5.

Q6.   (a) If the derivative is interpreted as slope, what does the formula $\dfrac{d}{dx}\sin x = \cos x$ say about the relation between the graphs of Fig. 3-7? Is the relation visible? (b) Graph $\cos x$ and $-\sin x$ together, and answer the same questions for that pair of functions.

Q7.   (a) Suppose $0 \le u < 2\pi$ and $0 \le v < 2\pi$. Prove that if $\sin u = \sin v$ and $\cos u = \cos v$ then $u = v$. Do this in two ways: (i) by geometry, (ii) by calculating $\cos(u - v)$. (b) If $\sin u = \sin v$ and $\cos u = \cos v$, but u and v are otherwise unrestricted, how are they related?

Q8.   An oscillating particle has momentum $10\cos t$ kg m/sec at any time t. What force acting on the particle would produce this momentum?

Q9.   (a) (§3.3.7) Establish the stated symmetries. (b) Show that the derivative of an odd function is even, and of an even function, odd. (Consider the difference quotient at x with increment Δx and at $-x$ with increment $-\Delta x$.)

Q10.   For a geometrical representation of the tangent function which explains its name, augment Fig. 3-6 by the tangent line to the circle at A and let T be the intersection of OP (extended if necessary) with that tangent (Fig. 3-10). (a) Show that $\tan x$ is the vertical coordinate of T. (b) Show that $\tan x$ is periodic and find its period. (c) With the aid of the geometrical representation, sketch the graph of $y = \tan x$. (d) Use this graph to sketch the graph of $y = \cot x$, which equals $1/\tan x$ except when x is a multiple of $\pi/2$.

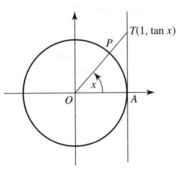

Fig. 3-10

Q11.   Prove directly from the geometrical definitions of $\sin x$ and $\cos x$ that both functions are continuous at $x = 0$.

Q12.   Show that the differentiation formulas for $\sin x$ and $\cos x$ are equivalent (each implies the other).

Q13.   Prove the *law of cosines*: if a triangle has sides a, b, c, with θ the angle opposite c, then
$$c^2 = a^2 + b^2 - 2ab\cos\theta.$$

To do this, calculate c^2 by the distance formula of analytic geometry applied to Fig. 3-11.

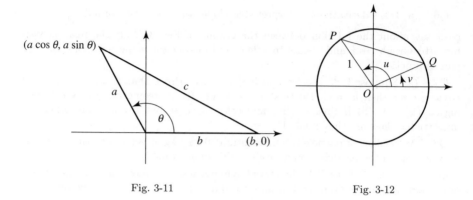

Fig. 3-11   Fig. 3-12

Q14.   (a) Deduce the case $0 \leq v \leq u \leq \pi$ of the addition formula for $\cos(u - v)$ from the law of cosines by calculating PQ^2 (Fig. 3-12) in two ways (note that $\angle QOP = u - v$). (b) Establish the general case as a consequence of this case (note that any u can be written as $u' + m\pi$, where $0 \leq u' \leq \pi$ and m is an integer). (c) Derive the remaining addition formulas from this one. (d) Derive the other identities of §3.3.7 from the addition formulas.

Q15 (§3.3.7).   Why does the graph of a polynomial have only finitely many hills and valleys? How many can there be?

Q16.   Let a curve have equation $r = f(\theta)$ in polar coordinates. Show by infinitesimals that the cotangent of the angle ζ from the radius vector to the curve is given by

$$\cot \zeta = \frac{1}{r} \frac{dr}{d\theta}$$

Fig. 3-13

(Fig. 3-13).

3.3.10   The exponential functions a^x and the logarithmic functions $\log_a x$, and their derivatives. A. The functions a^x.

Let a be a positive number. For any number α, the power a^α is defined, according to §3.3.2: it is the value at $x = a$ of the power function x^α. Let us now think of the *exponent* in a^α as variable; then we have a new function a^x, the **exponential function** with **base** a, defined for all x. Taking $a = 2$, for example, the function is 2^x; we have $2^0 = 1$, and since $2^{x+1} = 2 \cdot 2^x$ the value of the function doubles with every unit increase in x (Fig. 3-14). A similar remark applies to the exponential functions with other bases.

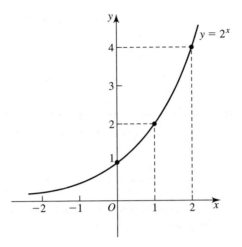

Fig. 3-14

The function a^x is decreasing if $a < 1$, constant if $a = 1$, and increasing if $a > 1$. It is always positive, and except in the case $a = 1$ assumes every positive number as a value. The rate of increase of a^x for $a > 1$ grows with a; that is, the larger a is, the more rapidly a^x increases as x increases. All of the functions a^x take the value 1 at $x = 0$, i.e. their graphs all pass through $(0, 1)$; and the slopes of the graphs at that point become larger and larger as a grows (Fig. 3-15). There is a certain value of a for which the slope at $(0, 1)$ is exactly 1; this number is called e. As Fig. 3-15 indicates, e is between 2 and 4. In fact its value is approximately 2.718 (as will be proved in §6.2.3 and §8.2.7).

Among the exponential functions—indeed, among all functions—e^x holds a distinguished place, and it is known simply as *the* exponential function. This is

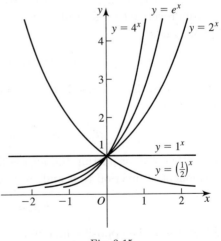

Fig. 3-15

partly because of its remarkable differentiation formula:

$$\frac{d}{dx}\,e^x = e^x$$

—it is its own derivative. In geometrical language, at each point its slope equals its ordinate.

The derivatives of the other exponential functions are proportional, rather than equal, to the respective functions. The general differentiation formula has the form

$$\frac{d}{dx}\,a^x = Ca^x\,,$$

where the constant C, which depends on a, is equal to the value of the derivative at $x = 0$ (the slope of the graph at $(0,1)$), because

$$\left.\frac{d}{dx}\,a^x\right|_0 = Ca^0 = C\,.$$

Only for $a = e$ is the constant 1. In a moment I will give its value in the general case.

3.3.11   B. The functions $\log_a x$. Derivatives.   Associated with a^x is its **inverse** function $\log_a x$, the **logarithm** or **logarithmic function** to the base a, which is defined for $x > 0$. By "inverse" function I mean that the logarithm reverses the effect of a^x; i.e., if the value of a^x at u is v, the value of $\log_a x$ at v is u:

$$a^u = v \text{ is the same as } \log_a v = u\,.$$

In other words, the two functions are related in the same way as the nth power and nth root:

$$u^n = v \text{ is the same as } \sqrt[n]{v} = u\,.$$

Still another way of putting the idea is that if starting with u we first apply the exponential function to it and then the logarithm we get u again,

$$\log_a a^u = u\,,$$

and likewise in the opposite direction,

$$a^{\log_a v} = v\,.$$

I will have more to say about inverse functions in Chapter 4 (Art. 4.3).

The fundamental law of exponents

$$a^{s+t} = a^s a^t$$

asserts that the function a^x applied to a *sum* $s + t$ yields a *product* $a^s a^t$. The inverse function $\log_a x$ reverses that law, so that applied to a product it yields a sum:

$$\log_a vw = \log_a v + \log_a w \,.$$

This is proved by applying the logarithm to both ends of

$$vw = a^{\log_a v} a^{\log_a w} = a^{\log_a v + \log_a w} \,.$$

The logarithm to the base e is called the **natural** logarithm, or simply *the* logarithm, and is denoted $\log$ (no subscript) or $\ln$. Its derivative is

$$\frac{d}{dx} \log x = \frac{1}{x}, \quad x > 0 \,.$$

Note that x^{-1} is the only power of x that was not the derivative of a monomial cx^α (§3.3.3.Q4).

With the aid of the natural logarithm I can now state the general formulas for the derivatives of the exponential and logarithmic functions. They are:

$$\frac{d}{dx} a^x = (\log a)\, a^x \,;$$

$$\frac{d}{dx} \log_a x = \left(\frac{1}{\log a} \right) \frac{1}{x}, \quad x > 0 \,.$$

I assume them here; in Chapter 4 (§4.1.5.Q4 and §4.3.6.Q3) they will be derived from the formula for the derivative of e^x, and that formula will be established in Chapter 6 when the functions are properly defined.

3.3.12   C. The logarithmic spiral. This curve is defined by an equation $r = r_0 a^\theta$, or equivalently $\theta = \log_a(r/r_0)$, in polar coordinates (§3.3.8), where a and $r_0 = r(0)$ are positive constants. To obtain a spiral we take $a \neq 1$: the curve spirals counterclockwise outward about the pole if $a > 1$ (Fig. 3-16), inward if $a < 1$; for $a = 1$ we get a circle of radius r_0 rather than a spiral. The rate of growth of the radius $OP = r$ of the spiral with respect to θ is proportional to the length of the radius: $dr/d\theta = (\log a)r$. Another property of the curve is stated in Q9.

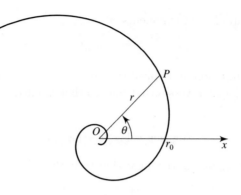

Fig. 3-16

3.3.13   QUESTIONS

Q1.   (a) Explain why the graph of $y = 2^x$ (Fig. 3-14) approaches the x-axis to within any given distance. (b) How are the graphs of $y = 2^x$ and $y = (\frac{1}{2})^x$ related (Fig. 3-15)? Explain.

Q2.   Evaluate. (a) $\log e^{2x}$ (b) $e^{2 \log x}$

Q3.   Show that (a) $\log (1/x) = - \log x$, (b) $\log x^{\alpha} = \alpha \log x$.

Q4.   Differentiate. (a) e^x (b) $2e^x + \sin x$ (c) 2^x (d) $e^x + \log x$ (e) $a^x + b^x$ (f) $2 \log_2 x + 4x^4$

Q5.   Find another non-zero function (besides e^x) equal to its own derivative.

Q6.   Show that $\dfrac{d}{dx} e^{\alpha x} = \alpha e^{\alpha x}$. (Notice that $e^{\alpha x} = (e^\alpha)^x$.)

Q7.   "Exponential growth" (Chap. 6) obeys an equation such as $y = 2e^{3t}$. (a) What is $y(0)$? (b) At what time does $y = 4$?

Q8.   Prove that if $a > 1$, a^x increases so rapidly that $\lim\limits_{n \to \infty} \dfrac{a^n}{n} = \infty$. (Let $\sqrt{a} = 1 + h$ and use §3.2.3.Q5 to show that $a^n > n^2 h^2$.)

Q9.   (a) Show that the angle ζ from the radius vector of the logarithmic spiral $r = r_0 a^\theta$ to the curve is determined by the relation $\cot \zeta = \log a$, and therefore is constant. For this reason the curve is also called an *equiangular* spiral. (Apply §3.3.9.Q16.) (b) Write the equation of the curve in terms of ζ rather than a.

3.4   RULES FOR DIFFERENTIATION CORRESPONDING TO ARITHMETIC OPERATIONS ON FUNCTIONS

3.4.1   Review of known rules.   In order to differentiate a polynomial (§3.2.4) it was necessary to introduce two differentiation rules, which can be expressed concisely as

$$(cf)' = cf'$$

and

$$(f + g)' = f' + g',$$

the latter extending to any number of summands. These apply to all differentiable functions, a fact which has already been appealed to in a number of questions. An example of their use together may be given here:

$$\frac{d}{dx}\left(\sqrt{x} + 3\cos x + \frac{1}{2}e^x\right) = \frac{d}{dx}(x^{1/2}) + \frac{d}{dx}(3\cos x) + \frac{d}{dx}\left(\frac{1}{2}e^x\right)$$

$$= \frac{d}{dx}(x^{1/2}) + 3\frac{d}{dx}(\cos x) + \frac{1}{2}\frac{d}{dx}(e^x)$$

$$= \frac{1}{2}x^{-1/2} + 3(-\sin x) + \frac{1}{2}e^x$$

$$= \frac{1}{2\sqrt{x}} - 3\sin x + \frac{1}{2}e^x.$$

Let us now obtain rules for differentiating the other arithmetic combinations of functions.

3.4.2 The difference, product, and quotient rules. Since $f - g = f + (-1)g$ we have

$$(f - g)' = (f + (-1)g)' = f' + ((-1)g)' = f' + (-1)g' = f' - g',$$

and the rule for subtraction is

$$(f - g)' = f' - g'.$$

The rule for differentiating a product cannot be $(fg)' = f'g'$, for if it were we would have

$$f' = (1 \cdot f)' = 1'f' = 0f' = 0$$

for every function f. To discover it we write out the difference quotient for the function fg,

$$\frac{f(x + \Delta x)g(x + \Delta x) - f(x)g(x)}{\Delta x},$$

and then use a trick of adding and subtracting the same quantity in the numerator, which without changing the numerator's value allows us to isolate the variations of the separate functions. The numerator equals

$$f(x + \Delta x)g(x + \Delta x) - f(x)g(x + \Delta x) + f(x)g(x + \Delta x) - f(x)g(x)$$

$$= [f(x + \Delta x) - f(x)]g(x + \Delta x) + f(x)[g(x + \Delta x) - g(x)],$$

so the difference quotient can be written

$$\left[\frac{f(x + \Delta x) - f(x)}{\Delta x}\right] g(x + \Delta x) + f(x) \left[\frac{g(x + \Delta x) - g(x)}{\Delta x}\right].$$

Here there are four quantities whose behavior as $\Delta x \to 0$ is to be determined. The difference quotients of f and g approach $f'(x)$ and $g'(x)$ respectively, and $f(x)$ remains fixed. As for $g(x+\Delta x)$, we know that $x+\Delta x \to x$, while g, being differentiable at x, is continuous there (§3.1.7); hence $g(x + \Delta x) \to g(x)$. By the rules for limits of products and sums (§3.1.9), therefore, the whole quantity approaches $f'(x)g(x) + f(x)g'(x)$. We have found the product rule:

$$(fg)' = f'g + fg'.$$

For example,

$$\frac{d}{dx}(x^2 \sin x) = \frac{d}{dx}(x^2) \sin x + x^2 \frac{d}{dx}(\sin x)$$

$$= 2x \sin x + x^2 \cos x.$$

The product rule extends to products of more than two functions, in a pattern sufficiently indicated by the case of three functions:

$$(fgh)' = f'gh + fg'h + fgh'$$

(Q4).

A similar argument leads to the quotient rule

$$\left(\frac{f}{g}\right)' = \frac{f'g - fg'}{g^2},$$

valid at any x for which $g(x) \neq 0$. This rule can also be deduced from the product rule and the special case

$$\left(\frac{1}{g}\right)' = -\frac{g'}{g^2}$$

(Q6). An example of its use follows:

$$\frac{d}{dx}\left(\frac{\sin x}{x^2+1}\right) = \frac{\left(\frac{d}{dx}\sin x\right)(x^2+1) - (\sin x)\frac{d}{dx}(x^2+1)}{(x^2+1)^2}$$

$$= \frac{(\cos x)(x^2+1) - (\sin x)2x}{(x^2+1)^2}$$

$$= \frac{(x^2+1)\cos x - 2x\sin x}{(x^2+1)^2}.$$

3.4.3 Summary. We now have the following rules (which should be committed to memory).

Sum: $(f+g)' = f' + g'$ (extends to more summands)

Difference: $(f-g)' = f' - g'$

Product: $(fg)' = f'g + fg'$ (extends to more factors),

and a special case: $(cf)' = cf'$

Quotient: $\left(\dfrac{f}{g}\right)' = \dfrac{f'g - fg'}{g^2}$, $g(x) \neq 0$,

and a special case: $\left(\dfrac{1}{g}\right)' = -\dfrac{g'}{g^2}.$

3.4.4 QUESTIONS

Q1. Differentiate. (a) $\sin t - \cos t$ (b) $x\log x - x$ (c) xe^x (d) $e^x \tan x$ (e) $2/\log x$ (f) $8x\sin x$ (g) $2^t/(t-1)$ (h) $(2x^2 - 1)e^x \cos x$

Q2. From the derivatives of $\sin x$ and $\cos x$ together with the differentiation rules obtain the differentiation formulas for the other four trigonometric functions (§3.3.7).

Q3. According to Boyle's law, at constant temperature the product of the pressure P and volume V of a given quantity of gas is constant, $PV = C$. Holding the temperature fixed, let the pressure and volume change with time. (a) Express the rate of change of volume in terms of the pressure and its rate of change. (b) Is the answer to (a) a new law of physics?

Q4 (§3.4.2). Prove the product rule for three factors.

Q5. Deduce $(x^n)' = nx^{n-1}$, n any integer > 1, from the product rule.

Q6 (§3.4.2). (a) Establish the special case by calculating the limit of the difference quotient. (See also §4.1.5.Q3.) (b) Prove the quotient rule from that case together with the product rule.

Q7. Use L'Hospital's rule (§3.1.11.Q8) to evaluate the following limits.

(a) $\lim\limits_{x \to 0} \dfrac{\sin x}{x}$ (b) $\lim\limits_{x \to 0} \dfrac{1 - \cos x}{x}$ (c) $\lim\limits_{x \to 1} \dfrac{x - 1}{\sqrt{x} - 1}$ (d) $\lim\limits_{x \to 0} \dfrac{a^x - 1}{x}$ $(a > 0)$

(e) $\lim\limits_{x \to 1} \dfrac{x^2 - 3x + 2}{\log x}$.

3.5   THE FUNDAMENTAL THEOREM OF CALCULUS

3.5.1   Integration and differentiation. From the beginning of our study we have worked with *pairs* of quantities or functions united by the operations that have been identified as integration and differentiation. The existence of these pairs is a consequence of the inverse relation between the two operations. To this relation, which constitutes the central fact of calculus, I now aim to give formal expression.

3.5.2   The correspondence between functions created by differentiation and integration. Differentiation and indefinite integration are operations which associate functions with functions. From a given function $F(x)$ differentiation produces a new function dF/dx; from a given function $f(x)$ indefinite integration produces a new function $\int_a^x f(x)\,dx$. The Fundamental Theorem of Calculus asserts, roughly speaking, that these operations are inverse to one another: that together they define a correspondence back and forth between functions, differentiation furnishing one direction of it and integration the other, so that each function F is paired with its derivative f, and each f with its integral F.

> Thus any one given function will belong to *two* pairs, the other member being in the one case its derivative (of which it is the integral), in the other its integral (of which it is the derivative). Cf. §§1.3.4, 1.5.8.

This simple picture needs correction in two respects. First, we recall that $\int_a^x f(x)\,dx$ depends on the choice of a, for the indefinite integrals resulting from different choices differ by constants (§2.5.2). For this reason "integration" cannot be said to assign to f a *single* function F, rather it assigns to it a *group* of functions $F + C$, which are all the same except for additive constants.[14]

---

[14]For certain functions f not all the functions $F + C$ are obtainable as integrals, as will be shown in §3.5.7.

Correspondingly, differentiation does not distinguish between functions that differ by a constant, but assigns them all the same derivative (§2.9.2(2)). So the correspondence set up by differentiation and integration is not quite one-to-one. This is the price we pay for preferring functions of points to functions of intervals.[15]

Secondly, not all functions participate in the correspondence. Integration applies to arbitrary continuous functions, and differentiation of an integral returns us to that class; differentiation requires differentiable functions, and indeed ones whose derivatives are continuous if the integrals of those derivatives are in turn to be defined. Therefore the correspondence is between the *continuous* and the *continuously differentiable* functions.

Although these two considerations cause asymmetry in the form of the theorem, its character as the expression of a correspondence is apparent.

3.5.3   The Fundamental Theorem.   Let a be a given number.

I. *Let f be a continuous function. Let*

$$F(x) = \int_a^x f(x)\, dx\,.$$

Then F is continuously differentiable and

$$\frac{d}{dx}\, F(x) = f(x)\,.$$

II. *Let F be a continuously differentiable function. Let*

$$f(x) = \frac{d}{dx}\, F(x)\,.$$

Then f is continuous and

$$\int_a^x f(x)\, dx = F(x) - F(a)\,.$$

In each part, if the hypothesis is fulfilled only on a given x-interval containing a, the conclusion will hold on the same interval.

Part I says that if we first integrate a function, then differentiate the integral, we return to the original function. Part II says that if we first differentiate a function, then integrate the derivative, the function we obtain is the same as the original function except for an additive constant (which may, of course, be zero). Taken together they give rise to the correspondence described in the preceding section (Q1).

---

[15]See the remark after §3.5.6(2).

3.5.4   Demonstration by infinitesimals.   The truth of the Fundamental Theorem is inherent in the very concepts of integral and derivative, and its demonstration by infinitesimals is straightforward. The essential ideas have in fact already been repeatedly presented in Chapter 2, in physical and geometrical terms; here they will be stated concisely in the language of functions. Afterwards I will move on to apply the theorem, only later returning to prove it by means of limits (Art. 3.7).

To establish Part I we must show that $dF/dx = f(x)$. For the sake of clarity let us specially designate the point at which the derivative is taken: we are to prove that for any x_0,

$$\left.\frac{dF}{dx}\right|_{x_0} = f(x_0)\,.$$

Here dF is the differential at x_0 of the function $F(x) = \int_a^x f(x)\,dx$, i.e. the infinitesimal change that occurs in that function's value when x changes from x_0 to $x_0 + dx$. But seeing that $F(x)$ is a sum of products $f(x)\,dx$, the change in it is merely to add one more product, namely $f(x_0)\,dx$. Therefore

$$dF = f(x_0)\,dx\,,$$

or

$$\left.\frac{dF}{dx}\right|_{x_0} = f(x_0)\,.$$

For Part II we again designate a selected point: the problem is to show for any x_0 that

$$\int_a^{x_0} f(x)\,dx = F(x_0) - F(a)\,.$$

Since by hypothesis $f(x) = dF/dx$, each summand $f(x)\,dx$ is equal to a differential dF, the infinitesimal change or increment corresponding to a change dx in x. If all the increments of F that occur as x increases (or decreases) from a to x_0 are added together, they constitute as their sum the total change in F over that x-interval. (It may happen that dF is positive at certain points and negative at others; then cancellation occurs in the sum.) But that total change is $F(x_0) - F(a)$, while the sum is $\int_a^{x_0} f(x)\,dx$. This completes the proof (such as it is) by infinitesimals.

3.5.5   QUESTIONS

Q1 (§3.5.3). Argue that the correspondence as described is assured by the Fundamental Theorem. (This point will be clarified in the remainder of Art. 3.5.)

Q2. Restate the proof of the Fundamental Theorem in terms of the various physical and geometrical pairs from Chapter 1, comparing the arguments to those in Chapter 2.

3.5.6   Reformulation of the theorem in terms of primitives.

A somewhat different statement of the Fundamental Theorem helps one grasp its implications. Given a function $f(x)$, let us call any function $F(x)$ whose derivative is $f(x)$ a **primitive** of $f(x)$.[16] With this terminology we can restate the theorem in the following way.

I. *An indefinite integral of a continuous function is a primitive of that function.*

II. *A primitive of a continuous function is, up to an additive constant, an indefinite integral of that function.*

In Part II the additive constant has not been specified, but its value is easily deduced, for in $F(x) = \int_a^x f(x)\,dx + C$ we can put $x = a$ to determine that $C = F(a)$.[17]

Since adding a constant to a function does not change its derivative, Part I implies that an indefinite integral with a constant added is also a primitive. Combining this with Part II, we have:

(1)   *The primitives of a continuous function $f(x)$ are the functions $F(x) = \int_a^x f(x)\,dx + C$, where a and C are arbitrary constants.*

Changing a will only change $F(x)$ by a constant. Therefore:

(2)   *If $F(x)$ is any primitive of $f(x)$, all the primitives of $f(x)$ are the functions $F(x) + C$, where C is an arbitrary constant.*

For example, since $(x^2)' = 2x$, the primitives of $2x$ are the functions $x^2 + C$.

---

[16]The name indicates that the "derivative" f is derived from F as from a source. "Antiderivative" is more widely used, and has twice as many syllables, too.

[17]Recall the comment immediately following the first statement of the theorem in §3.5.3.

In (2) I have omitted the proviso that the derivative function be continuous. Although the Fundamental Theorem as stated requires this, it is in fact unnecessary for the proposition at hand, as will be seen later (§3.7.9).

Note that while many primitives F correspond to a given function f, over any given interval $[a, b]$ they all have the same increment $F(b) - F(a) = \int_a^b f(x)\,dx$, by the original version of Part II. Thus we can regard the Fundamental Theorem as assigning a *unique* function of *intervals* to f, namely the function $\mathscr{S}$ defined by $\mathscr{S}([a, b]) = \int_a^b f(x)\,dx$ (§2.4.3; see also §2.10.3). The *one-to-one* correspondence $f \leftrightarrow \mathscr{S}$ is the correspondence of specific to total quantity on which Chapter 1 was based.

Stated in terms of derivatives, (2) amounts to the known fact that two differentiable functions which differ by a constant have the same derivative, together with its new converse:

(3)   *If two functions have the same derivative, they differ by a constant.*

In particular:

(4)   *If the derivative of a function is zero, the function is constant.*

For $0' = 0$; hence if $F' = 0$, F differs by a constant from 0.

3.5.7   The indefinite integral generalized.

In the general expression $F(x) = \int_a^x f(x)\,dx + C$ for a primitive of an arbitrary continuous function $f(x)$ we cannot dispense with the added **constant of integration** C, as the case $f(x) = 0$ shows—without the C we would not get all the primitives of 0, because $\int_a^x 0\,dx = 0$ for every a. (A more interesting example is suggested in Q4.) Thus a primitive is not, in general, an integral. Nevertheless, I now *extend* the term "indefinite integral," or simply "integral," to mean "indefinite integral (in the original sense) plus a constant," or equivalently "primitive," and employ the integral sign without upper and lower limits to denote it. The symbol

$$\int f(x)\,dx$$

accordingly signifies any primitive of $f(x)$.

This symbol is used in two slightly different ways. We may say, "Let $F(x) = \int f(x)\,dx$" ("Let $F(x)$ be an integral of $f(x)$"), which means, "Let $F(x)$ be any particular one of the primitives of $f(x)$"; or, given a particular primitive $F(x)$, we may say, "$\int f(x)\,dx = F(x) + C$" ("The integral of $f(x)$ is $F(x) + C$"), which means, "The primitives of $f(x)$ are the functions $F(x) + C$, where C is an arbitrary constant." To "integrate" $f(x)$ is to find $\int f(x)\, dx$.

3.5.8   A picture of the correspondence.[18]   A schematic portrait of the correspondence established by the Fundamental Theorem is shown in Fig. 3-17. Differentiation projects each vertical line, which represents by its points a class of functions differing from one another by constants, onto the point on the horizontal line which represents the common derivative of the functions of that class. Integration reverses this: each primitive, or integral (in the generalized sense of §3.5.7), of $f(x)$, and in particular each integral $\int_a^x f(x)\, dx$, is represented by a point on the vertical line above the point representing $f(x)$. (In each case pictured the general form of the integral is written to the left of the vertical line and a particular integral to its right.)

continuously differentiable functions

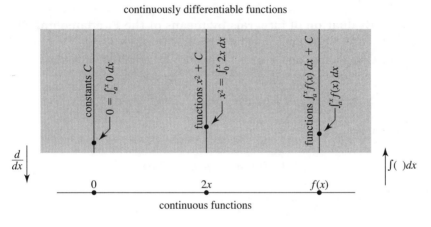

Fig. 3-17

While this diagram has some value as an uncomplicated depiction of the relation between differentiation and integration, in certain respects it is inaccurate and misleading. Among other errors, it represents the continuous and the continuously differentiable functions as distinct classes, whereas in fact the latter is contained in the former.

3.5.9   QUESTIONS

Q1.   Find the integrals (i.e. the primitives in their general form). (a) $\int 0\, dx$ (b) $\int 1\, dx$ (c) $\int 3t^2\, dt$ (d) $\int -\sin x\, dx$ (e) $\int e^t\, dt$ (f) $\int G'(x)\, dx$

---

[18]This section can be omitted.

Q2. What do you conclude if two functions $u(x)$ and $v(x)$ (a) have the same derivative $f(x)$? (b) have the same primitive $F(x)$?

Q3. (a) Give geometrical and physical interpretations of §3.5.6(4). (b) Give a geometrical argument for its truth, based on the interpretation of the derivative as the slope of a graph. (c) Deduce §3.5.6(3) from its special case (4).

Q4. Are all the primitives of e^x represented by integrals $\int_a^x e^x\,dx$?

Q5. Show that every primitive of $1/x\,(x>0)$ can be written in the form $\log ax$, a a positive constant.

Q6. Since all continuously differentiable functions are continuous, but not conversely, there are *more* continuous than continuously differentiable functions. Yet Fig. 3-17 indicates that each continuous function corresponds to many (in fact, infinitely many) continuously differentiable functions. How is this possible?

3.6   THE EVALUATION OF INTEGRALS

3.6.1   Evaluation of integrals by means of the Fundamental Theorem. Part II of the Fundamental Theorem provides a rule for evaluating integrals. Given a function $f(x)$, let $F(x)$ be any primitive of it; then

$$\int_a^x f(x)\,dx = F(x) - F(a),$$

or, if a definite integral is wanted,

$$\int_a^b f(x)\,dx = F(b) - F(a).$$

The rule can be briefly stated this way:

$$\text{If } F(x) = \int f(x)\,dx, \quad \int_a^b f(x)\,dx = F(b) - F(a).$$

Often the difference of values is abbreviated by use of a vertical line or square brackets:

$$\int_a^b f(x)\,dx = F(x)\Big|_a^b = \Big[F(x)\Big]_a^b.$$

How are we to find a primitive $F(x)$ of the given $f(x)$? We can look for $f(x)$ among the derivatives of functions already differentiated; if it is there, a primitive is given. To evaluate $\int_{-\pi/2}^{\pi/2} \cos x\,dx$, for example, we recall that $\cos x = \dfrac{d}{dx}\sin x$, i.e. $\sin x$ is a primitive of $\cos x$; the rule then directs the calculation

$$\int_{-\pi/2}^{\pi/2} \cos x \, dx = \sin x \Big|_{-\pi/2}^{\pi/2} = \sin \frac{\pi}{2} - \sin\left(-\frac{\pi}{2}\right) = 1 - (-1) = 2 \,.$$

Since this integral can be interpreted as the area under an arch of the cosine's graph, we have incidentally solved a very pretty geometrical problem.

When $f(x)$ is not a known derivative, it may be similar enough to one that a primitive can readily be discovered. Faced with $\int_0^1 x^2 \, dx$, I recall having found not x^2, but $3x^2$, as the derivative of x^3, and then realize that $x^3/3$ will be a primitive of x^2:

$$\frac{d}{dx}\left(\frac{x^3}{3}\right) = \frac{1}{3}\frac{d}{dx}x^3 = \frac{1}{3} \cdot 3x^2 = x^2 \,.$$

Consequently

$$\int_0^1 x^2 \, dx = \frac{x^3}{3}\Big|_0^1 = \frac{1}{3} - 0 = \frac{1}{3} \,.$$

3.6.2   Integration of powers of x. The essential step in the preceding example generalizes at once to the formula

$$\int x^\alpha \, dx = \frac{x^{\alpha+1}}{\alpha+1} + C \,, \quad \alpha \neq -1 \,.$$

For differentiation shows that $x^{\alpha+1}/(\alpha+1)$ is a primitive of x^α; then all primitives are obtained by adding constants to it (§3.5.6(2)). This fails if $\alpha = -1$, but we know that x^{-1} has a primitive in $\log x$, provided that $x > 0$ (§3.3.11). For $x < 0$ the function $\log(-x)$ serves as a primitive (Q10), so we have

$$\int \frac{1}{x} \, dx = \begin{cases} \log x + C \,, & x > 0 \,, \\ \log(-x) + C \,, & x < 0 \,, \end{cases}$$

or

$$\int \frac{1}{x} \, dx = \log|x| + C \,, \quad x \neq 0 \,.$$

Thus all powers of x can be integrated. One must remember, however, that negative powers are discontinuous at $x = 0$. While

$$\int_1^2 x^{-2} \, dx = \frac{x^{-1}}{-1}\Big|_1^2 = \frac{1}{2}$$

is correct,

$$\int_{-1}^{2} x^{-2}\, dx = \frac{x^{-1}}{-1}\Big|_{-1}^{2} = -\frac{3}{2}$$

is *incorrect*: the latter integral is not defined, because x^{-2} is not continuous over the interval $[-1, 2]$.[19]

3.6.3   Integrals of arithmetic combinations of functions.   To supplement the ways of finding a primitive that were proposed in §3.6.1 there are certain general methods of integration, of which the simplest will be given here. From the rules for differentiating a sum or difference of functions and a product of a function by a constant we obtain the following integration rules:

$$\int (f(x) + g(x))\, dx = \int f(x)\, dx + \int g(x)\, dx\,,$$

$$\int (f(x) - g(x))\, dx = \int f(x)\, dx - \int g(x)\, dx\,,$$

$$\int c f(x)\, dx = c \int f(x)\, dx\,.$$

The meaning of the first of these is: if $F' = f$ and $G' = g$, then $(F + G)' = f + g$—which we know to be true (§3.2.4). Similarly for the others. Each rule has an analogue for definite integrals, e.g.

$$\int_{a}^{b} (f(x) + g(x))\, dx = \int_{a}^{b} f(x)\, dx + \int_{a}^{b} g(x)\, dx$$

(Q7).

With these rules we can integrate any polynomial, as the following example shows.

$$\int (x^3 - 4x^2 + 5x + 1)\, dx = \int x^3\, dx - \int 4x^2\, dx + \int 5x\, dx + \int 1\, dx$$

$$= \int x^3\, dx - 4\int x^2\, dx + 5\int x\, dx + \int 1\, dx$$

$$= \frac{x^4}{4} - \frac{4x^3}{3} + \frac{5x^2}{2} + x + C\,,$$

---

[19]If the exponent is > -1, discontinuity does not prevent such an integral from existing; see §4.7.3.Q6(a).

the last step by §3.6.2. More generally, we can handle any **linear combination** (sum with numerical coefficients) of functions whose integrals are known, for example:

$$\int \left(\sqrt{2} \, \cos x + \frac{3}{x^2} - e^x \right) dx = \sqrt{2} \int \cos x \, dx + 3 \int \frac{1}{x^2} \, dx - \int e^x \, dx$$

$$= \sqrt{2} \, \sin x - \frac{3}{x} - e^x + C \, .$$

The integration of products of functions will be taken up in Chapter 4 (§4.6.6), together with other methods of integration.

3.6.4 The constant of integration.

The significance of the constant of integration C becomes clear when we solve problems requiring the determination of particular integrals (primitives) of given functions.

Suppose, for example, that a particle moves along the x-axis at the increasing velocity $v(t) = e^t$, and that it passes the origin at $t = 0$. What is its location $x(t)$ at any time t? We are tempted to say: Location is the integral of velocity; the integral of e^t is e^t; therefore $x(t) = e^t$. But this solution would make $x(0) = 1$, whereas by hypothesis we have $x(0) = 0$.

The error lay in asserting that e^t was the integral of e^t: e^t is *an* integral of e^t, but *the* integral of e^t is $e^t + C$, where C is indeterminate (§3.5.7); hence $x(t) = e^t + C$. To determine the value of the constant in the present circumstances we put $t = 0$ (because we know $x(0)$) and find $0 = e^0 + C$, or $C = -1$. Then $x(t) = e^t - 1$.

Another way to solve the problem is to represent $x(t)$ by a definite integral to begin with. Since $x(0) = 0$, $x(t)$ is the distance accumulated from time $t = 0$, hence

$$x(t) = \int_0^t v(t) \, dt = \int_0^t e^t \, dt = e^t \Big|_0^t = e^t - 1 \, .$$

3.6.5 Characterizations of uniformity. Linear functions.[20]

A polynomial function of degree ≤ 1 has the form $\alpha x + \beta$, where α and β are constants. Such a function is called **linear**, because its graph is a straight line.[21] As linear

---

[20]This section can be omitted at a first reading of the chapter. It is used in §3.8.6 and Chapter 6.

[21]The name is sometimes reserved for the case $\beta = 0$, when the graph is a line through the origin, and sometimes for the case $\alpha \neq 0$, when the degree of the polynomial is 1.

functions are everywhere it is useful to have alternative descriptions of them, and integration can supply an essential step in relating one description to another.

Let y be a differentiable function of x. I claim that the following statements are equivalent; that is, if any one of them is true, so are the others.[22]

(i) y is a linear function of x.
(ii) Change in y is proportional to[23] change in x.
(iii) Equal changes in x produce equal changes in y.
(iv) y' is constant.

As an illustration, consider a uniform wire, and let $y = m(x)$ be its cumulative mass measured from some point. Then (ii) says that the mass of an interval (the change in m) is proportional to its length (the change in x); (iii) says that equal intervals have equal masses; (iv) says that line density is constant; and (i) gives the form of the function $m(x)$ (cf. §1.6.3). In general, (i) and (iv) together can be interpreted to mean that *linearity of a cumulative quantity* (y) *is equivalent to constancy of a specific quantity* (y'); both conditions characterize *uniformity*, of which (ii) and (iii) are other descriptions.

In order to establish the equivalence of the four statements it suffices to prove the logical implications (i) $\Rightarrow$ (ii) $\Rightarrow$ (iii) $\Rightarrow$ (iv) $\Rightarrow$ (i).[24] Let $y = \alpha x + \beta$, then if $y_1 = y(x_1)$ and $y_2 = y(x_2)$ we have $y_2 - y_1 = \alpha(x_2 - x_1)$, which proves (ii). Given (ii), (iii) holds *a fortiori*. To show that (iii) $\Rightarrow$ (iv) we argue indirectly: suppose y' not constant, then for some x_1 and x_2, $y'(x_1) \neq y'(x_2)$. Since the difference quotients at x_1 and x_2 approach $y'(x_1)$ and $y'(x_2)$ respectively, for a small enough value of Δx we have

$$\frac{y(x_1 + \Delta x) - y(x_1)}{\Delta x} \neq \frac{y(x_2 + \Delta x) - y(x_2)}{\Delta x},$$

and consequently the numerators are unequal, in contradiction to (iii). Finally, the implication (iv) $\Rightarrow$ (i) is demonstrated by integration. Let y' have the constant value α, then $y = \int \alpha \, dx = \alpha x + \beta$ for some constant β, which is what (i) asserts.

---

[22]See also §3.8.5.Q3(b).

[23]The "constant of proportionality" may be zero: see the comment at the end of the section.

[24](i) implies (ii), i.e. if (i) then (ii); (ii) implies (iii); etc.

The constant α which in (i) is the *coefficient of* x is the *constant of proportionality* in (ii), the *change in* y *corresponding to unit change in* x in (iii), and the *value of* y' in (iv). In the trivial case $\alpha = 0$, y is constant, there is no true proportionality, and "uniformity" is present only in a manner of speaking.

3.6.6 QUESTIONS

Q1.   Integrate. (a) $\int_0^1 x^4\, dx$ (b) $\int \sqrt{t}\, dt$ (c) $\int_0^{\pi/2} \sin x\, dx$ (d) $\int (3t + 5e^t)\, dt$
(e) $\int_1^2 \left(\dfrac{2}{x} - \dfrac{3}{x^2} \right) dx$ (f) $\int (a \sin x + b \cos x)\, dx$ (a, b constants)

Q2.   Find the area of the segment of the parabola $y = 4 - x^2$ cut off by the x-axis. (Sketch the curve and express the area as an integral of y.)

Q3.   A water tank is being filled by two pipes. Through one of them a steady flow of 1 m^3/sec is maintained, while the flow through the other diminishes after $t = 1$, so that for $t \geq 1$ its rate is $1/t^2$ m^3/sec. How much water enters the tank from $t = 1$ to $t = 10$? (The total rate of flow is the sum of the two rates; integrate it.)

Q4.   In Fig. 2-36 (§2.10.2) the expansive force of the spring will be proportional to the displacement of its free end from position x_1 (Hooke's law). Let x be the location of the end of the spring, or of the left face of the block (once it has reached x_1); then the force is $f(x) = c(x_1 - x)$, where c is a positive constant. Show that the work done on the block as it moves from x_1 to x_2 is the negative quantity $-\frac{1}{2}c(x_1 - x_2)^2$. (Integrate the force as a difference of two terms, then use algebra to identify the result with the quantity named.)

Q5.   An oscillating particle has velocity $v(t) = 2 \cos t$. (a) Find the general form of its position function $x(t)$. (b) What is the meaning of the constant of integration?

Q6.   Discuss Q3 of §1.5.10 with reference to the indefinite integral.

Q7 (§3.6.3).   (a) Deduce the rules for definite integrals from the rules for indefinite integrals. (b) Prove the rules directly from the definition of the integral as a limit of sums.

Q8.   Interpret (i)–(iv) of §3.6.5 as they apply to the physical and geometrical situations of Chapter 1 other than that of the wire. In particular, carefully characterize uniformly accelerated motion (§1.3.3) in each of the four ways.

Q9.   Show directly that the function $y = x^2$ does not satisfy (ii), (iii), or (iv) of §3.6.5.

Q10.   In order to show that $\dfrac{d}{dx} \log (-x) = \dfrac{1}{x}$ for $x < 0$, proceed as follows. Let $g(x) = \log (-x)$. (a) Write down the difference quotient for $g(x)$ at x, with increment Δx. (b) Rewrite it in terms of log, and show that the result equals the negative of the difference quotient for log at $-x$ with increment $-\Delta x$. (c) Conclude that
$$g'(x) = -\frac{1}{-x} = \frac{1}{x}.$$

3.7 THE FUNDAMENTAL THEOREM AND THE MEAN VALUE THEOREMS[25]

3.7.1 Maximum and minimum values of a continuous function. The proof of the Fundamental Theorem rests upon a property of continuous functions whose demonstration would require closer attention to the foundations of the real number system than I have space to give. Fortunately, the property itself appears obvious to the intuition, so we may willingly grant it to the functions and go on to the proof. It is as follows.

> Maximum–Minimum Property. *A function $f(x)$ continuous on an interval $[a, b]$ has a maximum and a minimum value there.*

This means that there are numbers x_1 and x_2 in $[a, b]$ such that for all x in $[a, b]$,

$$f(x_1) \leq f(x) \leq f(x_2).$$

The **minimum** value m is $f(x_1)$ and the **maximum** value M is $f(x_2)$. Geometrically, the meaning is that a continuous graph drawn from a point A to a point B has a maximum height M and a minimum height m; of course, each may occur at more than one point (Fig. 3-18).

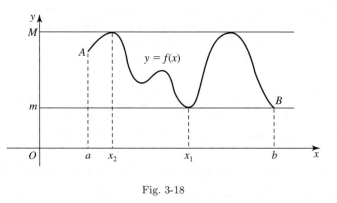

Fig. 3-18

---

[25] Avoiding theory and proof, one can find enough information to go on with in the following brief portions of this article: §3.7.1; §3.7.2(1); §3.7.5; §3.7.7, first paragraph; §3.7.11; §3.7.14. The proofs omitted in §§3.7.1, 3.7.12, and 3.7.14 can be found in the sources cited in §2.4.5, including Courant, pp. 63–67.

Examples show that there need not be a maximum or minimum if the function is not continuous, or if the interval on which the maximum and minimum are sought is one which does not include both endpoints, such as an interval $a < x < b$ (Q1).

3.7.2   Integral inequalities.   The immediate application to be made of the maximum–minimum property arises from the following fact about definite integrals.

(1)   *If $g(x) \leq h(x)$ everywhere in $[a, b]$, then*

$$\int_a^b g(x)\, dx \leq \int_a^b h(x)\, dx.$$

Its straightforward proof is the subject of Q3. From this integral inequality we reason that since by the definition of maximum and minimum

$$m \leq f(x) \leq M$$

everywhere in $[a, b]$, we have (regarding m and M as constant functions, Fig. 3-19)

$$\int_a^b m\, dx \leq \int_a^b f(x)\, dx \leq \int_a^b M\, dx.$$

If the integrals of the constants m and M are evaluated the inequalities become

$$m(b - a) \leq \int_a^b f(x)\, dx \leq M(b - a),$$

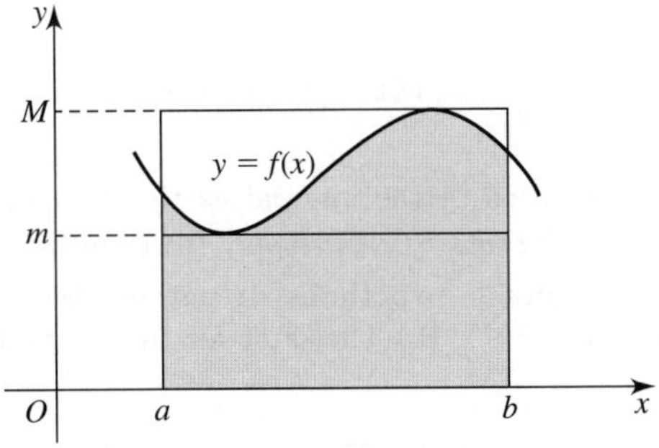

Fig. 3-19

or

(2) $$m \le \frac{1}{b-a} \int_a^b f(x)\,dx \le M\,,$$

or finally

(3) $$f(x_1) \le \frac{1}{b-a} \int_a^b f(x)\,dx \le f(x_2)\,,$$

a result we will need for the Fundamental Theorem.

Note that since

$$\int_b^a f(x)\,dx = -\int_a^b f(x)\,dx\,,$$

$$\frac{1}{a-b} \int_b^a f(x)\,dx = \frac{1}{b-a} \int_a^b f(x)\,dx$$

and hence we also have

$$m \le \frac{1}{a-b} \int_b^a f(x)\,dx \le M\,;$$

in other words, (2) and (3) hold whether the upper limit or the lower limit of the integral is the greater.

3.7.3   QUESTIONS

Q1 (§3.7.1).   Provide an example for each case.

Q2.   Deduce from (1) that

$$\int_a^b g(x)\,dx \le \int_a^b |g(x)|\,dx\,.$$

Q3.   (a) From the definition of the integral as the limit of sums argue that if $f(x) \ge 0$ on $[a, b]$, then $\int_a^b f(x)\,dx \ge 0$. (b) Apply (a) to prove (1).

Q4.   (a) By adding a suitable hypothesis strengthen the conclusion of (1) to a strict inequality ($<$ instead of $\le$). (b) The same for the inequalities in (2).

3.7.4   Proof of Part I of the Fundamental Theorem.   Let $f(x)$ be

a continuous function. The derivative at any x_0 of $F(x) = \int_a^x f(x)\,dx$, if it exists, is the limit as $\Delta x \to 0$ of

$$\frac{F(x_0 + \Delta x) - F(x_0)}{\Delta x} = \frac{\int_a^{x_0+\Delta x} f(x)\,dx - \int_a^{x_0} f(x)\,dx}{\Delta x}$$

$$= \frac{1}{\Delta x} \int_{x_0}^{x_0+\Delta x} f(x)\,dx$$

(the last step uses §2.4.4, with §2.10.9.Q3).

On the interval of integration between x_0 and $x_0 + \Delta x$, which is $[x_0, x_0 + \Delta x]$ if $\Delta x > 0$ or $[x_0 + \Delta x, x_0]$ if $\Delta x < 0$, $f(x)$ has a maximum and a minimum value. Let the minimum occur at x_1 and the maximum at x_2; then by §3.7.2(3), whether Δx is positive or negative we have

$$f(x_1) \le \frac{1}{\Delta x} \int_{x_0}^{x_0+\Delta x} f(x)\,dx \le f(x_2)$$

(Fig. 3-20). As $\Delta x \to 0$, both x_1 and x_2 approach x_0; by continuity, therefore, $f(x_1) \to f(x_0)$ and $f(x_2) \to f(x_0)$. Then $\dfrac{1}{\Delta x} \displaystyle\int_{x_0}^{x_0+\Delta x} f(x)\,dx$, which is trapped between $f(x_1)$ and $f(x_2)$, must approach $f(x_0)$ as well. Thus $F'(x_0)$ exists and equals $f(x_0)$, as Part I asserts.[26]

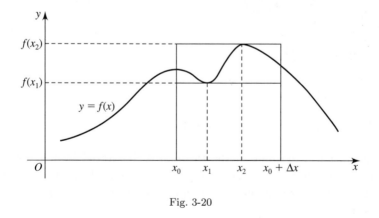

Fig. 3-20

---

[26] The same proof, in the special case of an increasing function, appeared as a geometrical argument at the end of §2.9.12. See also §3.7.6.Q5 and §3.7.17.Q7.

3.7.5   The mean value theorem.   Part II of the Fundamental Theorem is proved with the aid of the following theorem, which is important in its own right.

> The Mean Value Theorem. *If $f(x)$ is continuous on $[a, b]$ and differentiable for $a < x < b$ there is a number ξ, $a < \xi < b$, such that*
>
> $$f(b) - f(a) = f'(\xi)(b - a).$$
>
> *The same equation holds if $b < a$, provided that f is continuous on $[b, a]$ and differentiable for $b < x < a$; in that case $b < \xi < a$.*

The second part of the theorem follows from the first part upon interchanging a and b and reversing the sign of each side of the resulting equation.

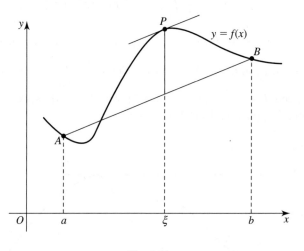

Fig. 3-21

The geometrical meaning of the theorem is illustrated by Fig. 3-21. The slope of secant AB is $[f(b) - f(a)]/(b - a)$; it is asserted that there is at least one point P on the graph between A and B at which the graph has the same slope as AB, i.e.

$$f'(\xi) = \frac{f(b) - f(a)}{b - a}.$$

As the figure suggests, such a point will be found where the graph is farthest from AB. Indeed, if we shift AB parallel to itself until it is as far as it can be from its original position without losing contact with the curve, it will evidently be tangent to the curve where it meets it. This idea of a maximum will form the basis of the proof of the theorem.

The difference quotient $[f(b) - f(a)]/(b - a) = \Delta f/\Delta x$ is called the **mean value** of the derivative function f' over the interval $[a, b]$, in agreement with §2.6.2, where the average or mean line density $\lambda = m'$ over $[a, b]$ was defined to be $[m(b) - m(a)]/(b - a)$. Thus in Fig. 3-21 the slope of the secant is the mean value of the slope of the curve between A and B. The theorem asserts that at some point ξ the derivative takes its mean value.[27]

As another illustration, consider the application of the mean value theorem to the location function $x(t)$ of a particle. Its derivative is $v(t)$, so there exists a time τ at which

$$\frac{\Delta x}{\Delta t} = v(\tau) \, .$$

If, for example, the particle travels 100 m in a given direction in 20 sec, there must be some instant at which its speed is exactly equal to its mean speed of 5 m/sec.

Often the theorem is applied the other way: from information about the value of the derivative a conclusion is drawn about the function. The first such applications will be made in §3.7.9.

3.7.6   QUESTIONS

Q1.   (a) Write down the equation of the mean value theorem for the function $f(x) = x^2$ and solve it for ξ. (b) Interpret the result geometrically.

Q2.   The same question for $f(x) = 1/x$, assuming a and $b > 0$.

Q3.   Explain the connection between the mean value theorem and the "matching motions" of §1.3.1.

Q4.   Does the following argument prove that all derivatives are continuous?

Let $f(x)$ be differentiable, and form the difference quotient at any x_0. We have $\Delta f/\Delta x \to f'(x_0)$, and by the mean value theorem, $\Delta f/\Delta x = f'(\xi)$ for ξ between x_0 and $x_0 + \Delta x$. Now as $\Delta x \to 0$, $\xi \to x_0$; hence as $\xi \to x_0$, $f'(\xi) \to f'(x_0)$, which means that f' is continuous at x_0.

---

[27]Here it can be assumed that f' exists at a and b. The connection between this notion of mean value and the familiar definition of a mean or average in terms of a sum will be made in §3.7.16.

Q5.   Use §3.1.3 to formalize the argument at the end of §3.7.4 about a "trapped" function having a certain limit.

3.7.7   The derivative at a maximum or minimum.   We remarked in §3.7.5 that a point ξ to satisfy the mean value theorem is found where a certain quantity assumes its maximum value. As a preliminary to the proof of the theorem, let us observe that where a function has a maximum or minimum, its graph has a horizontal tangent and its derivative is accordingly zero. More precisely, there is the following proposition.

> Proposition. *Let a function $f(x)$ be continuous on an interval $[a, b]$, and suppose that the maximum or minimum value of $f(x)$ occurs at an interior point of the interval. If $f(x)$ is differentiable at that point, its derivative is zero there.*

The restriction to interior points is explained by Fig. 3-22.

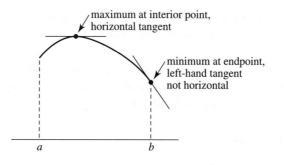

Fig. 3-22

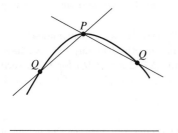

Fig. 3-23

The geometrical proof of this proposition is as follows. It is sufficient to treat the case of a maximum. Let the maximum occur at point P of the graph, and let Q be another point on the graph (Fig. 3-23). Q is not higher than P; therefore if Q is to the left of P the slope of secant PQ is ≥ 0, while if Q is to the right of P the slope of PQ is ≤ 0. Now the slope of PQ approaches that of the tangent at P whether Q approaches P from the left or from the right. If the approach is from the left the limit must be ≥ 0, while if it is from the right the limit must be ≤ 0 (§3.1.9(1)). Hence it can only be 0, and that is then the slope of the tangent, which equals the value of the derivative.

A proof independent of geometry is easily devised in imitation of the geometrical proof. Let $f(x)$ take its maximum at x. For any Δx small enough that $x + \Delta x$ is in the interval, $\Delta f = f(x + \Delta x) - f(x) \leq 0$. Then for $\Delta x < 0$, $\Delta f / \Delta x \geq 0$, while for $\Delta x > 0$, $\Delta f / \Delta x \leq 0$. But $f'(x)$ is the limit of $\Delta f / \Delta x$ whether Δx approaches 0 through negative or through positive values. The former case gives $f'(x) \geq 0$, the latter $f'(x) \leq 0$; hence $f'(x) = 0$.

3.7.8   Proof of the mean value theorem.   We are now ready to prove the mean value theorem. It is convenient (and traditional) to begin with the special case, known as Rolle's theorem, in which the value of the function at the endpoints of the interval is zero. Under this additional hypothesis the mean value theorem asserts the existence of an interior point at which the derivative vanishes (Fig. 3-24).

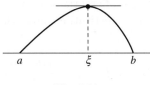

Fig. 3-24

Suppose, then, that $f(a) = f(b) = 0$. We know that $f(x)$ has a maximum and a minimum value on the interval. They cannot both occur *only* at endpoints, for in that case f would be zero everywhere (its greatest and least values both being zero) and *every* point of the interval would be a maximum and a minimum point—contrary to the assumption that only a and b could be such. Consequently there is a maximum or a minimum at an interior point ξ, and then $f'(\xi) = 0$, by the preceding proposition. This proves Rolle's theorem.

To establish the mean value theorem in the general case we apply Rolle's theorem to a function which represents the (signed) vertical distance from secant to graph (Fig. 3-25).[28]   Let $y = g(x)$ be the equation of line AB;

---

[28] The vertical distance of a point P on the graph from the secant differs only by a constant factor from the distance from P to the secant, as is easily verified.

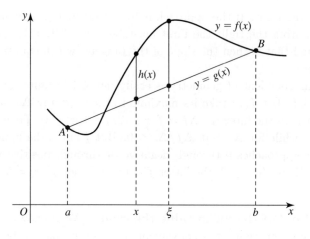

Fig. 3-25

that is, let $g(x) = \mu(x - a) + f(a)$, where $\mu = [f(b) - f(a)]/(b - a)$. Then
the distance function is $h(x) = f(x) - g(x)$, which satisfies the hypotheses of
Rolle's theorem. For as a difference of continuous functions h is continuous,
and as a difference of functions differentiable on the interior of the interval it is
differentiable there; moreover, substitution shows that $h(a) = h(b) = 0$. Hence
there is an interior point ξ such that $h'(\xi) = 0$. But $h'(\xi) = f'(\xi) - g'(\xi)$, so

$$f'(\xi) = g'(\xi) = \mu = \frac{f(b) - f(a)}{b - a} \,.$$

The proof of the mean value theorem is complete.

3.7.9   Proofs of Part II of the Fundamental Theorem.   A. First
proof.   An immediate corollary of the mean value theorem is the proposition
that a function whose derivative vanishes everywhere is constant (§3.5.6(4)).
For if $f'(x) = 0$ for all values of x then for any a and b

$$f(b) - f(a) = f'(\xi)(b - a) = 0 \,,$$

i.e. $f(a) = f(b)$, so that the value of the function is the same at all points. This
in turn implies that two primitives of the same function differ by a constant
(§3.5.6(3)): if $F' = G'$ then $(F - G)' = F' - G' = 0$, hence $F - G$ is constant.

The latter result enables us to deduce Part II of the Fundamental Theorem
from Part I. It is sufficient to prove Part II in the form given in §3.5.6: that any

primitive of a continuous function $f(x)$ differs only by a constant from some indefinite integral $\int_a^x f(x)\,dx$ of the function.[29] But Part I tells us that any such integral is a primitive, and we now know that any two primitives differ by a constant. Thus Part II is established.

3.7.10   B. Second proof.   Unlike the argument by infinitesimals in §3.5.4, the preceding proof does not demonstrate the all-important formula

$$\int_a^b f(x)\,dx = F(b) - F(a)$$

directly. We can imitate the earlier argument by a different application of the mean value theorem, in the following way.

Let $f(x)$ be continuous and let $F' = f$. Let the interval from a to b be divided into n subintervals by points $x_0 = a$, $x_1, x_2, \ldots, x_n = b$. For each i from 1 to n we write as usual $\Delta x_i = x_i - x_{i-1}$; also, let $\Delta F_i = F(x_i) - F(x_{i-1})$. By the mean value theorem, there is a point ξ_i in the ith subinterval such that

$$\Delta F_i = f(\xi_i)\Delta x_i\,.$$

The sum of the increments ΔF_i is the total increment of F over $[a, b]$:

$$\Delta F_1 + \Delta F_2 + \cdots + \Delta F_n$$
$$= F(x_1) - F(x_0) + F(x_2) - F(x_1) + \cdots + F(x_n) - F(x_{n-1})$$
$$= F(x_n) - F(x_0) = F(b) - F(a)\,.$$

Therefore

$$f(\xi_1)\Delta x_1 + f(\xi_2)\Delta x_2 + \cdots + f(\xi_n)\Delta x_n = F(b) - F(a)\,.$$

As the number of subintervals is increased in such a way that the length of the longest approaches zero, the sum on the left approaches $\int_a^b f(x)\,dx$, while $F(b) - F(a)$ remains fixed. This proves the formula.

---

[29]The term "indefinite integral" is here used in its original sense, not the extended sense introduced in §3.5.7.

3.7.11   Estimation of values.   At the beginning of §3.7.9 we applied the mean value theorem when $f' = 0$. A less drastic assumption is that f' is bounded in absolute value—that is, that $|f'|$ is nowhere greater than some fixed number B. Geometrically, this means that the steepness of the graph of f nowhere exceeds B. By the maximum-minimum property (§3.7.1) $|f'|$ will surely be bounded on an interval $[a, b]$ if f' is continuous there; for the continuity of f' implies that of $|f'|$ (§3.1.6.Q4).

Assuming that $|f'|$ is bounded by B on $[a, b]$, we can apply the mean value theorem to any subinterval of $[a, b]$, with endpoints x_0 and x_1, say, to obtain

$$|f(x_1) - f(x_0)| = |f'(\xi)(x_1 - x_0)| = |f'(\xi)|\, |x_1 - x_0|$$

and hence

(4) $$|f(x_1) - f(x_0)| \leq B|x_1 - x_0|\,.$$

If B is known, the inequality gives us an upper bound to the distance of $f(x_1)$ from $f(x_0)$ in terms of the distance of x_1 from x_0.

As an application of this idea, suppose that the value of $f(x)$ is known at x_0, but not at the nearby point x_1. We may be willing to accept $f(x_0)$ as an estimate of $f(x_1)$, if we can be assured that the error involved in the approximation is not large. Let us write

$$f(x_1) = f(x_0) + R\,,$$

where R is the remainder, or error; then our inequality says that in magnitude R does not exceed $B|x_1 - x_0|$, and the latter quantity is easily calculated.

To take an example, how well is $\sin 31°$ estimated by $\sin 30° = 0.5$? Here $f(x) = \sin x$, and $|f'(x)| = |\cos x| \leq 1$ for all x; therefore we can put $B = 1$. The difference in x-values is $1°$, or $\pi/180$ radians, which is less than 0.018. Thus the error in the approximation is less than 0.018 in absolute value.

Without any other knowledge of the function, we could say at this point that

$$0.482 < \sin 31° < 0.518\,.$$

If we recall that $\sin x$ increases between $x = 0$ and $x = \pi/2$, so that $\sin 30° < \sin 31°$, we can halve the range of possibility:

$$0.5 < \sin 31° < 0.518\,.$$

As a matter of fact $\sin 31°$ to three decimal places is 0.515.

With further refinement, reasoning such as we have done here is of great service to practical calculation—e.g., of the value 0.515 just cited. It also forms the basis of the study of functions by way of so-called infinite series. Further development of these matters will be undertaken in §3.8.6 and Chapter 8.

3.7.12   Uniform continuity. Inequality (4) has an interesting consequence pertaining to the idea of continuity. Continuity of f assures us that as $x_1 \to x_0$, $f(x_1) \to f(x_0)$, wherever x_0 may be; but how near to x_0 we must choose x_1 in order to get $f(x_1)$ within a prescribed distance of $f(x_0)$ may very well depend strongly on the location of x_0. Given the inequality, however, we know that $f(x_1)$ will necessarily be within any given distance ε of $f(x_0)$, provided only that x_1 is within ε/B of x_0; for if $|x_1 - x_0| \leq \varepsilon/B$,

$$|f(x_1) - f(x_0)| \leq B(\varepsilon/B) = \varepsilon.$$

In other words, *the size of the difference in values of f is controlled by the size of the difference in x-values, without regard to location in $[a, b]$*—controlled, then, *uniformly across the interval.*

When this uniformity obtains (whether or not $|f'|$ is bounded), f is said to be **uniformly continuous** on $[a, b]$. We have proved that f is uniformly continuous on $[a, b]$ if it is continuously differentiable; it is a remarkable fact that continuity alone is sufficient to ensure uniform continuity:

> Proposition. *If a function $f(x)$ is continuous on an interval $[a, b]$, it is uniformly continuous there.*

I omit the proof, which depends on the same kind of considerations as that of the maximum-minimum property.

On an interval which does not include both endpoints a function can be continuous without being uniformly continuous (Q4).

3.7.13   QUESTIONS

Q1.   Show that 0.99^{10} is within 0.1 of 1. (Consider the derivative of x^{10} on the interval $[0.99, 1]$.)

Q2.   Estimate $\tan 44°$.

Q3.   Suppose that $|f'(x)| \leq 2$ everywhere on $[a, b]$, and the straight line $y = \alpha x$ cuts the graph of $f(x)$ twice. What can you conclude about α?

Q4.   (a) Show that the derivative of $1/x$ is not bounded in absolute value on the interval $0 < x \leq b$. (b) Show that $1/x$ is not uniformly continuous on that interval; that is, given $\varepsilon > 0$, there is no distance so small that if x_0 and x_1 in the interval are within that distance of each other, regardless of their location, $1/x_1$ is within ε of $1/x_0$.

Q5.   Make the argument suggested in step 2 of §2.4.5.

3.7.14   Intermediate values of a continuous function.   A companion to the maximum-minimum property of §3.7.1 is a characteristic of continuous functions that is implied by the very name "continuous." It too will have to be accepted without proof.

> Intermediate Value Property. *A function $f(x)$ continuous on an interval $[a, b]$ assumes there as a value every number between its maximum and its minimum value.*

Let M be the maximum and m the minimum value of f on $[a, b]$, and let c be any number such that $m \leq c \leq M$; then by the intermediate value property there is some number x_0 in $[a, b]$ such that $f(x_0) = c$. The geometrical meaning is that a horizontal line $y = c$, drawn at a height between the maximum and the minimum, meets the graph at least once (Fig. 3-26).

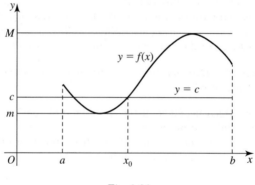

Fig. 3-26

If $f(x)$ assumes on $[a, b]$ every value between m and M, surely $f(x)$ *assumes there as a value every number between $f(a)$ and $f(b)$.* That statement is in fact equivalent to the intermediate value property, as is also this one: *if $f(a)$ and $f(b)$ are opposite in sign, there is a point x_0 between a and b such that $f(x_0) = 0$* (Q4).

3.7.15   The integral mean value theorem.   In conjunction with inequality (2) of §3.7.2 the intermediate value property assures us of the truth of the following proposition.

Integral Mean Value Theorem. *If $f(x)$ is continuous on $[a,b]$ there is a number ξ, $a \le \xi \le b$, such that*

$$\int_a^b f(x)\,dx = f(\xi)(b-a)\,.$$

(*The* mean value theorem of §3.7.5 can be called the *differential* mean value theorem if it is necessary to distinguish it from this one. As with the differential theorem, the equation here remains valid if a and b are interchanged.) The quantity $\mu = [1/(b-a)] \int_a^b f(x)\,dx$, which the theorem asserts to be a value of $f(x)$, is called the **mean value** of the function over the interval $[a,b]$. Geometrically, if $f(x)$ is non-negative its mean value represents the height of a rectangle based on the interval and equal in area to the figure under the graph of $y = f(x)$ (Fig. 3-27; see also Q1(c)).

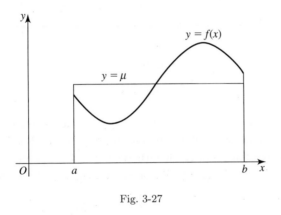

Fig. 3-27

3.7.16   The mean value of a function.   We now have two definitions of the mean value of a function $f(x)$ over an interval $[a,b]$. According to §3.7.5 the mean value is[30]

$$\frac{F(b)-F(a)}{b-a}\,,$$

where $F' = f$, while we have just now said that it is

$$\frac{1}{b-a}\int_a^b f(x)\,dx\,.$$

---

[30]Writing F in place of the f in §3.7.5.

The Fundamental Theorem tells us that these quantities are equal, so the definitions are not in conflict.

> Given the Fundamental Theorem, it is easy to deduce the integral mean value theorem from the differential one, and in fact with the slightly stronger conclusion that $a < \xi < b$ instead of $a \le \xi \le b$ (Q6(a)). This way of proceeding conceals the elementary character of the integral theorem. To reverse the deduction we must assume f continuously differentiable (Q6(b)).

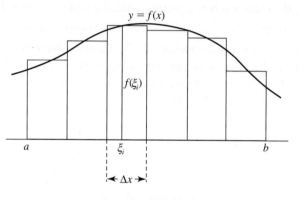

Fig. 3-28

The mean value of a function as I have defined it can be regarded as the limit of a mean or average of numbers in the ordinary sense. Let $[a, b]$ be divided into n *equal* subintervals, each of length $\Delta x = (b-a)/n$, and set up rectangles on them as in §2.4.1 (Fig. 3-28). The average height of the n rectangles is

$$\frac{f(\xi_1) + f(\xi_2) + \cdots + f(\xi_n)}{n} = \frac{[f(\xi_1) + f(\xi_2) + \cdots + f(\xi_n)]\Delta x}{n\Delta x}$$

$$= \frac{f(\xi_1)\Delta x + f(\xi_2)\Delta x + \cdots + f(\xi_n)\Delta x}{b - a},$$

the limit of which as $n \to \infty$ is $[1/(b - a)] \int_a^b f(x)\, dx$. Thus the mean value of $f(x)$, or average height of the graph, is the number approached as a limit by the average height of the rectangles as their number is increased without limit—a very reasonable notion of "mean."

Alternatively, without invoking geometry we can simply say that the mean is the limit of the average of the numbers $f(\xi_i)$, where ξ_i belongs to the ith subinterval. At each stage these numbers constitute a sample of n values of f, taken at x-values distributed pretty evenly across $[a, b]$; as the size of the sample increases, its average approaches the mean value of f.

More generally, if the subintervals are not required to be equal, the mean value is represented as the limit of a sequence of weighted averages of the $f(\xi_i)$. A **weighted average** of numbers $\alpha_1, \alpha_2, \ldots, \alpha_n$ is a sum of the form

$$w_1\alpha_1 + w_2\alpha_2 + \cdots + w_n\alpha_n \,,$$

where each $w_i \geq 0$ and $w_1 + w_2 + \cdots + w_n = 1$. If all the "weights" w_i are equal, each equals $1/n$ and the weighted average is the ordinary average; otherwise, the relative sizes of the w_i determine the relative contributions of the α_i to the sum (i.e. how much "weight" each α_i has). In the present case, for each i we let $w_i = \dfrac{\Delta x_i}{b-a}$; then it is easy to see that the w_i satisfy the conditions to form a set of weights, and that the mean is the limit of $w_1 f(\xi_1) + w_2 f(\xi_2) + \cdots + w_n f(\xi_n)$ as n is increased and the length of the longest subinterval approaches zero.

As an example let us calculate the mean value of $1 - x^2$ over the interval $[-1, 1]$. The length of the interval is 2, so the mean value is

$$\frac{1}{2}\int_{-1}^{1}(1-x^2)\,dx = \frac{1}{2}\left[x - \frac{x^3}{3}\right]_{-1}^{1} = \frac{2}{3}\,.$$

By the integral mean value theorem we know that this value is assumed by the function at a point ξ, i.e.

$$1 - \xi^2 = \frac{2}{3}\,;$$

solving, we find $\xi = \pm 1/\sqrt{3}$—the mean value is taken twice. The geometrical interpretation is shown in Fig. 3-29, where the area of the rectangle equals the area under the parabola.

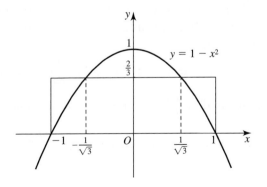

Fig. 3-29

3.7.17   QUESTIONS

Q1.   (a) Write down the equation of the integral mean value theorem for the function $f(x) = x$ and solve it for ξ. (b) Interpret the result geometrically. (c) Interpret the theorem geometrically for any $f(x)$, not necessarily non-negative.

Q2.   (a) Find the mean value of e^x over $[0, 1]$. (b) At what value of x is it assumed by the function?

Q3.   From $t = 0$ to $t = \pi$ a force $f(t)$ increases, then decreases, obeying the law $f(t) = \sin t$. Find the average force over the time interval.

Q4 (§3.7.14).   Prove the three statements equivalent.

Q5.   (a) Prove that if $\alpha^2 + \beta^2 = 1$ there is a number u, unique up to an additive multiple of 2π, such that $\cos u = \alpha$ and $\sin u = \beta$. (For the uniqueness use §3.3.9.Q7(b).) (b) Hence show analytically that the second polar coordinate θ of a point P (other than the pole) is determined up to a multiple of 2π by the Cartesian coordinates (x, y) of P.

Q6 (§3.7.16).   (a), (b): Perform the deductions.

Q7.   Use the integral mean value theorem to shorten the proof of Part I of the Fundamental Theorem given in §3.7.4.

3.8   HIGHER DERIVATIVES AND LINEAR APPROXIMATION

3.8.1   Higher derivatives of a function.   The derivative of a differentiable function $f(x)$ is another function $f'(x)$. If that function is differentiable its derivative is still another function $f''(x)$ ("f-double-prime of x"), and continuing in this way we generate a sequence of functions f', f'', f''', $f^{(4)}$, ... We call f'' the **second derivative** of f, f''' the third derivative, etc., and refer to f' as the **first derivative** if a distinction has to be made. The nth derivative is also called the derivative of **order** n, or the nth-order derivative.[31] The function f is **twice differentiable** if f'' exists, and in general n times differentiable if $f^{(n)}$ exists.

Instead of f'', f''', etc., we can use the notation $\dfrac{d^2f}{dx^2}$ ("d two f dx squared"

or "d squared f dx squared"), $\dfrac{d^3f}{dx^3}$, etc., and denote the value at x_0 by $\dfrac{d^2f}{dx^2}\Big|_{x_0}$,

etc. If the function is $y = f(x)$, we also write $y'' = \dfrac{d^2y}{dx^2}$, etc. I will not try

---

[31] Occasionally it is convenient to call f the derivative of order zero and denote it by $f^{(0)}$.

to assign meaning to d^2y (a "second differential"), so we will regard $\dfrac{d^2y}{dx^2}$ as an inseparable symbol, not a quotient; the notation is justified by observing that as the derivative of the first derivative $\dfrac{d^2y}{dx^2}$ is $\dfrac{d}{dx}\left(\dfrac{dy}{dx}\right)$. Similarly for the derivatives of order higher than two.

3.8.2　Higher derivatives of polynomials and other functions.

A polynomial of degree n has derivatives of all orders, of which only the first n are non-zero. Consider, for example, the fourth-degree polynomial

$$y = x^4 + 5x^2 - 3x + 4\,.$$

We calculate

$$y' = 4x^3 + 10x - 3\,,$$
$$y'' = 12x^2 + 10\,,$$
$$y''' = 24x\,,$$
$$y^{(4)} = 24\,,$$
$$y^{(5)} = 0\,,$$

and all subsequent derivatives are also zero.

Conversely, if $y^{(n+1)} = 0$, y is a polynomial of degree at most n (exactly n if $y^{(n)} \neq 0$), as we see by repeated integration (§3.6.3). This generalizes §3.5.6(4). To illustrate, suppose $y''' = 0$. Then y'' is constant,

$$y'' = c_1\,;$$

integrating c_1 gives

$$y' = c_1 x + c_2\,,$$

where c_2 is another constant, and a further integration gives finally

$$y = \frac{c_1}{2}x^2 + c_2 x + c_3\,.$$

In this expression for y the constants can only be determined if the given equation $y''' = 0$ is supplemented by additional information. Three values of y will suffice: given, say, $y(0) = 1$, $y(1) = 0$, and $y(-1) = 4$, we find that $c_1 = 2$, $c_2 = -2$, and $c_3 = 1$, so that $y = x^2 - 2x + 1$. The same conclusion can be obtained from other data, e.g. the values of y and its first two derivatives at a single point: given $y(0) = 1$, $y'(0) = -2$, and $y''(0) = 2$, we again find that $y = x^2 - 2x + 1$ (Q5).

With functions other than polynomials, successive differentiation produces a variety of patterns. All the derivatives of e^x are the same as the original function, but its inverse $\log x$ gives rise to the sequence

$$\frac{1}{x}, -\frac{1}{x^2}, \frac{2}{x^3}, -\frac{6}{x^4}, \cdots.$$

Again, starting with

$$y = \sin x$$

we obtain

$$y' = \cos x,$$

$$y'' = -\sin x,$$

$$y''' = -\cos x,$$

$$y^{(4)} = \sin x = y,$$

and the sequence repeats. As a rule, the formulas for derivatives of complex functions become more and more complicated as the order of the derivatives increases.

3.8.3 QUESTIONS

Q1. Calculate all the derivatives of $y = 2t^5 - t^4 - 3t^3 + 4t^2 + 4t - 1$.

Q2. Calculate the second derivative of $x^2 \sin x$.

Q3. Show by repeated application of the product rule that the nth derivative of xe^x is $(x + n)e^x$.

Q4. Find a formula for the nth derivative of (a) x^n; (b) x^α.

Q5 (§3.8.2). Verify that each of the two sets of data determines y as stated.

Q6. Given that $y'' = 2$ (the constant function), $y(0) = 0$, and $y'(0) = 1$, find y.

Q7. Show that a polynomial $p(x)$ of degree $\leq n$ is uniquely determined by its value and the values of its first n derivatives at any given point a.

Q8. Find a function $y(x)$ such that $y'' + y = 0$, $y(0) = 1$, and $y'(0) = 0$.

Q9. Let $f(x)$ be an odd or even function. (a) What kinds of functions are its successive derivatives? (See §3.3.9.Q9(b).) (b) What can be said about their values at $x = 0$?

Q10. Use Q6 of §3.2.3 to prove *Leibniz's rule* for differentiating a product:

$$(fg)^{(n)} = f^{(n)}g + \frac{n}{1} f^{(n-1)}g' + \frac{n(n-1)}{1 \cdot 2} f^{(n-2)}g''$$

$$+ \frac{n(n-1)(n-2)}{1 \cdot 2 \cdot 3} f^{(n-3)}g''' + \cdots + fg^{(n)}.$$

3.8.4   Interpretations of the second derivative. The first derivative dy/dx represents the rate of change of y with respect to x, or y per x (§2.9.1). The second derivative $\dfrac{d}{dx}\left(\dfrac{dy}{dx}\right) = \dfrac{d^2y}{dx^2}$ therefore represents the rate of change of that rate of change, with respect to x; i.e. dy/dx per x.

In certain contexts there are natural interpretations for this quantity. Geometry provides two of these.

1. Let a curve that lies above the x-axis have equation $y = f(x)$ and let $F(x) = \int_a^x f(x)\,dx$, the area under the curve from a to x. By the Fundamental Theorem $F'(x) = f(x)$, so $F''(x) = f'(x)$—the second derivative of the area under the curve is the slope of the curve (Fig. 3-30). For example, the triangular area under $y = 2x$ from 0 to x is x^2; as x increases, the area grows at the increasing rate $2x$, and that rate grows at the constant rate 2 (Fig. 3-31).

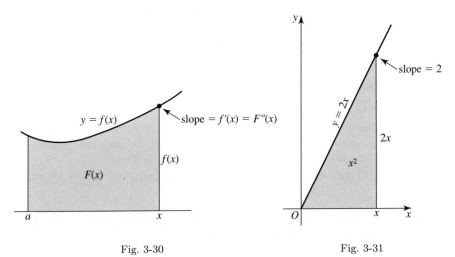

Fig. 3-30   Fig. 3-31

2. Again let $y = f(x)$ be the equation of a curve. The slope of the curve is $y' = f'(x)$, and $y'' = f''(x)$ is the rate of change of that slope with respect to

x. Assuming for simplicity that both are positive, y' is the rate at which the curve *rises* with increasing x, y'' the rate at which it *steepens*. For example, for $x > 0$ the parabola $y = x^2$ rises at the increasing rate $2x$ and steepens at the constant rate 2.

A basic physical interpretation of the second derivative is afforded by quantities associated with motion along a line (Art. 1.3). Velocity v is the derivative with respect to time of location x, and acceleration a the derivative of velocity; hence *acceleration is the second derivative of location*,

$$a = \frac{d^2x}{dt^2}.$$

A particle whose location at any t is $x = t^2$ m (to the right of 0) has velocity $2t$ m/sec, which increases at the constant rate of 2 m/sec^2 (2 m/sec per sec), the acceleration of the particle.

As by two differentiations we pass from location to acceleration, so by two integrations we can recover location from its second derivative acceleration, as in §3.8.2—at least when $a(t)$ is constant, or more generally when the integrals can be calculated. This is how the promise of §1.3.5 will be fulfilled when the subject of motion is developed in Chapter 5.

Derivatives of the third and higher orders lack such simple interpretations, and we will not be concerned with them until Chapter 8. There, however, they will be seen to play a fundamental role in the general theory of functions.

3.8.5   QUESTIONS

Q1.   Represent acceleration graphically.

Q2.   How does the graph of $y = \sin x$ steepen as x increases from $-\pi/2$ to $\pi/2$?

Q3.   (a) A particle has zero acceleration. What can you conclude about $x(t)$? (b) Add a fifth characterization of uniformity to the list in §3.6.5.

Q4.   The third derivative of location has been called **jerk**. Account for the name.

3.8.6   Linear approximation using the first derivative.[32]   A. Tangent line and linear function.   In §3.7.11 we considered the problem of

---

[32] This section and the two following are used in §4.5.7 and in Chapters 7 and 8.

estimating $f(x)$ by $f(a)$ for x near a.[33] In order to judge the accuracy of the approximation[34]

$$f(x) \approx f(a)$$

we wrote $f(x)$ in the form

$$f(x) = f(a) + R_0$$

and appealed to the mean value theorem for the calculation of R_0. According to that theorem, the magnitude of R_0 $(= f'(\xi)(x - a))$ was controlled by the length of the interval from a to x and the size of f' on that interval.

Such estimation has a simple description from the point of view of functions: for x in a small neighborhood of a, the function $y = f(x)$ is approximated by the constant function $y = f(a)$. This amounts to estimating the height of the graph of f near $x = a$ by the height of the horizontal line through $P(a, f(a))$ (Fig. 3-32).

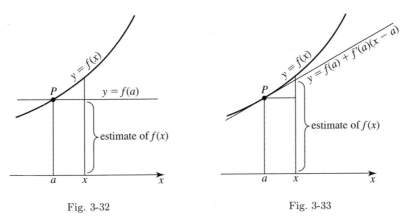

Fig. 3-32   Fig. 3-33

Plainly, we would in general do much better to use the *tangent* line at P to estimate $f(x)$ (Fig. 3-33). To do so is to approximate the function $y = f(x)$ in the vicinity of a by the *linear* function

$$y = f(a) + f'(a)(x - a)$$

---

[33] In the earlier discussion x_0 and x_1 were used instead of a and x, and R instead of the R_0 to be introduced below.

[34] The sign $\approx$ indicates approximate equality.

whose graph is the tangent.[35] Recalling from §3.6.5 that linearity is uniformity, you will recognize in this idea a recurrence of the theme of approximation of a non-uniform thing by the uniform thing that best matches it.

I therefore propose the approximation

$$f(x) \approx f(a) + f'(a)(x - a)$$

as an improvement over the earlier one. It is then appropriate to write

$$f(x) = f(a) + f'(a)(x - a) + R_1$$

and inquire about the size of R_1, the error by which $f(x)$ deviates from its linear estimate.

3.8.7 B. Second-order error. Satisfactory information about R_1 is provided by the following proposition, to be proved in Chapter 8 (§§8.2.1–3; see also §§7.3.2–4).

Second-order Mean Value Theorem. *If $f'(x)$ is continuous on $[a, b]$ and differentiable for $a < x < b$ there is a number ξ, $a < \xi < b$, such that*

$$f(b) = f(a) + f'(a)(b - a) + \frac{1}{2} f''(\xi)(b - a)^2 .$$

The same formula holds if $b < a$, provided that f' is continuous on $[b, a]$ and differentiable for $b < x < a$; in that case $b < \xi < a$.

Assuming f twice differentiable in a neighborhood of a, we can put x in place of b to obtain

$$f(x) = f(a) + f'(a)(x - a) + \frac{1}{2} f''(\xi)(x - a)^2$$

for x in that neighborhood. Here ξ, which depends on x, lies between a and x. Then

$$R_1 = \frac{1}{2} f''(\xi)(x - a)^2 .$$

---

[35]This function can easily be put in the form $\alpha x + \beta$ by which linear functions were defined in §3.6.5.

Comparing this to the earlier

$$R_0 = f'(\xi_0)(x - a)$$

(the subscript is added to distinguish the ξ's) we see that if $f''(\xi)$ is comparable in size to $f'(\xi_0)$, R_1 will be much smaller than R_0 when x is near a, because the square of the small quantity $x - a$ is much smaller than the quantity itself. This bears out the superiority of approximation by the tangent. Of course, advantage comes at a cost: the new method requires knowledge of $f'(a)$ as well as $f(a)$.

Let us now revisit the example presented in §3.7.11. The object is to compute $\sin 31°$. With $a = 30°$ and $f(x) = \sin x$ we have

$$\sin 31° \approx \sin 30° + (\cos 30°)(31° - 30°)$$

$$= \frac{1}{2} + \frac{\sqrt{3}}{2}\left(\frac{\pi}{180}\right) = 0.5151\,,$$

to four decimal places. Since $|f''(\xi)| = |\sin \xi| \le 1$, the magnitude of the error satisfies

$$|R_1| \le \frac{1}{2}\left(\frac{\pi}{180}\right)^2 < 0.0002\,.$$

Our estimate of $\sin 31°$ is therefore accurate to three places, and so is much better than the old estimate of 0.5.

More important than the calculation of any single value of $\sin x$ is the approximate representation of the function over a whole interval near $x = 30° = \pi/6$ as a linear function:

$$\sin x \approx \frac{1}{2} + \frac{\sqrt{3}}{2}\left(x - \frac{\pi}{6}\right)\,.$$

If we want the approximation to be accurate to within 0.0005, say, we have to make $|R_1| \le 0.0005$. Since

$$|R_1| \le \frac{1}{2}\left(x - \frac{\pi}{6}\right)^2$$

it will certainly suffice to take $|x - (\pi/6)| \le 0.03$; thus the approximation is good at least over the interval $[(\pi/6) - 0.03, (\pi/6) + 0.03]$, which includes $[0.50, 0.55]$.

3.8.8   C. Estimation of difference.   So far we have regarded the purpose of linear approximation as that of estimating $f(x)$. It also serves to estimate the difference between $f(x)$ and $f(a)$:

$$f(x) - f(a) \approx f'(a)(x - a),$$

or

$$\Delta f \approx f'(a)\Delta x,$$

the finite counterpart of the equation of infinitesimals

$$df = f'(a)\, dx.$$

This approximation is often applied to gauge the effect of inaccuracy in measurement. The following simple example will illustrate the method.

Suppose the edge of a cube is 10 m long. How much effect does an error in measuring this length have on the calculated volume of the cube? If x represents the length of an edge, the corresponding volume is $f(x) = x^3$, whose derivative is $f'(x) = 3x^2$. At $a = 10$ the derivative has the value 300; therefore if x varies from 10 by the small amount Δx, the corresponding change in calculated volume will be

$$\Delta f \approx 300\Delta x$$

—error is multiplied by 300. For example, a measurement error of 1 cm = 0.01 m results in an error of 3 m^3 in calculated volume.

Instead of the **absolute** error Δf caused by Δx we may wish to find the **relative** error $\dfrac{\Delta f}{f(a)}$, or the **percentage** error $100\dfrac{\Delta f}{f(a)}$%—change in $f(x)$ as a part or percent of its basic value. The preceding example used the formula

$$\Delta f \approx 3a^2\Delta x\,;$$

dividing both sides by $f(a) = a^3$ gives

$$\frac{\Delta f}{a^3} \approx 3\frac{\Delta x}{a},$$

or

$$100\frac{\Delta f}{a^3} \approx 3\left(100\frac{\Delta x}{a}\right)$$

—relative error and percentage error are multiplied by 3. The measurement error of 0.01 m when $a = 10$ is a percentage error of $100 \left(\dfrac{0.01}{10} \right) \% = 0.1\%$ in x, and results in a percentage error of 0.3% in the volume.

> The "differential" of a variable, which I have so far defined only as an infinitesimal (§§2.2.7, 2.7.2), has also a different definition as a finite quantity. The differential of the independent variable is defined to be the same as its increment, $dx = \Delta x$. Then the differential of the dependent variable at a is defined by the equation $dy = y'(a)dx$, or $df = f'(a)dx$. That is, df is the change, not in $f(x)$, but in the linear function by which $f(x)$ is approximated at a—the rise of the tangent line to the graph rather than the graph itself. With this definition the approximate equation $\Delta f \approx f'(a)\Delta x$ can be written $\Delta f \approx df$: the increment of f is approximated by its (finite) differential.

3.8.9　QUESTIONS

Q1.　Show that $(0.99)^{10}$ is within 0.0045 of 0.9 (cf. §3.7.13.Q1).

Q2.　Find the linear function that best approximates $y = \tan x$ at $x = \pi/4$.

Q3.　The earth is approximately a sphere of radius 6.4×10^6 m. How does a small error in measurement of the radius affect the calculated (a) surface area, (b) volume? Consider absolute, relative, and percentage error. (The area of a sphere is $4\pi r^2$, the volume $\frac{4}{3}\pi r^3$.)

3.9　BEHAVIOR OF FUNCTIONS AS REVEALED BY DERIVATIVES

3.9.1　The first derivative.　A. Increase and decrease.　From this section through §3.9.5 functions will be assumed smooth (continuously differentiable), although it will be apparent that continuity of the derivative is not always necessary.

Recall that a function $f(x)$ is said to be *increasing* if its values increase as x increases, *decreasing* if its values decrease as x increases (§1.6.8). It was pointed out in §2.10.8 that the sign of the derivative indicates increase or decrease. In particular, we can say that *if $f'(x)$ is positive throughout an interval $f(x)$ is increasing there, and if $f'(x)$ is negative $f(x)$ is decreasing.* This is geometrically evident—positive and negative slopes respectively imply increase and decrease. A proof follows from the mean value theorem: if $a < b$, the equation

$$f(b) - f(a) = f'(\xi)(b - a),$$

where $a < \xi < b$, shows that the change in f
resulting from the increase in x from a to b
has the same sign as the derivative.

The converse proposition does not quite
hold: it is possible for the derivative of a
monotonic function to vanish at one or more
individual points, as the example $f(x) = x^3$
illustrates (Fig. 3-34). Although this function is increasing everywhere, its derivative
$3x^2$ is zero at $x = 0$, where the graph has
a horizontal tangent. Everywhere else the
derivative is positive. The derivative of an
increasing function cannot vanish along an
interval, for if it did the function would
be constant there (§3.5.6(4)); nor can it be

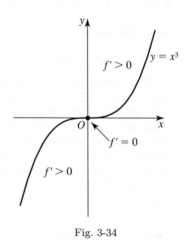

Fig. 3-34

negative at any point, for then continuity would make it negative in a neighborhood of the point and the function would decrease. Similar remarks apply
to a decreasing function.

In the case of x^3 the derivative does not change sign from one side of the
point where it vanishes to the other. With $f(x) = x^2$ the situation is different,
since $f'(x) = 2x$ has the same sign as x. The derivative passes from negative
values through 0 to positive values; the function is decreasing for $x \leq 0$ and
increasing for $x \geq 0$ (Fig. 3-35). At $x = 0$ it takes its minimum value. Similarly,
the function $f(x) = -x^2$ has a maximum value at $x = 0$, where its derivative
changes sign from positive to negative (Fig. 3-36).

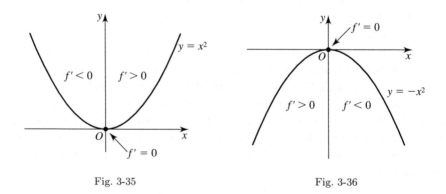

Fig. 3-35   Fig. 3-36

3.9.2 B. Local maxima and minima. In general, suppose the derivative of a function $f(x)$ vanishes at a point $x = x_0$, but is non-zero at all points nearby. Then if $f'(x)$ changes from negative to positive at x_0 the function has a **local minimum** there, a value smaller than its values at *nearby points on both sides*—there may be still smaller values at a distance (Fig. 3-37). In symbols: for all $x \neq x_0$ in some neighborhood of x_0, $f(x_0) < f(x)$. Note that a local minimum is defined by strict inequality, whereas in §3.7.1 a minimum value only had to be $\leq$ other values (Q1). If $f'(x)$ changes from positive to negative a **local maximum** (analogously defined) occurs at x_0, and if it does not change sign there is neither a local maximum nor a local minimum.

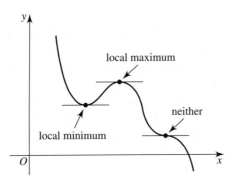

Fig. 3-37

These facts are again consequences of the mean value theorem. Suppose that $f'(x)$ changes from negative to positive at x_0; it is claimed that there is a local minimum at x_0, i.e. $f(x_0) < f(x)$ for all x near x_0. For such x we have

$$f(x) - f(x_0) = f'(\xi)(x - x_0),$$

where ξ lies between x_0 and x—in particular, $\xi \neq x_0$, so $f'(\xi) \neq 0$. If $x < x_0$ both $f'(\xi)$ and $x - x_0$ are negative; if $x > x_0$ both are positive; in either case $f'(\xi)(x - x_0)$ is positive, i.e. $f(x) - f(x_0) > 0$, or $f(x_0) < f(x)$. Similar arguments prove the other assertions.

Thus the vanishing of $f'(x_0)$ is not alone *sufficient* to ensure that there is a local maximum or minimum at x_0. It is of course *necessary*—*if* there is a local maximum or minimum at x_0 *then* $f'(x_0) = 0$; this is a geometrically obvious corollary of the proposition of §3.7.7.

When referring to a local maximum or minimum value we frequently omit the adjective "local." Besides the caution about strict inequality just given one should keep in mind that such a value (i) need not be the greatest or least value of a function (*the* maximum or minimum in the sense of §3.7.1), and (ii) must occur at an *interior* point x_0 of the interval of definition of the function (because it must be comparable to values taken on both sides of x_0). A term that covers both a (local) maximum and a (local) minimum is **extremum**, or **extreme** value.

3.9.3   QUESTION

Q1. How should the foregoing account be modified if it is to deal with functions monotone in the weaker sense (§1.6.8) and local extrema defined by weak inequalities?

3.9.4   Procedure for finding maxima and minima.
Assuming that the derivative of a function $f(x)$ vanishes only at points that are **isolated**, i.e. each contained in a neighborhood that contains none of the others (as is ordinarily the case[36]), we can lay out a practical procedure for determining the local maxima and minima of $f(x)$ and its regions of increase and decrease. This much information is often sufficient for us to form a good idea of the function's overall behavior and produce a rough sketch of its graph. The importance of the specific problem of locating extrema can hardly be exaggerated, in view of the interest that attaches to greatest and least values of all kinds of mathematical and physical quantities.

The following steps compose the basic method.

1. Differentiate $f(x)$ to obtain $f'(x)$.

2. Solve the equation $f'(x) = 0$. Its solutions (if there are any) are called **critical points** of f.[37]

3. Determine the sign of f' on each of the intervals marked off by the critical points. Since f' is continuous, it cannot change sign without taking the value 0 (by the intermediate value property); hence its sign is constant over such an interval, and can be found by evaluating f' at any x-value in the interval. A positive sign indicates increase along the interval, a negative sign decrease.

---

[36] A non-constant smooth function which does not have this property is defined in §4.2.9 (Example 4).

[37] Some ambiguity is found in the use of such terms as critical point, maximum, and minimum. Instead of saying that x_0 is a critical point of f we may say that $(x_0, f(x_0))$ is one, in effect substituting a point of the graph for a point of the x-axis. Instead of saying that f has a maximum at x_0 we may say that it has one at $(x_0, f(x_0))$; or that x_0, or $(x_0, f(x_0))$, is a maximum of f.

4. At a critical point x_0, $f(x)$ has a local maximum if f' changes from positive to negative there, a local minimum if it changes from negative to positive. When it does not change sign the graph of f has a horizontal tangent but no extremum. To find the local extreme value, or the height of the horizontal tangent, calculate $f(x_0)$.

A monotone function may have critical points, but will not have local extrema. Of course, any function, if restricted to an interval $[a, b]$, will have there a maximum and a minimum value in the sense of §3.7.1; but these may occur at endpoints, in which case they are not "local" as defined in §3.9.2, even if they are strictly greater or less than the values taken nearby in the interval. Several possibilities are illustrated in the examples that follow.

3.9.5 Examples. EXAMPLE 1. Let $f(x) = \alpha x$, $\alpha > 0$. Then $f'(x) = \alpha$, so there are no critical points, hence no local maxima or minima. The function is monotone increasing. On any interval $[a, b]$ there is a minimum value αa at a and a maximum value αb at b.

EXAMPLE 2. Let $f(x) = 3x^4 - 4x^3$. We have $f'(x) = 12x^3 - 12x^2 = 12x^2(x - 1)$, so the critical points are 0 and 1. At other points the sign of $f'(x)$ is the same as that of $x - 1$ ($12x^2$ being positive), i.e. negative for $x < 1$ and positive for $x > 1$ (Fig. 3-38). This means that the function is decreasing for $x \leq 1$ and increasing for $x \geq 1$. There is a local minimum at $x = 1$, but neither maximum nor minimum at $x = 0$. Since $f(0) = 0$ and $f(1) = -1$ the graph resembles Fig. 3-39. To the left of 0 and to the right of 1 the function becomes arbitrarily large, because $3x^4$ becomes much larger in absolute value than $4x^3$ as $|x|$ increases.

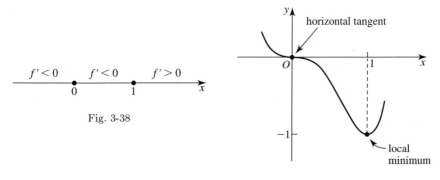

Fig. 3-38

Fig. 3-39

EXAMPLE 3. Of all rectangles of given perimeter, which has the greatest area? Let the given perimeter be p, and let x be the length of a side. An adjacent side is then $\frac{p}{2} - x$ and the area is $A(x) = x\left(\frac{p}{2} - x\right) = \frac{p}{2}x - x^2$. Then $A'(x) = \frac{p}{2} - 2x$ vanishes when $x = p/4$: this is the sole critical point. Since $A'(x) > 0$ if $x < p/4$ and $A'(x) < 0$ if $x > p/4$, at $p/4$ there is a local maximum, which must also be the greatest value assumed by the function. When $x = p/4$ the other side is $\frac{p}{2} - \frac{p}{4} = \frac{p}{4}$ also; the rectangle is therefore a square.

In this problem $A(x)$ is defined only for $0 < x < p/2$. On that interval the function has no minimum: its value can be arbitrarily close to 0. If we allow a "rectangle" to have one side of length 0, $A(x)$ is defined on $[0, p/2]$ and takes its minimum value of 0 at both endpoints (Fig. 3-40).

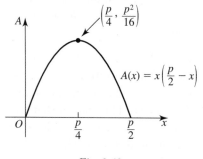

Fig. 3-40

EXAMPLE 4. (a) If the location of a particle is given by $x(t) = \sin t$, $t \geq 0$, the particle oscillates about $x = 0$ between the limits -1 and 1. Its maximum coordinate 1 obviously occurs at $t = \frac{\pi}{2}, \frac{\pi}{2} + 2\pi, \frac{\pi}{2} + 4\pi, \ldots$, its minimum coordinate -1 at $t = \frac{3\pi}{2}, \frac{3\pi}{2} + 2\pi, \frac{3\pi}{2} + 4\pi, \ldots$. This is confirmed by noting that the critical points occur when $x'(t) = \cos t = 0$ and checking the sign of $\cos t$ between them. Since $x'(t)$ is the velocity of the particle, the critical points are the times at which velocity vanishes, which happens when the particle is at the limits of oscillation.

(b) Suppose now that the particle's location is $x(t) = \dfrac{\sin t}{e^t} = e^{-t}\sin t$. The rapidly decreasing positive factor e^{-t} forces the particle closer and closer to the origin as t increases; in fact, only the first movement away from $x = 0$ attains

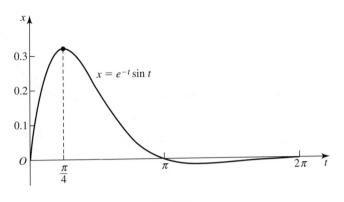

Fig. 3-41

any significant distance (Fig. 3-41[38]). At what time does the first maximum occur? Using the quotient rule we find $x'(t) = e^{-t}(\cos t - \sin t)$. Since e^{-t} is never 0, the critical points are where $\cos t = \sin t$. This happens for the first time at $t = \pi/4$, and $\cos t - \sin t$ does change sign from positive to negative at that value of t (because $\cos t$ is decreasing and $\sin t$ increasing at $\pi/4$). Thus the first maximum is at $t = \pi/4 \approx 0.79$; the value of the function at that time is $(1/\sqrt{2})e^{-\pi/4} \approx 0.32$.

Additional examples will be presented in Chapter 4 (§4.2.6), after more techniques of differentiation have been developed.

3.9.6   The second derivative: concavity and inflection. Let us now assume that $f''(x)$ exists and is continuous. If $f'' > 0$ on an interval the first derivative f' is increasing, because *its* derivative is positive. Geometrically, this means that the slope of the graph of f is increasing; that is, the graph is **concave upward** (or just concave "up")—it lies above every tangent line, except at the point of tangency (Fig. 3-42; Q10). When this is the case in a neighborhood of a critical point x_0, the graph lies above its horizontal tangent and a local minimum occurs. (A proof by the mean value theorem is suggested in Q9.) Since we are supposing f'' continuous, if it is positive at x_0 it will be positive nearby; a sufficient condition for a local minimum to occur at x_0 is therefore that $f''(x_0) > 0$.[39] Similarly, where $f'' < 0$ the graph of f is **concave downward** and a critical point is a maximum.

---

[38]The vertical scale is exaggerated.

[39]The continuity of f'' is not necessary; see Q13.

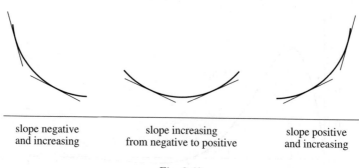

| slope negative and increasing | slope increasing from negative to positive | slope positive and increasing |

Fig. 3-42

Instead of checking the sign of f' on both sides of a critical point x_0, then, we can check the sign of $f''(x_0)$ to determine what kind of critical point it is. If f'' is easy to calculate the latter method may be preferable. However, it gives no information when $f''(x_0)$ vanishes. In such a situation there are several possibilities, as we are about to see.

Suppose that f'' vanishes at an isolated point x_0, not necessarily a critical point. Then either the sign of f'' changes at x_0, or it does not. In the former case we have an **inflection point**, at which the graph of f switches from concave upward to concave downward, or vice versa; if x_0 is a critical point, there is a horizontal tangent but no extremum at x_0 (Q11). As Fig. 3-43 shows, the origin is an inflection point for x^3, and also a critical point; for $\sin x$ it is a non-critical inflection point.

If on the other hand the sign of f'' does not change at x_0, neither does the direction of concavity; then if x_0 is a critical point, there is a minimum at x_0 if

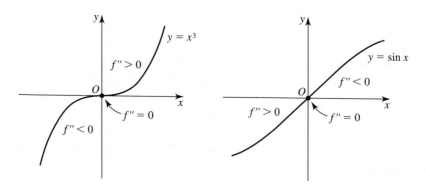

Fig. 3-43

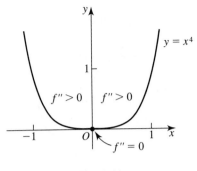

Fig. 3-44

that direction is upward, as with x^4 (Fig. 3-44), or a maximum in the opposite case. Thus when x_0 is a critical point there is either an inflection point, a maximum, or a minimum.

Let us now reconsider the examples of §3.9.5 with the aid of the second derivative. The first illustrates the vanishing of f'' on an interval rather than at isolated points.

3.9.7 Examples. EXAMPLE 1. $f(x) = \alpha x$, $\alpha > 0$. Then $f'(x) = \alpha$, $f''(x) = 0$. There are no critical points, and there is no concavity, consequently there are no inflection points.

EXAMPLE 2. $f(x) = 3x^4 - 4x^3$. Then $f'(x) = 12x^3 - 12x^2 = 12x^2(x - 1)$, $f''(x) = 36x^2 - 24x = 36x(x - \frac{2}{3})$. At the critical points 0 and 1 we find $f''(0) = 0$, $f''(1) > 0$. Hence the latter is a minimum; as for the former, it is an inflection point, either by the argument given before or because f'' changes sign there. The graph is concave up for $x \leq 0$ and $x \geq \frac{2}{3}$, concave down for $0 \leq x \leq \frac{2}{3}$. A second inflection point occurs at $x = \frac{2}{3}$ (Fig. 3-45).

EXAMPLE 3. $A(x) = \dfrac{p}{2}x - x^2$. Then $A'(x) = \dfrac{p}{2} - 2x$, $A''(x) = -2 < 0$. The critical point $p/4$ is a maximum; there are no inflection points, and the curve is concave down everywhere.

EXAMPLE 4. (a) $x(t) = \sin t$, $t \geq 0$. Then $x'(t) = \cos t$, $x''(t) = -\sin t$. Critical points are at $(\pi/2) + n\pi$; there $x''(t)$ is negative if n is even, positive if n is odd, so the former points are maxima, the latter minima. Inflection points occur wherever $\sin t$ vanishes, i.e. at points $n\pi$. Since $x''(t)$ is the acceleration,

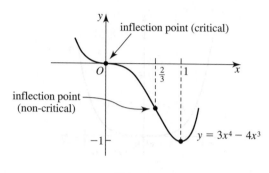

inflection point (critical)

inflection point
(non-critical)

$y = 3x^4 - 4x^3$

Fig. 3-45

positive acceleration corresponds to concave up, negative to concave down. The inflection points are where acceleration vanishes—at the center of the oscillation, where the direction of acceleration changes (cf. §5.2.2).

(b) $x(t) = e^{-t}\sin t$, $x'(t) = e^{-t}(\cos t - \sin t)$, $x''(t) = -2e^{-t}\cos t$. Since e^{-t} is always positive, the sign of $x''(t)$ is the same as that of $-\cos t$. The first critical point is at $\pi/4$, where $x''(t)$ is negative; hence it is a maximum (Fig. 3-46). The first point at which $x''(t)$ vanishes is $\pi/2$; this is an inflection point, where the graph turns from concave down to concave up. What happens at $t = \pi/2$ is that the leftward-moving particle begins to decelerate, well before it reaches the origin.

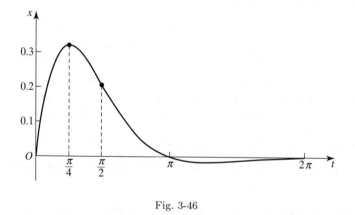

Fig. 3-46

3.9.8   Other methods. The behavior of functions can be investigated by other means, which are needed when derivatives do not exist, or are hard to calculate, or when an equation $f'(x) = 0$ or $f''(x) = 0$ cannot be solved.

Even when the methods of this article are applicable, a given question about a function may be more easily answered without them. Although I cannot discuss alternative methods at any length, I will mention some of the principal ones.

1. It goes without saying that we may want to *evaluate* $f(x)$ itself at one or more points besides critical and inflection points, and perhaps find out where it *vanishes* (its graph meets the x-axis) by solving the equation $f(x) = 0$. Evaluating f' at points of interest will give slopes there.

2. The *behavior* of $f(x)$ *in the limit* as x approaches ∞, or $-\infty$, or a point where f is not continuous, can reveal much. (See the end of Example 2, §3.9.5.) It may also be useful to look at the behavior of f'. For example, examination of $1/x$ and its derivative as $x \to \infty$, $x \to -\infty$, and $x \to 0$ shows that the axes are asymptotes.[40]

3. Considerations of *symmetry* save duplication of effort, as with odd and even functions (§3.3.7).

4. General *properties of a class of functions* should be kept in mind, e.g. that a polynomial of degree n can vanish at most n times (cf. §3.3.9.Q15), or that trigonometric functions are periodic.

5. Simple *transformations of familiar functions* may affect behavior in a predictable way; e.g., from knowledge of $\sin x$ we can at once deduce certain properties of $\sin^2 x$.

This partial list should make it clear that ingenuity and resourcefulness are by no means rendered obsolete by our general methods, effective as they are. Such tasks as graphing ordinary functions can of course be carried out very well by machine; in order to understand what has been graphed, however, one has to know in detail the whole range of principles that govern the visible forms.

3.9.9 QUESTIONS

Q1.   Investigate the behavior of the following functions and sketch their graphs.
(a) x^6 (b) $2x^3 - 9x^2 + 12x - 3$ (c) $\frac{1}{4}x^4 - \frac{3}{2}x^2 + 2x$ (to factor f' notice that $f'(1) = 0$)

(d) $\tan x$ (e) $\dfrac{1}{x^2 + 1}$ (f) $\dfrac{x}{x^2 + 1}$

---

[40] The positive x-axis is an **asymptote** of the graph (the graph is asymptotic to it) because as $x \to \infty$ the distance from the curve (the graph) to the straight line (the axis) approaches 0 and the slope of the curve approaches that of the line (in this case also 0). This indicates the meaning of the term in general. (For a vertical asymptote the slope of the curve approaches ∞ or $-\infty$.)

Q2. Prove that $e^x > x$ everywhere. (Consider the function $e^x - x$.) Hence $e^x \to \infty$ as $x \to \infty$ (also a consequence of §3.3.3.Q6(a)).

Q3. Investigate the behavior of x/e^x and sketch its graph. (Use §3.3.13.Q8 to determine the limit of the function as $x \to \infty$.)

Q4. The cumulative mass of a wire is $m(x) = 1 - \cos x$, $0 \le x \le \pi/2$. At what point in the interval $0 \le x \le 1$ is its line density greatest?

Q5. Of all rectangles with a given area, find the one with the least perimeter.

Q6. The velocity of a particle is $v(t) = t^2 e^{-t}$, $t \ge 0$. When is it greatest? What happens to the acceleration at that moment?

Q7. Find the right circular cylinder of unit volume having the least surface area, including its top and bottom faces. (This describes the can most economical in material, if thickness of material is uniform.) (Express height h in terms of radius r, then minimize area as a function of r.)

Q8. Improve §3.2.3.Q5 by investigating the truth of $(1 + x)^n > 1 + nx$, where n is an integer > 1, for $x < 0$. (Let $t = 1 + x$ and consider $f(t) = t^n - 1 - n(t - 1)$.)

Q9 (§3.9.6). For the proof, apply the mean value theorem to f' and reason as in the mean value argument of §3.9.1.

Q10 (§3.9.6). Use the second-order mean value theorem (§3.8.7) to show that the graph lies above every tangent line.

Q11. Show that an inflection point cannot be a maximum or minimum.

Q12. Let A and B be points of a graph which is concave upward. Show that every point of the graph between A and B lies below secant AB. (Cf. the definition of convexity in §1.5.5.Q2(b).)

Q13 (§3.9.6). Argue from the difference quotient of f' at x_0 that a minimum must occur if $f''(x_0) > 0$, regardless of whether f'' is continuous.

Q14. (a) Let $f(x)$ be continuous for $0 \le x \le b$ and differentiable for $0 < x < b$, and suppose that $f(0) = 0$ and $f'(x) > 0$ for $0 < x < b$. Prove that $f(x) > 0$ for $0 < x \le b$. (Use the mean value theorem.) (b) Show that $x > \sin x$ for $x > 0$.

Q15. The celebrated *brachistochrone* (shortest-time) problem seeks the curve down which a particle will slide without friction from one given point to another lower one in the least time. Explain why this is *not* the kind of problem of finding a minimum we have considered.

CHAPTER 4

Differentiation and Integration: II

INTRODUCTION

There are two operations on functions that are more fundamental in character than the arithmetic ones, and extend the class of differentiable functions far beyond the members we have already encountered. While the arithmetic operations take their origin in *number* and apply to functions $f(x)$ only because they are numerical-valued, the operations of "composition" and "inversion" to be introduced in this chapter arise from the idea of *function* and apply to numerical-valued functions just because they are functions. These operations, their relation to differentiation, and a number of applications of them, are the main business of the first four articles of the chapter. The fifth article is concerned with the extension of differentiation to functions of more than one independent variable.

The remaining three articles deal with integration. Article 4.6 takes up the problem of integrating combinations of elementary functions in terms of elementary functions. Although no universal solution is to be had, several methods make it possible to calculate many integrals of interest. In Art. 4.7 the integral is extended to infinite intervals and discontinuous functions, and in Art. 4.8 to regions of the plane and regions of space.

4.1 COMPOSITE FUNCTIONS AND THEIR DERIVATIVES

4.1.1 Composition of functions. Let us recall from Chapter 1 (§1.6.5) the representation of a function as an input–output machine as in Fig. 4-1, in which the hopper and chute can be thought of as a "reader" and "writer"

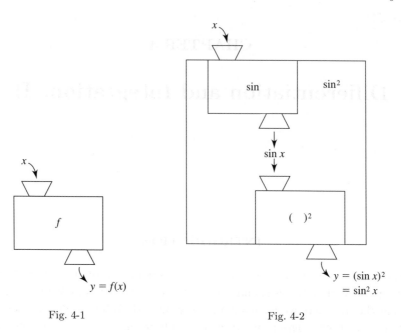

Fig. 4-1   Fig. 4-2

respectively. To each x is assigned one $y = f(x)$; the same y may be assigned to more than one x.

Certain function-machines are usually viewed as operating in successive steps. To calculate a value of the function $y = \sin^2 x$, for example, we would proceed as follows: given x, find $\sin x$; then square the result. There are two steps, because $\sin^2$ is constructed from two simpler functions, $\sin$ and $(\ )^2$. Its machine can be assembled out of their machines (Fig. 4-2). Such a combination is often dealt with by introducing another letter to represent the output of the first function and the input of the second:

$$u = \sin x,$$

$$y = u^2.$$

With this notation we speak of y as being a function of x by **substitution** of $\sin x$ for u in $y = u^2$, and also refer to y as a **function of a function** of x. Since $\sin x$ is substituted *into* $y = u^2$ and appears *inside* the parentheses in $(\sin x)^2$, we call it the **inner** function and $y = u^2$ the **outer** function of $y = \sin^2 x$.

In general, given two functions

$$u = g(x)$$

and

$$y = f(u)$$

we can express y as a function of x by substitution,

$$y = f(g(x)),$$

in this way defining a *new function*. The operation by which the new function is made up out of f and g is called **composition** or **compounding**, and the composite or compound function is denoted $f \circ g$ ("f of g" or "f circle g"). Thus by definition

$$(f \circ g)(x) = f(g(x)).$$

The function-machine for $f \circ g$ performs the actions of g and f sequentially (Fig. 4-3). Note carefully that although the outer function f comes first in the notations $f \circ g$ and $f(g(x))$, it is the inner function g that is evaluated first.

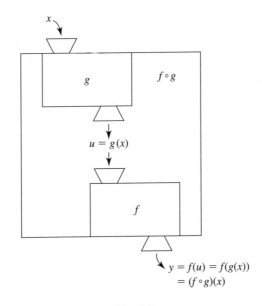

Fig. 4-3

The function $y = \sin^2 x$ can now be written as $(\ )^2 \circ \sin$ (first take sine, then square):

$$((\ )^2 \circ \sin)(x) = (\ )^2(\sin x) = (\sin x)^2 = \sin^2 x\,.$$

We may, if we wish, compound the sine and $(\ )^2$ in the reverse order, obtaining $\sin \circ (\ )^2$ (first square, then take sine):

$$(\sin \circ (\ )^2)(x) = \sin(x^2) = \sin x^2\,.$$

The two functions are quite different; e.g.,

$$\sin^2(\pi/2) = 1^2 = 1\,, \quad \sin(\pi/2)^2 = \sin(\pi^2/4) \approx 0.62\,.$$

As another example, consider $y = (x-1)^3$. It can be analyzed as the composite $(\ )^3 \circ ((\ ) - 1)$ (first subtract 1, then cube), the inner function being $u = x - 1$ and the outer $y = u^3$. Compounding in the opposite order produces

$$y = [((\ ) - 1) \circ (\ )^3](x) = ((\ ) - 1)(x^3) = x^3 - 1\,,$$

where the inner function is $u = x^3$ and the outer $y = u - 1$.

There is one restriction on forming composite functions $f \circ g$: f *must be defined on the values taken by* g. Values of x such that f is not defined at $g(x)$ will be values where $f \circ g$ is not defined. Of course, $f \circ g$ will also fail to be defined wherever g is not defined.[1] Consider, for example, the composites $\sin \sqrt{x}$ and $\sqrt{\sin x}$. The former is defined for $x \geq 0$, because that is where $g(x) = \sqrt{x}$ is defined. In the latter, $g(x) = \sin x$ is defined everywhere, but $f(u) = \sqrt{u}$ is defined only where $u = g(x) = \sin x \geq 0$; therefore $f \circ g$ is defined between 0 and π but not between π and 2π, etc.

The composite of continuous functions is continuous. For if $g(x)$ is continuous and $x \to a$, then $g(x) \to g(a)$, and if $f(u)$ is also continuous then $f(g(x)) \to f(g(a))$, i.e. $(f \circ g)(x) \to (f \circ g)(a)$. (See also Q4.)

Three or more functions can be compounded in the same way as two—the chain of functions joining original input to ultimate output will simply have more links. If

$$v = h(x)\,, \quad u = g(v)\,, \quad \text{and} \quad y = f(u)\,,$$

---

[1] This is a consequence of our definition of $(f \circ g)(x)$ as $f(g(x))$. See Q3.

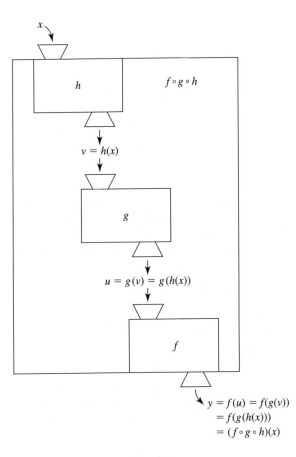

Fig. 4-4

then

$$y = f(g(v)) = f(g(h(x))) = (f \circ g \circ h)(x)$$

(Fig. 4-4). Thus, e.g., $y = \sin^2(x-1)$ is

$$(\ )^2 \circ \sin \circ ((\ ) - 1),$$

or the composite of

$$v = x - 1, \quad u = \sin v, \quad \text{and} \quad y = u^2.$$

There are no longer an inner and an outer function, but we can call $y = f(u)$ the *outermost* function, $u = g(v)$ the *next function in* from f, and so on, to the *innermost* function h.

4.1.2 QUESTIONS

Q1. Express y as the indicated function of x. For example, given $\sin \circ (\ )^2$ you would write $y = \sin x^2$. (a) $\tan \circ \log$ (b) $e^{(\ )} \circ (\ )^2$ (c) First take cosine, then cube, then double. (d) $y = \cos u$, $u = 2v$, $v = x^3$ (e) $e^{(\ )} \circ (\ )^{1/2} \circ \sec$ (f) $y = 1/u$, $u = v^2 + 3v$, $v = \tan x$

Q2. Analyze each function as a composite, expressing the result in two or three ways. For example, given $\sin x^2$ you would write $\sin \circ (\ )^2$ and $y = \sin u$, $u = x^2$, and perhaps draw a composite function-machine in the style of Fig. 4-2. (a) $2^{1/x}$ (b) $|\log x|$ (c) $1/\tan^4 x$ (d) $\sin e^x$ (e) $\cos \sqrt{x^2 + a^2}$ (f) $\log \log x$

Q3. Express each of the following functions in a simpler way, and tell where each is defined. (a) $(\sqrt{x})^2$ (b) $\sqrt{x^2}$

Q4 (§4.1.1). (a) Extend this proposition and argument to the case $x \to \infty$. (b) Apply (a) to show $\lim\limits_{x \to \infty} \sqrt{\dfrac{1}{x^2} + 1} = 1$. (c) Apply (b) to find $\lim\limits_{x \to \infty} \dfrac{x}{\sqrt{1 + x^2}}$.

4.1.3 The chain rule.

It will be proved in §4.1.6 that *the composite of differentiable functions is differentiable*. Can the derivative of a composite function be expressed in terms of the derivatives of its components? From the point of view of infinitesimals, nothing could be easier. If

$$y = f(u) \quad \text{and} \quad u = g(x)$$

then

$$\frac{dy}{dx} = \frac{dy}{du}\frac{du}{dx},$$

because the du's cancel when the fractions are multiplied. *The derivative of a composite function is the product of the derivatives of its component functions.* This is called the **chain rule**. Its derivation by limits will be given in §4.1.6; here I will explain how to put it into practice.

The first point to realize is that the derivative of y with respect to u is to be evaluated at the u-value *corresponding* to the x-value at which the other two derivatives are evaluated. A more detailed statement of the rule makes this clear:

$$\frac{dy}{dx}\bigg|_x = \frac{dy}{du}\bigg|_{u=g(x)} \frac{du}{dx}\bigg|_x.$$

We may be tempted to express the chain rule in functional notation as

$$(f \circ g)' = f'g',$$

but this is misleading because $(f \circ g)'$ and g' are functions of x while f' is a function of u. The correct expression is

$$(f \circ g)' = (f' \circ g)\, g',$$

i.e.

$$(f \circ g)'(x) = f'(g(x))\, g'(x).$$

In words:
  To find the derivative at x of a composite function $f \circ g$,

  (i) differentiate the outer function f,
  (ii) evaluate its derivative at $g(x)$, the value of the inner function at x,
  (iii) differentiate the inner function g at x, and
  (iv) multiply the results of (ii) and (iii).

4.1.4   Examples and extension. As a first example I again take $y = \sin^2 x$. Setting $u = \sin x$, $y = u^2$, we have

$$\frac{dy}{dx} = \frac{dy}{du}\frac{du}{dx} = 2u \cdot \cos x = 2\sin x \cos x.$$

At the particular value $x = \pi/6$ the derivative is then $2\sin(\pi/6)\cos(\pi/6) = \sqrt{3}/2$. This value of dy/dx can be calculated directly by applying the chain rule at $x = \pi/6$, if we first note that when x has that value, $u = \sin(\pi/6) = \frac{1}{2}$. Then

$$\frac{dy}{dx}\bigg|_{\pi/6} = \frac{dy}{du}\bigg|_{1/2}\frac{du}{dx}\bigg|_{\pi/6} = 2\left(\frac{1}{2}\right) \cdot \cos\frac{\pi}{6} = 1 \cdot \frac{\sqrt{3}}{2} = \frac{\sqrt{3}}{2}.$$

With $y = \sin x^2$, on the other hand, we put $u = x^2$, $y = \sin u$, and find

$$\frac{dy}{dx} = \frac{dy}{du}\frac{du}{dx} = (\cos u)2x = (\cos x^2)2x,$$

or $2x \cos x^2$. At $x = 3$ the rule takes the form

$$\left.\frac{dy}{dx}\right|_3 = \left.\frac{dy}{du}\right|_9 \left.\frac{du}{dx}\right|_3 = (\cos 9)\, 2(3) = 6 \cos 9 \,.$$

When the chain rule is used in this way to differentiate a given function we do not ordinarily introduce u explicitly, but rather employ the form

$$(f \circ g)'(x) = f'(g(x))\, g'(x) \,,$$

thinking something like this:

> "First differentiate the outer function, leaving the inner function alone; then differentiate the inner function."

Just as on seeing x^2 I immediately write down $2x$ as its derivative, so on seeing $\sin^2 x$ I immediately write down $2\sin x$ to begin with ("leaving the inner function alone"), then append $\cos x$ (the derivative of the inner function $\sin x$):

$$\frac{d}{dx}\left(\sin^2 x\right) = 2\sin x \cos x \,.$$

Similarly, given $\sin x^2$ I first write $\cos x^2$, then append $2x$:

$$\frac{d}{dx}\left(\sin x^2\right) = (\cos x^2) 2x \,.$$

In the same way,

$$\frac{d}{dx}(x-1)^3 = 3(x-1)^2(1) = 3(x-1)^2 \,,$$

where the "(1)" is the derivative of $x - 1$.

The chain rule extends to composites of more than two functions. Suppose

$$v = h(x), \quad u = g(v), \quad \text{and} \quad y = f(u) \,,$$

then

$$\frac{dy}{dx} = \frac{dy}{du}\frac{du}{dv}\frac{dv}{dx} \,,$$

or

$$(f \circ g \circ h)'(x) = f'(g(h(x)))\, g'(h(x))\, h'(x) \,.$$

We work from the outermost function inward, at each stage leaving untouched the functions not yet reached.

In the following examples the successive derivatives are separated by dots indicating multiplication, and square brackets surround the parts that are left alone.

EXAMPLE 1. $\dfrac{d}{dx}\sqrt{x^2+1} = \dfrac{1}{2}[x^2+1]^{-1/2}\cdot 2x = \dfrac{x}{\sqrt{x^2+1}}.$

EXAMPLE 2. $\dfrac{d}{dx}\sin^4(2x-3) = 4\left[\sin(2x-3)\right]^3\cdot\cos\left[2x-3\right]\cdot 2$

$$= 8\sin^3(2x-3)\cos(2x-3).$$

EXAMPLE 3. $\dfrac{d}{dx}\log\left(1+e^{x^2}\right) = \dfrac{1}{[1+e^{x^2}]}\cdot e^{[x^2]}\cdot 2x = \dfrac{2xe^{x^2}}{1+e^{x^2}}.$

4.1.5   QUESTIONS

Q1.   Differentiate. (a) $\cos^3 x$ (b) $(2x+1)^4$ (c) e^{-x} (d) $(1-x)^{1/3}$ (e) $1/(1+x)$
(f) $t\sin(1/t)$ (g) $\log\log x$ (h) $(1+4\sin^2 t)^{3/2}$ (i) $e^{\sqrt{1-x^2}}$ (j) $\tan\dfrac{1+t}{1-t}$

Q2.   (a) Find dy/dx if $u = \sqrt{x}$ and $y = 1-u$. (b) Find dy/du if $u = \cos(2x-1)$ and $dy/dx = 2\sin(2x-1)$.

Q3.   Use the chain rule to prove the special case of the quotient rule $(1/g)' = -g'/g^2$, where $g(x)\neq 0$ (§3.4.2).

Q4.   Use the chain rule to deduce from $(e^x)' = e^x$ that $(a^x)' = (\log a)a^x$. (Express a^x as a power of e.)

Q5.   Differentiate $\log f(x)$.

Q6.   If $y = f(g(x))$, find y''.

Q7.   **Hyperbolic functions.** The hyperbolic sine, cosine, and tangent are defined by the following formulas:

$$\sinh x = \frac{e^x - e^{-x}}{2}, \quad \cosh x = \frac{e^x + e^{-x}}{2}, \quad \tanh x = \frac{\sinh x}{\cosh x} = \frac{e^x - e^{-x}}{e^x + e^{-x}}.$$

(*Sinh* can be pronounced *cinch*, *cosh* to rhyme with *Josh*, and *tanh* to rhyme with *ranch*.) (a) Establish addition formulas for the hyperbolic sine and cosine (i.e., express $\sinh(x\pm y)$ and $\cosh(x\pm y)$ in terms of $\sinh x$, $\sinh y$, $\cosh x$, and $\cosh y$). Derive from them the following identities:

$$\cosh^2 x - \sinh^2 x = 1, \quad 1 - \tanh^2 x = \frac{1}{\cosh^2 x},$$

$$\sinh 2x = 2\sinh x \cosh x, \quad \sinh^2\frac{x}{2} = \frac{1}{2}(\cosh x - 1).$$

(b) Differentiate $\sinh x$ and $\cosh x$. (c) Use (b) and the quotient rule to differentiate $\tanh x$, and simplify the result by (a). (d) Investigate the behavior of (i) $\sinh x$, (ii) $\cosh x$, and (iii) $\tanh x$, and sketch their graphs.

4.1.6   Proof of the chain rule.[2]   Let $y = f(u)$ and $u = g(x)$ be differentiable functions. In order to prove that $f \circ g$ is differentiable and find its derivative we consider the limit of the difference quotient

$$\frac{\Delta y}{\Delta x} = \frac{(f \circ g)(x + \Delta x) - (f \circ g)(x)}{\Delta x}.$$

Here Δy is the change in y corresponding to the change Δx in x. There is also a change in u corresponding to Δx, $\Delta u = g(x + \Delta x) - g(x)$, and this change in turn causes a change in y, namely $f(u + \Delta u) - f(u)$. Let us verify that this latter change is the same as Δy. Since $u + \Delta u = g(x + \Delta x)$, we have

$$f(u + \Delta u) - f(u) = f(g(x + \Delta x)) - f(g(x))$$

$$= (f \circ g)(x + \Delta x) - (f \circ g)(x)$$

$$= \Delta y.$$

It would seem that we can now write $\Delta y / \Delta x$ as a product of difference quotients,

$$\frac{\Delta y}{\Delta x} = \frac{\Delta y}{\Delta u} \frac{\Delta u}{\Delta x},$$

and reason as follows.

Let $\Delta x \to 0$. Then $\dfrac{\Delta u}{\Delta x} \to \dfrac{du}{dx}$. Furthermore, $\Delta u \to 0$ (because g, being differentiable, is continuous); then also $\dfrac{\Delta y}{\Delta u} \to \dfrac{dy}{du}$. Consequently

$$\frac{\Delta y}{\Delta x} = \frac{\Delta y}{\Delta u} \frac{\Delta u}{\Delta x} \to \frac{dy}{du} \frac{du}{dx},$$

by the rule for the limit of a product, and our proposition is proved.

In fact, however, this straightforward translation from the language of infinitesimals is an instructive failure. The problem is that while we can and do choose Δx to be non-zero, the corresponding Δu determined by that choice

---

[2]This section can be omitted.

and the function g may turn out to be zero, in which case $\Delta y/\Delta u$ will be undefined. This obviously occurs for *any* Δx if g is a constant function, but it happens for *arbitrarily small* Δx in less trivial cases as well.[3]

To correct the argument we observe that whether Δu is zero or not we can write

(1)
$$\Delta y = \left(\frac{dy}{du} + \varepsilon \right) \Delta u \,,$$

where ε is a quantity which depends on Δu and approaches zero as $\Delta u \to 0$. The definition of ε is different for $\Delta u \neq 0$ and for $\Delta u = 0$:

(i) Provided that $\Delta u \neq 0$, $\Delta y/\Delta u$ becomes arbitrarily close to dy/du as $\Delta u \to 0$. Hence for non-zero Δu we can satisfy (1) and the condition $\varepsilon \to 0$ by putting $\varepsilon = \dfrac{\Delta y}{\Delta u} - \dfrac{dy}{du}$.

(ii) When $\Delta u = 0$, $\Delta y = 0$ too; then (1) is satisfied by any number ε. We put $\varepsilon = 0$ so as to have $\varepsilon \to 0$.

Now
$$\frac{\Delta y}{\Delta x} = \left(\frac{dy}{du} + \varepsilon \right) \frac{\Delta u}{\Delta x} \,,$$

and passage to the limit as Δx, hence Δu, approaches zero yields
$$\frac{dy}{dx} = \frac{dy}{du} \frac{du}{dx} \,.$$

4.2 APPLICATIONS OF THE CHAIN RULE

4.2.1 The chain rule and rates of change. To grasp the meaning of the chain rule it helps to recall the description of the derivative as the *rate of change* of the dependent variable with respect to the independent one (§2.9.1). If y per u is p, and u per x is q, surely y per x is pq; and that is what the rule asserts.

Suppose by way of illustration that a particle moves at velocity $v(t) = dx/dt$ along a wire on the x-axis of line density $\lambda(x) = dm/dx$ (Fig. 4-5). We understand dm/dx to be mass per distance, or kg/m along the wire, and dx/dt to

---

[3]E.g., if g is the function of Example 3, §4.2.9.

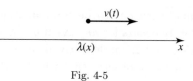

Fig. 4-5

be distance per time, or m/sec of the particle's motion. At what rate is the particle traversing *mass*? How many *kilograms* of wire does it pass per second? The answer is mass per time,

$$\frac{dm}{dt} = \frac{dm}{dx}\frac{dx}{dt} = \lambda v \text{ kg/sec}.$$

More precisely, if at time t the particle's location is $x(t)$, then at that instant and place its velocity is $\left.\dfrac{dx}{dt}\right|_t$ and the line density of the wire is $\left.\dfrac{dm}{dx}\right|_{x(t)}$, so the particle is moving past

$$\left.\frac{dm}{dt}\right|_t = \left.\frac{dm}{dx}\right|_{x(t)}\left.\frac{dx}{dt}\right|_t = \lambda(x(t))\,v(t)$$

kilograms per second.

As an example, let the cumulative mass from $x = 0$ be $m(x) = \log(1+x)$ and let the location of the particle at t sec be $x(t) = t^2$ m from $x = 0$; then $\lambda = 1/(1+x)$ and $v = 2t$, so the particle traverses mass at the rate of $2t/(1+t^2)$ kg/sec.

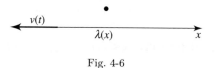

Fig. 4-6

If we change our viewpoint and regard the particle as fixed and the wire as moving (Fig. 4-6), the meaning of dm/dt will be the rate at which mass is passing a fixed point—how many kilograms go by per second. Such a quantity is called a **current** of mass, and has application to fluid flow as well as analogues in electrical, thermal, and other kinds of transport.

4.2.2   QUESTIONS

Q1. Writing dm/dt implies that m is a function of t. Explain the meaning of this function.

Q2. Let a wire have line density $\lambda = 2 + \sin x$ kg/m. Find dm/dt if the particle, initially at rest at $x = 0$, has velocity $v = \alpha t$ m/sec for $t \geq 0$, where α is a constant. (Integrate v to find x.)

Q3. The line density of a liquid flowing in a pipe is the same everywhere at any given moment, but increases as time passes in such a way that $\lambda = t$ kg/m. At the same time the velocity of flow is diminishing: $v = e^{-t^2}$ m/sec. When is the current a maximum?

4.2.3 Implicit differentiation.

Often the functional dependence of a variable y on a variable x is stated not *explicitly*, as in the expression

$$y = \frac{1}{x},$$

but *implicitly*, in an equation such as

$$xy = 1,$$

in which y does not stand alone. In this particular case it is easy to solve the equation for y, but in many other cases solution is difficult or impossible, or produces more than one function $y = y(x)$. We would be at a loss to deduce an explicit formula for y from

$$y + \sin y = x,$$

for example, while from

$$x^2 + y^2 = 1$$

we obtain both $y = \sqrt{1 - x^2}$ and $y = -\sqrt{1 - x^2}$. The similar equation $x^2 + y^2 = -1$ is not satisfied by any values of x and y, so cannot possibly define y as a function of x.

Bearing in mind these complications, let us nevertheless suppose that y is a differentiable **implicit function** of x: that we have an equation which relates the variables in such a way as implicitly to express y as a differentiable function of x. Then we may be able to *find y' in terms of x and y* without having to determine y as a function of x. The method for doing this, known as **implicit differentiation**, depends essentially on the chain rule. An example will explain it well enough.

Suppose that $y = y(x)$, $-2 \leq x \leq 2$, is expressed by the equation

(1) $$x^2 + y^2 = 4.$$

We differentiate both sides with respect to x, not forgetting that y is a function of x:

$$2x + 2yy' = 0,$$

where the chain rule has been used to differentiate y^2. This can be solved for y':

(2) $$y' = -\frac{x}{y}, \quad y \neq 0.$$

To obtain y' as a function of x we could now solve equation (1) for y, decide which of the two solutions was $y(x)$, and substitute (Q1). But if all we want is a particular value of y', we need only have particular values of x and y. For example, the slope of the tangent to the circle $x^2 + y^2 = 4$ at $(1, \sqrt{3}\,)$ is given at once by our formula: $y' = -1/\sqrt{3}$.

In order to justify this kind of calculation, notice that (1) equates two functions of x, namely $g(x) = x^2 + (y(x))^2$ and $h(x) = 4$. As equal functions g and h will have equal derivatives. (It is irrelevant that in this example $h(x)$ happens to be constant.)

4.2.4 Related rates of change. The same technique can be used when x and y are both dependent on a third variable t, which in many applications is time, in order to relate the derivatives dx/dt and dy/dt to one another. An equation in x and y is differentiated with respect to t by means of the chain rule, and a connection between dx/dt and dy/dt is thereby established.

Suppose, for example, that a particle moves around the circle just considered, so that its coordinates x and y are functions of time t. Without knowing the functions $x(t)$ and $y(t)$, we nevertheless can say that at all times $x^2 + y^2 = 4$, just because the particle is located on the circle. Differentiation with respect to t then yields

$$2x\frac{dx}{dt} + 2y\frac{dy}{dt} = 0,$$

or

(3) $$y\frac{dy}{dt} = -x\frac{dx}{dt}.$$

At the point $P(1, \sqrt{3}\,)$ this becomes

$$\sqrt{3}\,\frac{dy}{dt} = -\frac{dx}{dt}.$$

If we somehow know that at P the particle is moving towards the left at 2 m/sec, i.e. $dx/dt = -2$, we can conclude that it is simultaneously rising at $\dfrac{dy}{dt} = -\dfrac{1}{\sqrt{3}}(-2) = \dfrac{2}{\sqrt{3}}$ m/sec (Fig. 4-7). By the chain rule $\dfrac{dy}{dx} = \dfrac{dy}{dt} \Big/ \dfrac{dx}{dt}$, so equation (3) is consistent with equation (2).

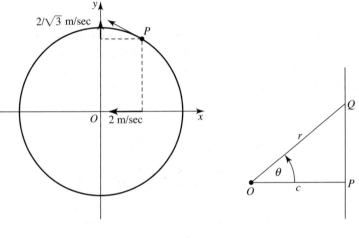

Fig. 4-7   Fig. 4-8

For another example let a straight line be at distance $c = OP$ from point O (Fig. 4-8), and let Q move along the line. What is the relation between the rates of change of angle $\theta = \angle POQ$ and distance $r = OQ$? We have

$$r \cos \theta = c \,,$$

hence

$$\frac{dr}{dt} \cos \theta - r \sin \theta \frac{d\theta}{dt} = 0 \,,$$

or

$$\frac{dr}{dt} = r \tan \theta \frac{d\theta}{dt} \,.$$

We will come back to implicit differentiation in §4.5.10.

4.2.5 QUESTIONS

Q1 (§4.2.3). (a) Carry this out, choosing the function $y(x)$ so that its value at 1 is $\sqrt{3}$. (b) Instead of differentiating implicitly, differentiate $y(x)$ directly. The result should agree with that of (a).

Q2. Find the equations of the lines tangent to the ellipse $\dfrac{x^2}{4} + \dfrac{y^2}{9} = 1$ at the points with x-coordinate 1.

Q3. In §3.4.4.Q3 Boyle's law $PV = C$ was used to find dV/dt in terms of P and dP/dt. Find dV/dt by differentiating $PV = C$ implicitly. Is the result the same?

Q4. The volume of a leaky round balloon diminishes at the constant rate of 0.2 m^3/sec. How rapidly is its radius shrinking when its volume is 36 m^3?

4.2.6 The principle of least time. Laws of reflection and refraction.[4] One reason for the usefulness of the chain rule in applications is that it enables us to differentiate the square roots of sums of squares which measure distance. I will illustrate by deriving two laws of optics.

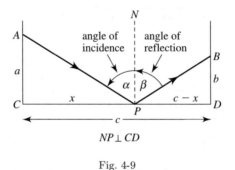

Fig. 4-9

1. A light ray is to travel in vacuum[5] from A to B via reflection in mirror CD, as illustrated in Fig. 4-9. Given these conditions on the ray, we may ask at what point P it must strike the mirror. The answer provided by physics is that P is so determined that the path is the *shortest* of all such paths from A to B.

In order to discover what this rule entails, let us undertake to minimize the variable path-length $AP + PB$. We first express it in terms of $CP = x$ as

$$f(x) = \sqrt{x^2 + a^2} + \sqrt{(c - x)^2 + b^2}\,.$$

Differentiation by means of the chain rule then yields

$$f'(x) = \frac{x}{\sqrt{x^2 + a^2}} - \frac{c - x}{\sqrt{(c - x)^2 + b^2}} = \sin\alpha - \sin\beta\,.$$

---

[4]This section can be omitted.

[5]Or in any homogeneous, isotropic, transparent medium.

Hence a value x_0 at which the **angle of incidence** α equals the **angle of reflection** β will be a critical point of f.

We can show that this condition actually makes $f(x)$ a minimum by examining the behavior of f'. As x increases from 0 to c, α increases from 0 to some value $< \pi/2$, while β decreases from some positive value $< \pi/2$ to 0. Therefore $\sin \alpha$ increases from 0, while $\sin \beta$ decreases to 0; consequently $\sin \alpha - \sin \beta = f'(x)$ increases from a negative to a positive value, hence (by the intermediate value property) vanishes at a point x_0, and is negative to the left of x_0 and positive to the right of it. This means that (i) x_0 is the only critical point, (ii) $f(x)$ has a local minimum there, and (iii) that minimum is actually the least value of $f(x)$, because the function decreases for $x \leq x_0$ and increases for $x \geq x_0$. Furthermore, $\alpha = \beta$ at x_0, because as a function monotonic between 0 and $\pi/2$ $\sin x$ does not take the same value twice on that interval.

Another way to prove the same thing is to show $f'' > 0$ (Q1). The foregoing reasoning has the advantages of avoiding additional calculation and allowing some insight into why the minimum occurs.

The value x_0 can be found by solving $f'(x) = 0$ (Q2), but it is of less interest than the *law of reflection* $\alpha = \beta$ we have deduced from the shortest-path hypothesis.[6] In this deduction the variable x had the virtue that the path length admitted a manageable expression in terms of it.

2. The hypothesis of the shortest path is only a special case of a more general principle, that light will take the path that requires *least time*. This may not be the shortest path if the light is not traveling in vacuum, because the speed of light depends on the medium through which it moves.[7]

Suppose that a light ray is to travel through one medium—air, say—from A to boundary CD, and then through another—water—to B, as in Fig. 4-10. Let its velocity in air be v_1, and in water v_2. It is clear that the least-time path will consist of two straight segments AP and PB (Q3). Then the time the light ray takes to go from A to B is the sum of the times it takes along AP and PB, or

$$\frac{AP}{v_1} + \frac{PB}{v_2}.$$

---

[6]The law has been celebrated by poets: Lucretius, 4.322-323 (line numbers as in Loeb Classical Library edition, 1975); Dante, *Purg.* 15.16-21.

[7]In certain ("anisotropic") mediums the speed also depends on the direction taken by the light; these are excluded here.

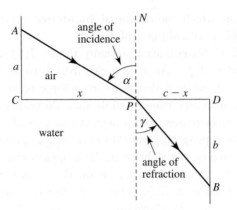

Fig. 4-10

In terms of x this is

$$f(x) = \frac{1}{v_1}\sqrt{x^2 + a^2} + \frac{1}{v_2}\sqrt{(c-x)^2 + b^2}\,.$$

As before, we find

$$f'(x) = \frac{1}{v_1}\frac{x}{\sqrt{x^2 + a^2}} - \frac{1}{v_2}\frac{c-x}{\sqrt{(c-x)^2 + b^2}} = \frac{\sin\alpha}{v_1} - \frac{\sin\gamma}{v_2}\,.$$

There is a critical point where

$$\frac{\sin\alpha}{\sin\gamma} = \frac{v_1}{v_2}\,.$$

The *law of refraction* states that this relation between angle of incidence and **angle of refraction** determines the path.[8] That the minimum actually occurs as the law asserts is proved exactly as in the preceding derivation (Q4).

4.2.7 QUESTIONS

Q1 (§4.2.6, no. 1). Calculate f'' to prove this.
Q2 (§4.2.6, no. 1). Find x_0, and also $c - x_0$.
Q3 (§4.2.6, no. 2). Give the argument.
Q4 (§4.2.6, no. 2). Give the argument.

---

[8]Poets have sung of refractive phenomena; have they acknowledged the role of the sine?

Q5. (a) In the case of refraction, show that if $v_1 > v_2$ then $\alpha > \gamma$. (b) Explain directly from the physical situation why this is so when $a = b$.

Q6. (a) Show that the law of refraction implies that an observer at A looking at an underwater object at B will see it displaced from its true position. (b) Find an expression for the apparent horizontal displacement in terms of the depth b and the angles of incidence and refraction. (c) When sodium light is used, the ratio v_1/v_2 for air and water is about $4/3$. If $\alpha = 60°$, find the apparent displacement of an object 1 meter deep. (You need not find γ; rather, use trigonometric identities.)

Q7. Prove the law of reflection by plane geometry, without calculus. (Extend AC to AA' so that $A'C = AC$ and join $A'P$.)

Q8. What makes these problems of determining paths solvable by our methods, in view of the distinction pointed out in §3.9.9.Q15?

Q9. Let $P(x_0, y_0)$ be a fixed point in the first quadrant of the xy-plane, not lying on an axis. Find the equation of the shortest straight line through P cut off by the axes. (Let the slope μ of the line be the independent variable.)

Q10. A lies on a straight road, B is 1 mile away from the road. It is 1 mile from A to the point on the road nearest B. To string wire along the road costs half as much per mile as to do so overland. What is the cheapest route for stringing wire from A to B? Compare the cost along this route with the cost along the straight route and the cost along the right-angle route which begins on the road.

Q11. Let $P(x_0, y_0)$ be a fixed point not on the graph of a differentiable function $y = f(x)$. Show that *if* the distance from P to the graph has a local extremum, it is attained along a straight line *perpendicular* to the graph. Give examples. (Let $Q(x, y)$ be a variable point on the graph and consider PQ.)

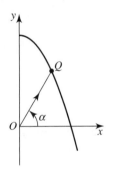

Q12. Let $Q(x, y)$ be a variable point on the parabola $y = 1 - 3x^2$, $x \geq 0$ (Fig. 4-11). Suppose that for any given position of Q, a constant force moves a particle from O to Q along the straight line OQ; and further, that the magnitude of the force is determined by the angle of inclination α of OQ according to the law $f = \cos^2 \alpha$. For what location of Q is the greatest work done on the particle?

Fig. 4-11

4.2.8 Derivatives of momentum and kinetic energy.[9] By means of the chain rule we can establish the equivalence asserted in §1.4.8 between two fundamental propositions of physics.

---

[9]This section can be omitted.

Let a particle move along the x-axis under the sole influence of a force f. Then (§2.8.4) (i) the force is the derivative with respect to time of the particle's momentum $p = mv$,

$$f = \frac{dp}{dt} \, ;$$

and (ii) the force is the derivative with respect to location of the particle's kinetic energy $k = \frac{1}{2}mv^2$,

$$f = \frac{dk}{dx} \, .$$

The two statements are equivalent because

$$\frac{dp}{dt} = \frac{dk}{dx} \, .$$

For

$$\frac{dk}{dx} = \frac{d}{dx}\left(\frac{1}{2}mv^2\right) = mv\frac{dv}{dx} \qquad \text{(by the chain rule)}$$

$$= m\frac{dx}{dt}\frac{dv}{dx} = m\frac{dv}{dt} \qquad \text{(by the chain rule)}$$

$$= \frac{d}{dt}(mv) = \frac{dp}{dt} \, .$$

4.2.9   A family of functions ill-behaved at a point.[10]   The extension of the range of differentiation afforded by the chain rule makes it possible to explore types of functions whose behavior may appear exotic or pathological. Although such functions will not be of practical use to us, on two grounds it is worth our while to examine some of them. In the first place, they provide examples to justify certain distinctions, such as that between differentiability and smoothness. In the second, they serve to chasten the untutored intuition, which, well acquainted with the graphs of elementary functions and encouraged by the simplicity of the *definitions* of function and graph, is likely to believe that it knows in a general way what to expect of functional behavior. From the examples given here one can get an inkling of the complexity that exists,

---

[10]This section can be omitted.

even though their pathology is limited to the vicinity of a single point. I will give only rather brief accounts of the functions, leaving details to you.

In each case the point in question will be $x = 0$, and each function is either odd or even. Because of the symmetry it suffices to consider non-negative x-values.

EXAMPLE 1. $y = f(x) = \sin \dfrac{1}{x}, \quad x \neq 0.$

Our interest is in the behavior of this function in the vicinity of $x = 0$. Let x decrease from some large value toward 0; then $1/x$ increases toward ∞. The

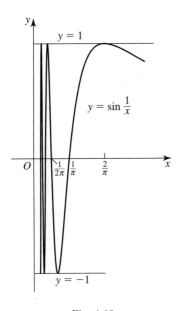

function increases from small positive values to the value $+1$ at $x = \dfrac{2}{\pi}$, then falls to -1 at $x = \dfrac{2}{3\pi} = \dfrac{1}{3}\left(\dfrac{2}{\pi}\right)$, rises to $+1$ at $x = \dfrac{1}{5}\left(\dfrac{2}{\pi}\right)$, falls to -1 at $x = \dfrac{1}{7}\left(\dfrac{2}{\pi}\right)$, and so on (Fig. 4-12). All the infinity of oscillations undergone by $\sin x$ for $\pi/2 \leq x < \infty$ are here crowded into $0 < x \leq 2/\pi$. Further, any neighborhood of 0, no matter how small, contains all but a finite number of these oscillations.

Although continuous (and even differentiable as many times as we wish) for $x \neq 0$, the function clearly has no limit as $x \to 0$.

Fig. 4-12

EXAMPLE 2. $y = f(x) = \begin{cases} x \sin \dfrac{1}{x}, & x \neq 0, \\ 0, & x = 0. \end{cases}$

The value assigned to y at $x = 0$ makes the function continuous there, for as $x \to 0$ we have

$$|f(x)| = \left| x \sin \frac{1}{x} \right| = |x| \left| \sin \frac{1}{x} \right| \leq |x| \to 0 = f(0).$$

Nevertheless, the function oscillates in value just as many times as $\sin \dfrac{1}{x}$; the difference is that the factor x tames or **damps** the oscillations, making them smaller and smaller in amplitude as $x \to 0$ (Fig. 4-13). Any neighborhood of 0 contains infinitely many alternating intervals of increase and decrease.

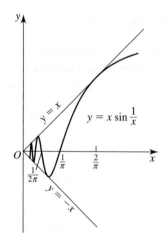

Fig. 4-13

As a continuous function, $f(x)$ can be integrated; but the existence of its integral cannot be demonstrated by the method of §2.4.2, owing to the hypothesis introduced at the beginning of that section.

While sufficient to produce continuity at 0, the damping does not control the slope of a secant drawn from the origin to a variable point on the curve well enough to make the function differentiable at 0. As $x \to 0$ the secant swings back and forth between the lines $y = x$ and $y = -x$, so that its slope has no limit. Indeed, the difference quotient at 0 is

$$\frac{f(\Delta x) - f(0)}{\Delta x} = \sin \frac{1}{\Delta x} \, ,$$

whose behavior as $\Delta x \to 0$ we have already seen.

EXAMPLE 3. $y = f(x) = \begin{cases} x^2 \sin \dfrac{1}{x}, & x \neq 0, \\[2mm] 0, & x = 0. \end{cases}$

The damping is now more severe, because x^2 approaches 0 more rapidly than x (Fig. 4-14). As a result the function is differentiable at 0: the slope of a secant drawn from the origin is forced toward 0 as $x \to 0$. The analytic expression of this fact is that the difference quotient at 0 equals $\Delta x \sin \dfrac{1}{\Delta x}$. Thus $f'(0)$ exists and has the value 0.

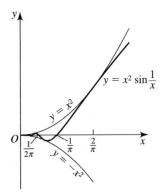

Fig. 4-14

Although $f(x)$ is differentiable everywhere, its derivative is discontinuous at 0. For when $x \neq 0$,

$$f'(x) = 2x \sin \frac{1}{x} - \cos \frac{1}{x}.$$

As $x \to 0$, the first term $\to 0$ while the second oscillates between $+1$ and -1. The function of Example 2 was continuous, but not differentiable; this one is differentiable but not continuously differentiable (smooth). Its graph has a tangent everywhere, but the tangent does not turn continuously as we pass 0. Instead, the tangent at x swings more and more rapidly between slope $+1$ and slope -1 as $x \to 0$, even though at 0 itself the tangent is the x-axis. The assumptions made in §2.9.6, where we showed that the secant approaches the tangent, are not fulfilled here.

EXAMPLE 4. $y = f(x) = \begin{cases} x^3 \sin \dfrac{1}{x}, & x \neq 0, \\ 0, & x = 0. \end{cases}$

Now the damping is effective enough to make the derivative continuous: we have

$$f'(0) = \lim_{\Delta x \to 0} \frac{\Delta y}{\Delta x} = \lim_{\Delta x \to 0} \left[(\Delta x)^2 \sin \frac{1}{\Delta x} \right] = 0 \,,$$

and for $x \neq 0$,

$$f'(x) = 3x^2 \sin \frac{1}{x} - x \cos \frac{1}{x} \to 0$$

as $x \to 0$. Yet the function is not twice differentiable, for the difference quotient of f' at 0 is

$$\frac{\Delta y'}{\Delta x} = 3\Delta x \sin \frac{1}{\Delta x} - \cos \frac{1}{\Delta x} \,.$$

Among the critical points of this smooth function there is one, namely $x = 0$, which is not *isolated* (§3.9.4). At that value of x the function has neither an extreme value nor an inflection point (Q4).

There is no need to continue with the series, as its pattern is plain by now. In any case, granted the possibility of containing an infinitude of oscillation in an arbitrarily small space,[11] a variety of behaviors at a point hardly astonishes. Much more wonderful is the fact that such an infinitude can exist at all points at once, so that, for example, there is a function continuous everywhere but nowhere differentiable. An example of that kind is less easily constructed than the ones I have given here.

4.2.10   QUESTIONS

Q1.   Supply any details omitted from the examples. In particular, verify the derivatives in Examples 3 and 4.

Q2.   In Example 3, find x-values arbitrarily near 0 such that $f'(x)$ is equal to $+1$; to -1.

Q3.   Sketch the graph for Example 4.

Q4 (§4.2.9, Example 4).   Prove the assertions made about the critical point $x = 0$. (Use the intermediate value property to locate critical points.)

Q5.   Discuss functions $x^\alpha \sin(1/x)$, where α is positive but not an integer.

---

[11] Cf. "infinite riches in a little room" (Marlowe).

4.3   INVERSE FUNCTIONS AND THEIR DERIVATIVES

4.3.1   Inverse of a function. The notion of the **inverse** of a function was introduced in §3.3.11, in connection with the exponential functions and their corresponding logarithmic functions. Recall that one function is the inverse of another if it performs the opposite operation. Thus the functions

$$(\ )^3 , \quad \text{or } y = x^3 ,$$

and

$$(\ )^{1/3}, \quad \text{or } y = x^{1/3} ,$$

are inverses of one another, since for any numbers u and v,

$$v = u^3$$

and

$$u = v^{1/3}$$

are equivalent. Similarly, e^x and $\log x$ are inverses; in fact, the latter was defined as the inverse of the former.

The inverse of a function f is denoted f^{-1} ("f inverse"). (Here the -1 is not an exponent.) To obtain the function machine for f^{-1} we turn the machine for f upside down (invert it) and run it in reverse (Fig. 4-15). The x's are now put in where the y's formerly came out. This is important: in order to have the inverse, like the original function, expressed as a function of x, we must *interchange the variables* after inverting. From $y = f(x)$ (e.g. $y = x^3$) follows

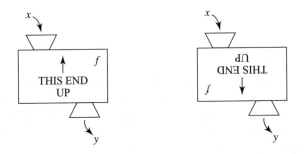

Fig. 4-15

$x = f^{-1}(y)$ $(x = y^{1/3})$; only after interchanging x and y do we get $y = f^{-1}(x)$ $(y = x^{1/3})$.

The values of x for which f^{-1} is defined are the numbers which are values of f, i.e. the outputs of the f-machine. For example, the values taken by e^x consist of all the positive numbers; therefore $\log x$ is defined for $x > 0$. Symmetrically, the values f^{-1} takes are the numbers on which f is defined, the inputs to the f-machine. Thus every number is a value of $\log x$, because e^x is defined for all x.

As Fig. 4-16 suggests, the graph of f^{-1} is obtained from that of f by **reflection** in the line $y = x$ as though in a mirror: to each point P on the graph of f corresponds the point P' such that $y = x$ is the perpendicular bisector of PP'. The reason is not hard to discover (Q5). Of course, reflection back again restores the graph of f, which is to say that the inverse of the inverse is the original function, $(f^{-1})^{-1} = f$.

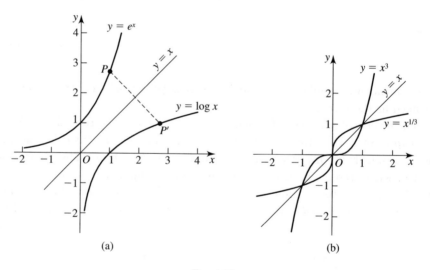

(a)   (b)

Fig. 4-16

4.3.2 Invertible functions. The arc sine. Not all functions have inverses, but only those whose function-machines can be run backwards as function-machines. By definition, a function f assigns a unique output to each input. If the reverse operation is to be a function as well, the original function f must also have a unique *input* for each *output*. In other words, f must define a *one-to-one correspondence* between its inputs and its outputs.[12] But many functions do not have this property; e.g., the constant function $f(x) = 2$ makes *every* number as input correspond to the single output 2, and (as a less extreme example) there are two inputs corresponding to each non-zero output of $f(x) = x^2$. Again, $f(x) = \sin x$ has infinitely many inputs for each output (in fact, all the numbers $n\pi + x$ (n even) and $n\pi - x$ (n odd) go to the single output $\sin x$).

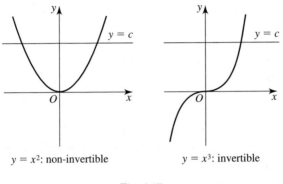

$y = x^2$: non-invertible $y = x^3$: invertible

Fig. 4-17

Geometrically speaking, a function is **invertible** (has an inverse) if its graph does not meet any horizontal line $y = c$ more than once; for in that case no output c can result from more than one input of an x-value. Provided that the graph is a continuous line, this means that it always rises or always falls (Fig. 4-17). We would therefore expect that *a continuous monotone function is invertible, and its inverse is again a continuous monotone function.* It is indeed almost obvious from the definition of monotonicity that a monotone function is invertible, and further that *the inverse of an increasing function is increasing, and of a decreasing function, decreasing* (Q6). While it is also true that the inverse of a continuous monotone function is continuous, the proof requires some care; it will be given in §4.3.7.

Continuous functions which are not monotone are, as a rule, nevertheless monotone over certain intervals of x-values. Thus, e.g., $y = x^2$ is monotone for $x \geq 0$, and also for $x \leq 0$ (Fig. 4-17); $y = \sin x$ is monotone for $-\pi/2 \leq x \leq \pi/2$, and also for $\pi/2 \leq x \leq 3\pi/2$, and over infinitely many other intervals. If

---

[12]Between its domain and its range (§1.6.6).

we restrict a non-invertible function to a chosen
interval on which it is monotone, it will become
invertible—as a function defined on that interval.

That is how the ordinary square-root function is defined: it is the inverse of the squaring
function on the interval $x \geq 0$ (Fig. 4-18). We
agree to think of squares as arising from non-
negative numbers only; then each square has only
one square root. If we wished, we could confine
the squaring function to the interval $x \leq 0$ instead, and have non-positive square roots. But
the other inverse of $y = x^2$ is more useful, so it
has become standard.

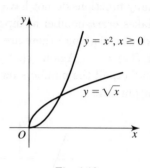

Fig. 4-18

In the case of $y = \sin x$ it is usual to choose the interval $[-\pi/2, \pi/2]$, over
which the function increases from -1 to 1. Its inverse is then defined on $[-1, 1]$
and increases there from $-\pi/2$ to $\pi/2$ (Fig. 4-19). This function is called the
arc sine, and written $y = \arcsin x$ (the arc [of a unit circle]—that is, the
number [of radians] between $-\pi/2$ and $\pi/2$—whose $sine$ is x). The notation
$\sin^{-1} x$ is also used, despite its potential to be confused with $1/\sin x$.

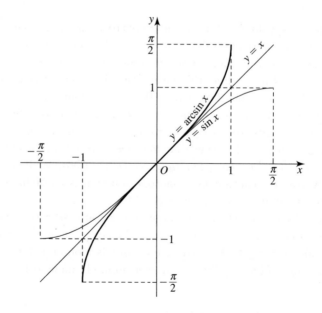

Fig. 4-19

The inverses of other trigonometric functions will be defined in §4.4.7.

4.3.3 QUESTIONS

Q1. If q is a positive integer, show that the function $y = x^q$ is invertible for all x if q is odd, and for $x \geq 0$ if q is even. Hence justify (iii) of §3.3.2.

Q2. Find the following values of the arc sine function. (a) arcsin 1 (b) arcsin $\left(-\frac{1}{2}\right)$ (c) arcsin $(\sqrt{3}/2)$

Q3. Is it true for all x that arcsin sin $x = x$?

Q4. Prove that arcsin x is an odd function.

Q5 (§4.3.1). Explain this.

Q6 (§4.3.2). Give the argument.

Q7. Let f be invertible. What are the functions $f^{-1} \circ f$ and $f \circ f^{-1}$? (Cf. §4.1.2.Q3.)

Q8. (a) Suppose that $y = f(u)$ and $u = g(x)$ are invertible functions, and that every number u for which f is defined is a value of g. Show that $f \circ g$ is invertible, and find its inverse. (b) Apply (a) to show that $x^{3/5}$ is invertible and find its inverse (recall (iv) of §3.3.2). (c) Is $x^{2/3}$ invertible?

Q9. (a) Show that if $f(x)$ is a continuous increasing function and $\lim_{x \to \infty} f(x) = \infty$, then $\lim_{x \to \infty} f^{-1}(x) = \infty$. (b) Apply this with $f(x) = a^x$, $a > 1$.

Q10. What connection is there between the idea of an inverse function and the Fundamental Theorem of Calculus?

4.3.4 Derivative of the inverse.

Let $f(x)$ be a differentiable monotone function. Differentiation of f^{-1} is easy in principle, but in practice somewhat confusing because of the interchange of variables. To simplify matters I am not going to interchange variables until the very end: I will at first be writing $x = f^{-1}(y)$ rather than $y = f^{-1}(x)$.

The derivative of $f^{-1}(y)$, if it exists, is then dx/dy. Under its interpretation as a quotient we obviously have

(1)
$$\frac{dx}{dy} = \frac{1}{\dfrac{dy}{dx}},$$

as long as $dy/dx \neq 0$. The two derivatives are to be evaluated at *corresponding* values of y and x: if $y_0 = f(x_0)$,

$$\left.\frac{dx}{dy}\right|_{y_0} = \frac{1}{\left.\dfrac{dy}{dx}\right|_{x_0}}.$$

Note that in formula (1) the right side is a function of x, while the left side, as the derivative of $f^{-1}(y)$, ought to be a function of y. In order to make the right side into a function of y we substitute $f^{-1}(y)$ for x.

As a final step we interchange the variables to obtain $(f^{-1})'(x)$, the derivative of $y = f^{-1}(x)$.

All this will be clarified in a moment by examples. Let me first state the essential fact in functional notation:

If $f(x_0) = y_0$ and $f'(x_0) \neq 0$, then f^{-1} is differentiable at y_0 and

$$(f^{-1})'(y_0) = \frac{1}{f'(x_0)} \, .$$

The proof will be given in §4.3.7.

Since f is monotone, f' can vanish only at individual points, ordinarily isolated (§§3.9.1, 3.9.4). Except at such points, $x = f^{-1}(y)$ is a differentiable function, and its derivative is given by

$$(f^{-1})'(y) = \frac{1}{f'(x)} = \frac{1}{f'(x(y))} \, ,$$

where $x(y) = f^{-1}(y)$ has been substituted for x as explained above. Moreover, if f' is continuous so is $(f^{-1})'$ (Q5).

If we interchange the variables and write $y = f^{-1}(x)$, the formula becomes

$$(f^{-1})'(x) = \frac{1}{f'(y(x))} \, .$$

4.3.5 Examples. For the first example, let $y = f(x) = x^3$. The inverse function is $x = f^{-1}(y) = y^{1/3}$. To determine its derivative we proceed as follows. From $y = x^3$ we obtain $dy/dx = 3x^2$; then the derivative of $x = y^{1/3}$ is

$$\frac{dx}{dy} = \frac{1}{\dfrac{dy}{dx}} = \frac{1}{3x^2}, \quad x \neq 0 \, .$$

Here we substitute $y^{1/3}$ for x to express dx/dy as a function of y:

$$\frac{dx}{dy} = \frac{1}{3(y^{1/3})^2} = \frac{1}{3y^{2/3}} = \frac{1}{3} y^{-2/3}, \quad y \neq 0 \, .$$

We have now found the derivative of $x = y^{1/3}$. The derivative of $y = x^{1/3}$ is obtained by interchanging the variables:

$$\frac{d}{dx}(x^{1/3}) = \frac{dy}{dx} = \frac{1}{3}x^{-2/3}, \quad x \neq 0,$$

in agreement with the rule for differentiating a power. At the origin the graph of $y = x^{1/3}$ has a vertical tangent, corresponding to the horizontal tangent to the graph of $y = x^3$ there (Fig. 4-16(b)). This explains the non-differentiability of the inverse function at $x = 0$.

Although we have as yet only postulated the existence of the exponential function and assumed without proof that the derivative of e^x is e^x (§3.3.10), as a second example we can now deduce from that assumption that the derivative of the inverse function $\log x$ is $1/x$. Let $y = e^x$, then $dy/dx = e^x$; hence the derivative of $x = \log y$ is

$$\frac{dx}{dy} = \frac{1}{e^x} = \frac{1}{y}.$$

Interchanging the variables then gives the derivative of $y = \log x$:

$$\frac{d}{dx}(\log x) = \frac{dy}{dx} = \frac{1}{x}.$$

Here $x > 0$, since the logarithm is defined only for positive x. In the same way the derivative of $\log_a x$ is found from that of a^x (Q3).

We can similarly differentiate $\arcsin x$, but I will defer the calculation until the other inverse trigonometric functions have been introduced (§4.4.7).

4.3.6　QUESTIONS

Q1.　(a) Show that $f(x) = x^2 + 2x + 3$ is invertible on the interval $x > -1$. (b) Let f^{-1} be its inverse there. Find $(f^{-1})'(3)$. (First find x_0 such that $f(x_0) = 3$.)

Q2.　Carry out the differentiation of the inverse of $y = x^{3/2}$, $x \geq 0$, following the model of $y = x^3$ in the text.

Q3 (§4.3.5).　Perform the differentiation.

Q4.　Show that $(f^{-1})' = \dfrac{1}{f' \circ f^{-1}}$.

Q5 (§4.3.4).　Prove this. (Apply Q4.)

4.3.7   Continuity and differentiability of the inverse.[13]   Let $y = f(x)$ be a continuous monotone function—to be definite, an increasing function. In order to prove f^{-1} continuous a preliminary remark is necessary.

Let $x_1 < x_2$, and let $y_1 = f(x_1)$, $y_2 = f(x_2)$; then $y_1 < y_2$. Let $y_1 < y < y_2$. By the intermediate value property (§3.7.14) f assumes the value y for some x in $[x_1, x_2]$; for y lies between y_1 and y_2, both of which are values assumed by f on $[x_1, x_2]$. Thus every y in $[y_1, y_2]$ is $f(x)$ for some x in $[x_1, x_2]$ (Fig. 4-20).

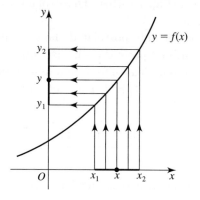

Fig. 4-20

Now we can show that the function $x = f^{-1}(y)$ is continuous at any $y_0 = f(x_0)$. Continuity of f^{-1} at y_0 can be defined in this way (§3.1.6.Q6): if S is any neighborhood of x_0, there is some neighborhood T of y_0 such that if y is in T, $f^{-1}(y)$ is in S. Suppose given a neighborhood $S = [x_1, x_2]$ of x_0, so that $x_1 < x_0 < x_2$. We have $f(x_1) < f(x_0) < f(x_2)$, so if we let $y_1 = f(x_1)$ and $y_2 = f(x_2)$ then $[y_1, y_2] = T$ is a neighborhood of y_0. By our initial remark, each y in that neighborhood is $f(x)$ for some x in $[x_1, x_2]$; that is, for each y in T, $f^{-1}(y)$ is in S. This establishes that f^{-1} is continuous.

Suppose now that the monotone function f is differentiable, and assume $f'(x_0) \neq 0$. Let $y_0 = f(x_0)$; I will show that f^{-1} is differentiable at y_0 and $(f^{-1})'(y_0) = 1/f'(x_0)$.

Let Δy be a non-zero increment added to y_0, and Δx the *corresponding* increment added to x_0; Δx is then necessarily non-zero. The difference quotient

---

[13]This section can be omitted.

for f^{-1} at y_0 is

$$\frac{\Delta x}{\Delta y} = \frac{1}{\dfrac{\Delta y}{\Delta x}} .$$

Because f^{-1} is continuous, as $\Delta y \to 0$ also $\Delta x \to 0$; and as $\Delta x \to 0$, $\Delta y/\Delta x$ approaches the non-zero limit $f'(x_0)$. Therefore as $\Delta y \to 0$, $\Delta x/\Delta y$ approaches $1/f'(x_0)$, q.e.d.

These arguments apply essentially unchanged to a function $f(x)$ defined only in some neighborhood of x_0. With obvious modifications they apply also if $f(x)$ is defined only on an interval of which x_0 is an endpoint (cf. §3.1.6.Q6(b)).

4.4 DIFFERENTIATION OF THE TRIGONOMETRIC FUNCTIONS AND THEIR INVERSES

4.4.1 Differentiation of the sine and cosine. Once either $\sin x$ or $\cos x$ has been differentiated, the other can immediately be differentiated as well. For suppose we know that $\sin x$ is differentiable and $(\sin x)' = \cos x$; then since $\cos x = \sin\left(x + \dfrac{\pi}{2}\right)$, the chain rule tells us that $\cos x$ is differentiable and

$$(\cos x)' = \cos\left(x + \frac{\pi}{2}\right)(1) = -\sin x .$$

The argument in the other direction is similar.[14]

To complete the differentiation of the trigonometric functions, then, we have to differentiate one of these two functions. Let it be the sine. The difference quotient is

$$\frac{\sin(x + \Delta x) - \sin x}{\Delta x} = \frac{\sin x \cos \Delta x + \cos x \sin \Delta x - \sin x}{\Delta x} ,$$

by application of an addition formula, or

$$\left[\frac{\sin \Delta x}{\Delta x}\right] \cos x - \left[\frac{1 - \cos \Delta x}{\Delta x}\right] \sin x .$$

---

[14] A direct proof using the difference quotient was expected in §3.3.9.Q12. For the identities used here and below, see §3.3.7.

I am about to prove that as $\Delta x \to 0$,

$$\frac{\sin \Delta x}{\Delta x} \to 1 \quad \text{and} \quad \frac{1 - \cos \Delta x}{\Delta x} \to 0.$$

It will then follow that the difference quotient for $\sin x$ approaches $\cos x$, as expected.

4.4.2   Limit of $(\sin h)/h$.

In considering these two limits let us replace the unwieldy Δx by the single letter h. A geometrical argument will demonstrate that

(1)
$$\lim_{h \to 0} \frac{\sin h}{h} = 1,$$

and then

(2)
$$\lim_{h \to 0} \frac{1 - \cos h}{h} = 0$$

will be derived as a consequence of (1).[15]

To begin with let $0 < h < \pi/2$, and place an angle POQ of h radians at the center of a unit circle, as shown in Fig. 4-21. QS is perpendicular to OP, and PR is the tangent at P cut off by OQ extended. By definition $QS = \sin h$ and $OS = \cos h$, and by similar triangles (since $OP = 1$) $PR = \sin h/\cos h = \tan h$.[16] Hence the inequalities of areas

$$\triangle OPQ < \text{sector } OPQ < \triangle OPR$$

are the same as

$$\frac{1}{2}\sin h < \frac{1}{2} h < \frac{1}{2}\tan h \, ;$$

here the area of the sector has been calculated from the proportion[17]

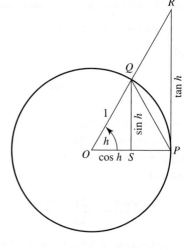

Fig. 4-21

---

[15] Cf. Newton's *Principia*, Book 1, Lemma 8. The demonstration of (1) will be reconsidered in the next section.

[16] Cf. §3.3.9.Q10.

[17] Cf. §3.3.6.Q7.

$$\frac{\text{sector } OPQ}{\text{area of circle}} = \frac{\text{arc } PQ}{\text{circumference of circle}}$$

or

$$\frac{\text{sector } OPQ}{\pi} = \frac{h}{2\pi}.$$

Dividing by the positive number $\frac{1}{2}\sin h$ gives

$$1 < \frac{h}{\sin h} < \frac{1}{\cos h},$$

or

$$\cos h < \frac{\sin h}{h} < 1.$$

These inequalities do not change if $-h$ is substituted for h, because $\cos(-h) = \cos h$ and $\dfrac{\sin(-h)}{-h} = \dfrac{-\sin h}{-h} = \dfrac{\sin h}{h}$. Therefore they hold for all non-zero h, $-\pi/2 < h < \pi/2$—that is, for all h near 0. If now we let $h \to 0$, $\cos h \to \cos 0 = 1$[18] and it follows that $(\sin h)/h$, trapped between $\cos h$ and 1, also $\to 1$. This establishes limit (1).

For limit (2), we apply the half-angle identity $\sin^2 \dfrac{h}{2} = \dfrac{1}{2}(1 - \cos h)$ to obtain

$$\frac{1 - \cos h}{h} = \frac{2\sin^2 \dfrac{h}{2}}{h} = \frac{\sin \dfrac{h}{2}}{\dfrac{h}{2}} \cdot \sin \frac{h}{2}.$$

As $h \to 0$, also $h/2 \to 0$, hence the first factor $\to 1$ and the second $\to 0$;[19] consequently the product $\to 0$, so that (2) is established.

4.4.3   QUESTIONS

Q1.   For any number x let SIN x and COS x be defined as the sine and cosine of the angle of x *degrees*. Express SIN x and COS x in terms of $\sin x$ and $\cos x$, and thereby determine the derivative of SIN x in terms of COS x. The answer indicates one good reason for preferring radian measure.

Q2.   Show that the limit $\dfrac{\sin \Delta x}{\Delta x} \to 1$ is an immediate consequence of $(\sin x)' = \cos x$.

---

[18]Continuity of $\sin x$ and $\cos x$ at $x = 0$ was proved in §3.3.9.Q11.

[19]See preceding footnote.

Q3. Find the limits of the functions considered in §4.2.9 as $x \to \infty$ and as $x \to -\infty$.

4.4.4   Arc and chord. The area and circumference of a circle.[20]

The limit of $(\sin h)/h$, §4.4.2(1), is of special interest geometrically (Fig. 4-22). It expresses the fact that the ratio of the length of the *straight* line QS to that of the *curved* line arc PQ approaches unity as $h \to 0$ (I am considering only positive values of h). If the angle is doubled as shown and QS is extended to chord QQ' we have

$$\frac{QQ'}{\operatorname{arc} QQ'} = \frac{2 \sin h}{2h} = \frac{\sin h}{h} \to 1.$$

Equivalently, $\dfrac{\operatorname{arc} QQ'}{QQ'} \to 1$, or simply

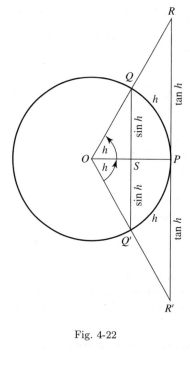

Fig. 4-22

(3) $$\frac{\operatorname{arc}}{\operatorname{chord}} \to 1:$$

the ratio of arc to chord approaches unity as the arc length approaches zero.

The ratio of arc to tangent—$\dfrac{\operatorname{arc} PQ}{PR}$, or $\dfrac{\operatorname{arc} QQ'}{RR'}$—also approaches unity, because

$$\frac{\tan h}{h} = \frac{\sin h}{h} \cdot \frac{1}{\cos h} \to 1.$$

Both limits are valid for a circle of any size (Q1).

It must not be supposed that we have really *proved* these geometrical propositions, in the sense of deducing them from more elementary ones. On the contrary, the demonstration of (1) amounted to no more than the reduction of the unfamiliar to its familiar equivalent. The reason is that it depended on the formulas $A = \pi r^2$ and $C = 2\pi r$ for the area and circumference of

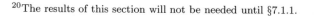

a circle, both implicitly in the use of radian measure and explicitly for the calculation of the area of a sector. Now if the number π is defined by the first of these formulas,[21] the second is equivalent to the limit in question. For $2\pi r$ is the limit of the perimeters of inscribed regular polygons as the number of their sides increases (I will give details in a moment), and these perimeters approach C if and only if (3) holds.

As the connection to approximating polygons may suggest, the limit of arc : chord is inseparable from the very concept of measurement of arc length— the idea that a curved line bears a ratio to a straight one. Property (3) of the circle is shared by the generality of curves, and can be used as the foundation of a theory of arc length akin to the theory of area we have already developed. These matters will be considered in Chapter 7 (Arts. 7.1–7.2).

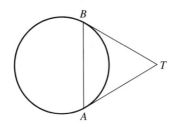

Fig. 4-23

It is worth mentioning also that (3) for the circle is equivalent to the following very simple fact: if AT and BT are intersecting tangents to a circle (Fig. 4-23), then in length

(4) $$AB < \text{arc } AB < ATB.$$

These are essentially the inequalities used in §4.4.2 (Q2(a)). Their significance with respect to polygons will be pointed out in the next section.

There follows an outline of several approaches to the arc length of a circle.

4.4.5　Arc length of a circle. Let A_n be the area, P_n the perimeter, and r_n the apothem (perpendicular from center to side) of a regular polygon of n sides inscribed in a circle of radius r and area A. Euclid proves $A_n \to A \propto r^2$ (*Elements*, Book 12, Prop. 2); let π be defined by $A = \pi r^2$. We can prove

---

[21] Euclid proves A proportional to r^2 (see the next section).

without difficulty that $A_n = \frac{1}{2}r_n P_n$ and $r_n \to r$; hence $P_n = 2A_n/r_n \to 2A/r = 2\pi r$.

1. If we disclaim any prior knowledge about the length of a circle, we can *define* it to be the limit of the perimeter of an inscribed regular polygon. This seems reasonable, because of the easily proved fact that if $\bar{P}_n$ is the perimeter of the *circumscribed* regular n-gon, $\bar{P}_n - P_n \to 0$. The length of a circular arc is then defined by means of the ratio (in the sense of Euclid[22]) which the arc bears to the whole circumference. With this definition the preceding remark implies $C = 2\pi r$, hence (as we have already proved) (4) holds and therefore (3).

2. Alternatively, we can argue directly that circular arcs can be assigned numerical sizes, or "lengths" (again following Euclid to define ratios of arcs), but make no prior assumption as to the ratio of an arc to a straight line.

(i) If then we accept (4), it obviously follows that $P_n < C < \bar{P}_n$. Then since P_n and $\bar{P}_n \to 2\pi r$, $C = 2\pi r$, etc.

(ii) If instead we assume (3), we reason as follows. Let p_n be a side of the inscribed n-gon and c_n the arc it subtends. Since $P_n = np_n$ and $C = nc_n$, we have $\dfrac{C}{P_n} = \dfrac{c_n}{p_n}$ and therefore $\dfrac{C}{2\pi r} = \lim \dfrac{c_n}{p_n} = 1$, or $C = 2\pi r$.

4.4.6 QUESTIONS

Q1 (§4.4.4).   Prove this.

Q2.   (a) (§4.4.4) Verify this. (b) Deduce the inequalities from §3.9.9.Q14. (c) Does (b) give a new basis for differentiating $\sin x$?

4.4.7 The inverse trigonometric functions and their derivatives.

The arc sine function, introduced in §4.3.2, is differentiated by the method of §§4.3.4–5. Starting with

$$y = \sin x, \quad -\pi/2 \le x \le \pi/2,$$

we have

$$x = \arcsin y, \quad -1 \le y \le 1,$$

and then

$$\frac{dx}{dy} = \frac{1}{\dfrac{dy}{dx}} = \frac{1}{\cos x}, \quad -\pi/2 < x < \pi/2$$

---

[22] *Elements*, Book 5, Def. 5; applied to circular arcs in Book 6, Prop. 33.

(excluding the endpoints, where $\cos x = 0$). On this x-interval $\cos x$ is positive, so from $\sin^2 x + \cos^2 x = 1$ we obtain

$$\cos x = \sqrt{1 - \sin^2 x} = \sqrt{1 - y^2}\,,$$

the positive square root giving the correct sign. Then

$$\frac{dx}{dy} = \frac{1}{\sqrt{1 - y^2}}\,, \quad -1 < y < 1\,,$$

or after interchanging variables

$$\frac{d}{dx}\left(\arcsin x\right) = \frac{1}{\sqrt{1 - x^2}}\,, \quad -1 < x < 1\,.$$

To invert the cosine function we must choose an interval on which it is monotonic. The standard choice is $[0, \pi]$, on which $y = \cos x$ *decreases* from 1 to -1. Its inverse there is the arc cosine,

$$y = \arccos x\,, \quad -1 \le x \le 1\,,$$

which decreases from π to 0 (Fig. 4-24).

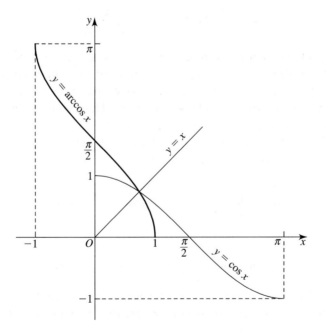

Fig. 4-24

The arc cosine differs by a constant from the negative of the arc sine: we have

$$\arccos x = \frac{\pi}{2} - \arcsin x .$$

For let $y = \arcsin x$, where $-1 \le x \le 1$ and $-\pi/2 \le y \le \pi/2$; then

$$x = \sin y = \cos \left(\frac{\pi}{2} - y \right), \quad \text{where } 0 \le \frac{\pi}{2} - y \le \pi ,$$

hence

$$\arccos x = \frac{\pi}{2} - y = \frac{\pi}{2} - \arcsin x .$$

As x increases from -1 to 1, $\arcsin x$ increases from $-\pi/2$ to $\pi/2$ and (I repeat) $\arccos x$ decreases from π to 0.

The derivative of $\arccos x$ can be calculated in exactly the same way as that of the arc sine (Q6), but it is easier to differentiate the preceding equation:

$$\frac{d}{dx} (\arccos x) = \frac{d}{dx} \left(\frac{\pi}{2} - \arcsin x \right)$$

$$= -\frac{d}{dx} (\arcsin x) = -\frac{1}{\sqrt{1 - x^2}}, \quad -1 < x < 1 .$$

The tangent function is not defined at $x = \pm\pi/2$, but for $-\pi/2 < x < \pi/2$ it is increasing, like the sine; unlike it, the tangent takes on every value there (see §3.3.9.Q10 and §3.9.9.Q1(d)). Its inverse on that interval, the arc tangent $y = \arctan x$, is accordingly defined for all x. The graph of the function is asymptotic to $y = \pi/2$ and $y = -\pi/2$ (Fig. 4-25).

We find the derivative by the method used before. Let $y = \tan x$, $-\pi/2 < x < \pi/2$. Then $x = \arctan y$ and

$$\frac{dx}{dy} = \frac{1}{\dfrac{dy}{dx}} = \frac{1}{\sec^2 x} = \frac{1}{1 + \tan^2 x} = \frac{1}{1 + y^2} ,$$

so

$$\frac{d}{dx} (\arctan x) = \frac{1}{1 + x^2}, \quad -\infty < x < \infty .$$

Inverses of the other trigonometric functions can be similarly defined and differentiated. For the arc cotangent see Q8; the other two are slightly more complicated.

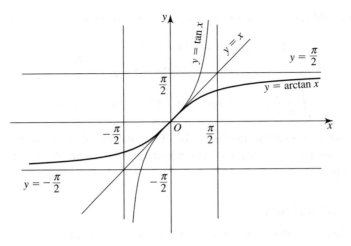

Fig. 4-25

4.4.8   Summary.   In summary:

The inverse of $\sin x$, $-\pi/2 \le x \le \pi/2$, is the increasing function

$$y = \arcsin x, \quad -1 \le x \le 1, \quad -\pi/2 \le y \le \pi/2,$$

and

$$\frac{d}{dx}(\arcsin x) = \frac{1}{\sqrt{1-x^2}}, \quad -1 < x < 1.$$

The inverse of $\cos x$, $0 \le x \le \pi$, is the decreasing function

$$y = \arccos x, \quad -1 \le x \le 1, \quad 0 \le y \le \pi,$$

and

$$\frac{d}{dx}(\arccos x) = -\frac{1}{\sqrt{1-x^2}}, \quad -1 < x < 1.$$

The inverse of $\tan x$, $-\pi/2 < x < \pi/2$, is the increasing function

$$y = \arctan x, \quad -\infty < x < \infty, \quad -\pi/2 < y < \pi/2,$$

and

$$\frac{d}{dx}(\arctan x) = \frac{1}{1+x^2}, \quad -\infty < x < \infty.$$

To these differentiation formulas correspond formulas for integration:

$$\int \frac{1}{\sqrt{1-x^2}}\,dx = \left\{ \begin{array}{l} \arcsin x + C\,, \\ -\arccos x + C\,, \end{array} \right\} \quad -1 < x < 1\,,$$

$$\int \frac{1}{1+x^2}\,dx = \arctan x + C\,, \quad -\infty < x < \infty\,.$$

In the first of these the necessity for the constant of integration (§3.5.7) is evident: $\arcsin x$ and $-\arccos x$ are primitives of the same function that differ by the constant $\pi/2$.

4.4.9   QUESTIONS

Q1.   Find the values. (a) $\arccos \frac{1}{2}$ (b) $\arccos\left(-\frac{1}{2}\right)$ (c) $\arctan 1$ (d) $\arctan(-1)$ (e) $\tan \arctan 0$ (f) $\arctan \tan \pi$

Q2.   Differentiate. (a) $\arcsin(2x-1)$ (b) $\arccos x^2$ (c) $5\arctan\sqrt{x}$

Q3.   Evaluate the integrals. (a) $\displaystyle\int_0^1 \frac{1}{1+x^2}\,dx$ (b) $\displaystyle\int_{-1}^1 \frac{1}{1+x^2}\,dx$

(c) $\displaystyle\int_{1/\sqrt{2}}^{\sqrt{3}/2} \frac{1}{\sqrt{1-x^2}}\,dx$

Q4.   (a) Express the function $\arctan x$ as an integral of the form $\int_a^x f(x)\,dx$. (b) Can $\arcsin x$ and $\arccos x$ be so expressed?

Q5.   Interpreting the derivative of $\arcsin x$ as the slope of its graph, explain why the derivative is not defined at $x = \pm 1$.

Q6 (§4.4.7).   Carry out the differentiation of $\arccos x$ by the method used for $\arcsin x$.

Q7.   Graph $\arcsin x$ and $\arccos x$ together to verify $\arccos x = (\pi/2) - \arcsin x$.

Q8.   (a) Show that $0 < x < \pi$ is a suitable interval on which to invert $\cot x$. (b) Hence define $\operatorname{arccot} x$, and show that $\operatorname{arccot} x = (\pi/2) - \arctan x$. (c) Differentiate $\operatorname{arccot} x$. (d) Sketch the graph of $\operatorname{arccot} x$. (e) Show that

$$\operatorname{arccot} x = \left\{ \begin{array}{ll} \arctan(1/x), & x > 0\,, \\ \pi + \arctan(1/x), & x < 0\,. \end{array} \right.$$

Do this in two different ways: (i) directly from the definition of the tangent and cotangent; (ii) by comparing the derivatives of $\operatorname{arccot} x$ and $\arctan(1/x)$.

Q9.   Let $f(x)$ be a differentiable function, and let $\alpha(x)$ be the angle of inclination of the tangent to its graph at x (§3.3.7). (a) Express $\alpha(x)$ in terms of $f(x)$. (b) Prove that if f is continuously differentiable, $\alpha(x)$ is continuous, and conversely. (This is a first answer to a question raised in §3.1.7; see also the remark at the end of §7.4.4.)

(c) Supposing that f'' exists, prove that α is differentiable and find $d\alpha/dx$. (d) The result of (c) shows that f'' and $d\alpha/dx$ have the same sign. Explain this in the light of §3.9.6.

Q10. *Inverse hyperbolic functions.* (a) Show that $\sinh x$ (§4.1.5.Q7) has an increasing inverse defined for all x, $\cosh x$ has a non-negative increasing inverse defined for $x \geq 1$, and $\tanh x$ has an increasing inverse defined for $-1 < x < 1$. The inverses are called[23] $\operatorname{arsinh} x$ or $\sinh^{-1} x$, $\operatorname{arcosh} x$ or $\cosh^{-1} x$, and $\operatorname{artanh} x$ or $\tanh^{-1} x$. (b) Show that

$$\frac{d}{dx}\operatorname{arsinh} x = \frac{1}{\sqrt{x^2+1}}, \quad \frac{d}{dx}\operatorname{arcosh} x = \frac{1}{\sqrt{x^2-1}}, \quad \frac{d}{dx}\operatorname{artanh} x = \frac{1}{1-x^2}.$$

Hence

$$\int \frac{1}{\sqrt{x^2+1}}\, dx = \operatorname{arsinh} x + C\,,$$

$$\int \frac{1}{\sqrt{x^2-1}}\, dx = \operatorname{arcosh} x + C\,, \quad x > 1\,,$$

$$\int \frac{1}{1-x^2}\, dx = \operatorname{artanh} x + C\,, \quad |x| < 1\,.$$

(c) By solving the equation $x = (e^y - e^{-y})/2$ for y, show that $\operatorname{arsinh} x = \log(x + \sqrt{x^2+1})$. Similarly show that $\operatorname{arcosh} x = \log(x + \sqrt{x^2-1})$ and $\operatorname{artanh} x = \frac{1}{2}\log\frac{1+x}{1-x}$.

4.5　DIFFERENTIATION OF FUNCTIONS OF SEVERAL VARIABLES[24]

4.5.1　Functions of several variables. The functions we have been studying express the dependence of one variable quantity on another. It can very well happen, however, that a quantity depends not just on one, but on two or more other quantities. The instantaneous rate of growth to be expected of a given kind of plant, for example, depends not only on time, but also on temperature, humidity, nutrient level, and so forth. Again, while one number

---

[23]The "ar" stands for "area": the hyperbolic functions can be associated with a certain area defined by the graph of a hyperbola, as the trigonometric functions are associated with an arc of a circle.

[24]This article can be omitted, except for its first section (used in Art. 4.8).

suffices to identify the location along a wire at which its line density is measured, to specify the **surface density**, or kilograms per square meter,[25] of a
plane sheet of metal we must refer to a point by two coordinate numbers.

The idea of function, and the functional notation, are readily expanded to
cover such situations. Let a variable z depend upon two other variables x and
y, both of which we call independent—y does not, as formerly, represent a
variable dependent on x; then we say that z is a function of x and y, and write

$$z = f(x, y)$$

("f of x, y") if we are using f to denote the function, or else simply $z = z(x, y)$.
An example is the function

$$z = f(x, y) = x^2 + 3xy,$$

defined for all values of x and y. At $x = 2$ and $y = -1$ it has the value

$$z = f(2, -1) = 2^2 + 3(2)(-1) = -2.$$

In the same way are defined functions of more than two variables, such as

$$w = f(x, y, z) = xyz^2 - y \sin z.$$

The properties of functions of several variables that will concern us are already
present in functions of two variables, so we can concentrate our attention on
the two-variable case.

As the example of surface density suggests, a function $z = f(x, y)$ can be
thought of as assigning a number z to each point (x, y) of the plane. If z
is used to specify a (signed) height above (x, y), a "graph" in the form of a
curved surface (rather than a curve) is created in the space of xyz-coordinates.
Figure 4-26 displays a **paraboloid of revolution**, which can be obtained
geometrically by revolving a parabola about its axis, as the graph of a function
(Q2). The continuity of the surface reflects the continuity of the function of two
variables, which we can define in geometrical terms as follows: as the variable
point (x, y) of the plane approaches a given point (x_0, y_0) (that is, as the
distance between them approaches zero), the height of the graph above (x, y)
approaches its height above (x_0, y_0) (that is, the value $z = f(x, y)$ approaches
$z_0 = f(x_0, y_0)$).

---

[25]The definition is analogous to that of line density in §1.1.1.

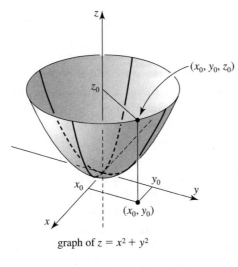

graph of $z = x^2 + y^2$

Fig. 4-26

4.5.2   Restriction to one variable.   From a function of two variables $z = f(x, y)$ functions of one of the variables, either x or y, can be derived by keeping the other fixed. Let us choose any definite value y_0 of y, and hold y at that value; then $z = f(x, y_0)$ expresses z as a function of x alone, under the condition $y = y_0$. If e.g. f represents surface density, then as x varies this function specifies the surface density at points along the line $y = y_0$ only (Fig. 4-27). Clearly each possible choice of y_0 gives rise to a function of x, and

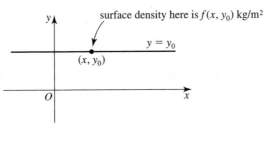

Fig. 4-27

these functions are in general different from one another. In the case of the function $z = x^2 + y^2$ graphed in Fig. 4-26, taking $y_0 = 0$ produces

$$z = f(x, 0) = x^2 \, ,$$

whereas if we choose $y_0 = 1$ we find instead

$$z = f(x, 1) = x^2 + 1 \, .$$

The graphical interpretation of these functions is dealt with in Q3.

In exactly the same way, by fixing x at a value x_0 we obtain $f(x_0, y)$, a function of y alone. For a function of more than two variables we hold *all but one* of the variables fixed in order to get a function of the remaining variable; e.g., given $w = f(x, y, z)$ we express w as a function of x alone under the conditions $y = y_0$ and $z = z_0$ by $w = f(x, y_0, z_0)$.

These one-variable functions may be of interest in themselves, as for example when we desire to know how fast a plant will grow at *specified* temperature, humidity, etc. But they are important also because, as it turns out, their derivatives can be said to govern the behavior of the original function. This means that in certain respects the theory of functions of several variables is reduced to the one-variable theory we have developed. I will be able to indicate how this is so in §4.5.7.

4.5.3   QUESTIONS

Q1.   Find the functions $f(x, 2)$ and $f(-1, y)$ if $f(x, y) = xy^2$.

Q2 (§4.5.1).   Verify that the graph is as described and depicted. (Consider its horizontal cross-sections defined by $z = $ constant.)

Q3 (§4.5.2).   Explain the relation between the given paraboloid and the parabolas $z = f(x, y_0)$ in the xz-plane, in the cases $y_0 = 0$ and $y_0 = 1$ and in general.

Q4.   Express the definition of continuity (§4.5.1) without geometrical language.

4.5.4   Partial derivatives.   The derivative of a function of one variable is the rate of change of the dependent variable with respect to the independent one. For a function of several variables we can consider the rates of change of the dependent variable with respect to each of the independent variables separately; in this way we define the several "partial derivatives" of the function, one for each independent variable.

To give formal expression to this idea let $z = f(x, y)$, choose a value y_0 of y, and consider the function of x defined by $z = f(x, y_0)$, as in §4.5.2. The derivative of that function is the **partial derivative**, or "partial" for short, **of** $f(x, y)$ **with respect to** x **at** $y = y_0$. Like any other derivative of a function of x it is itself a function of x; at any x it represents the rate of change of z with respect to x, given that y has the fixed value y_0. We denote it by $\left. \dfrac{\partial f}{\partial x} \right|_{y=y_0}$

or $f_x(x, y_0)$; its value at $x = x_0$ is $\left.\dfrac{\partial f}{\partial x}\right|_{(x_0, y_0)}$ or $f_x(x_0, y_0)$. In each symbol, f can be replaced by z.

If we now permit y_0 to vary, the partial derivative with respect to x becomes a function of two variables, like the original function f: $\left.\dfrac{\partial f}{\partial x}\right|_{(x, y)}$, or just $\dfrac{\partial f}{\partial x}$. The value of this function at (x_0, y_0) is the rate of change of z with respect to x at $x = x_0$, when y is fixed at the value y_0. (The symbol $\partial f/\partial x$ can be read "the partial derivative of f with respect to x" or "the partial of f with respect to x," or just "$df\, dx$"; f_x is "f sub x" or "$f\, x$.") Similarly we define $\partial f/\partial y$, the partial derivative with respect to y.

For a function of more than two variables the partial derivative with respect to each variable is defined by holding all the others constant and differentiating the resulting function of one variable.

4.5.5   Partial differentiation.   Higher-order derivatives. Since a partial derivative is nothing but an ordinary derivative with respect to a selected variable, the rules for differentiation of sums, products, etc., go over without change. *To calculate a partial derivative one simply differentiates as usual, treating as constants the variables held fixed.* For example, the partials of

$$z = x^2 y + \sin y$$

are

$$\frac{\partial z}{\partial x} = \frac{\partial}{\partial x}\left(x^2 y + \sin y\right)$$

$$= \frac{\partial}{\partial x}\left(x^2 y\right) + \frac{\partial}{\partial x}\left(\sin y\right)$$

$$= y\frac{\partial}{\partial x}\left(x^2\right) + 0 \quad \text{(treating } y \text{ as constant)}$$

$$= 2xy$$

and

$$\frac{\partial z}{\partial y} = \frac{\partial}{\partial y}\left(x^2 y + \sin y\right)$$

$$= \frac{\partial}{\partial y}\left(x^2 y\right) + \frac{\partial}{\partial y}\left(\sin y\right)$$

$$= x^2 \frac{\partial}{\partial y} (y) + \frac{\partial}{\partial y} (\sin y) \quad \text{(treating } x \text{ as constant)}$$

$$= x^2 + \cos y .$$

By differentiating the partial derivatives of a function we obtain higher-order partial derivatives. In particular, the definitions of the partials of the second order, and their values for the function of the preceding example, are as follows:

$$\frac{\partial^2 z}{\partial x^2} = \frac{\partial}{\partial x} \left(\frac{\partial z}{\partial x} \right) = 2y ,$$

$$\frac{\partial^2 z}{\partial x \partial y} = \frac{\partial}{\partial x} \left(\frac{\partial z}{\partial y} \right) = 2x ,$$

$$\frac{\partial^2 z}{\partial y \partial x} = \frac{\partial}{\partial y} \left(\frac{\partial z}{\partial x} \right) = 2x ,$$

$$\frac{\partial^2 z}{\partial y^2} = \frac{\partial}{\partial y} \left(\frac{\partial z}{\partial y} \right) = - \sin y .$$

The middle two, formed by differentiating first with respect to one variable, then with respect to the other, are called **mixed** partials. Their equality in this example represents a general phenomenon, that

$$\frac{\partial^2 z}{\partial x \partial y} = \frac{\partial^2 z}{\partial y \partial x}$$

—as a rule, the order of differentiation does not matter.[26] Hence a function of two variables normally has only three distinct second-order partial derivatives. More variables means more second-order partials, but the equality of mixed partials remains valid.

4.5.6   QUESTIONS

Q1.   Find $\partial z/\partial x$ and $\partial z/\partial y$ for the following functions $z(x,y)$. (a) $z = (x+3y^2)^3$ (b) $z = x/y^2$ (c) $z = x \sin xy$

Q2.   Find the first- and second-order partial derivatives of $z = e^{x^2-y}$ and verify the equality of the mixed partials.

---

[26]This is the case if the mixed partials are continuous.

Q3. How many distinct second-order partial derivatives does a function of three variables have?

Q4. The pressure of a given quantity of gas depends on its temperature and volume according to an equation $P = \alpha T/V$, where α is a constant. At what rate does P vary with respect to (a) T, (b) V, when the other quantity is held constant?

Q5. Write expressions for $\dfrac{\partial f}{\partial x}\bigg|_{y=y_0}$ and $\dfrac{\partial f}{\partial y}\bigg|_{x=x_0}$ as limits of difference quotients.

Q6. Find a geometrical meaning for the partial derivatives $\dfrac{\partial f}{\partial x}\bigg|_{y=y_0}$ and $\dfrac{\partial f}{\partial y}\bigg|_{x=x_0}$, and apply it to Q3 of §4.5.3.

Q7. (a) Find the extrema of $g(x) = x^3 - x^2 y_0 + 3$ (the answer will depend on the value of y_0). (b) Explain how (a) is related to partial differentiation.

Q8. (Cf. §3.6.5, (i) & (iv).) (a) Show that for a "linear" function $z = \alpha x + \beta y + \gamma$ (α, β, γ constants) the coefficients of the variables are the partials with respect to those variables, i.e. $\alpha = \partial z/\partial x$ and $\beta = \partial z/\partial y$. (b) Conversely, prove that if $z = f(x,y)$ and f_x and f_y have the constant values α and β respectively, then $f(x,y) = \alpha x + \beta y + \gamma$ for some constant γ. (First integrate $f_x = \alpha$ with respect to x, holding y constant. Then differentiate the result with respect to y.)

Q9. (Cf. §3.6.5, (ii).) Prove that the function of Q8(a) has the following property: if starting at point (x,y) we move in a given direction to any point $(x+\Delta x, y+\Delta y)$, the change in z is proportional to the distance we have moved. (Let θ be the angle from the positive x-direction to the direction of motion, r the distance moved. Then $\Delta x = r\cos\theta$, $\Delta y = r\sin\theta$.)

Q10. Define "local minimum" and "local maximum" for a function of two variables, and show that at a local minimum or maximum its partial derivatives vanish.

Q11. *Method of least squares.* Let x be a "controlled" physical variable, one whose values are somehow specified, and y an "observed" variable assumed to be a linear function of x, $y = a + bx$. To determine the function as well as possible from data points (x_i, y_i), $i = 1, 2, \ldots, n$—as we say, to find the line $y = a + bx$ which **best fits** the data—one chooses a and b so as to minimize the sum Q of the squares of the **errors** e_i, which are the differences between the y-values y_i *observed* to correspond to the x-values x_i, and the y-values $y(x_i) = a + bx_i$ that will be *predicted* by the equation $y = a + bx$ to correspond to those x-values x_i (Fig. 4-28). Let S_1 be the

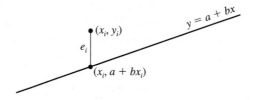

Fig. 4-28

sum of the x_i, S_2 the sum of the x_i^2, T the sum of the y_i, and U the sum of the $x_i y_i$. Show that a and b satisfy the following linear equations:

$$na + S_1 b = T,$$

$$S_1 a + S_2 b = U.$$

(Regard Q as a function of a and b and apply Q10.)

4.5.7   The differential of a function.   Equation of the tangent plane.

In this section I will sketch an argument to answer a basic question: given a function $z = f(x, y)$, if x changes by the infinitesimal quantity dx and y by the infinitesimal quantity dy, what is the corresponding infinitesimal change dz in z?

Let us first review the one-variable situation. In the vicinity of $x = x_0$ a differentiable function $y = f(x)$ can be compared to the linear function

$$y = f'(x_0)(x - x_0) + y_0,$$

where $y_0 = f(x_0)$ (§3.8.6); the graph of this function is the line tangent to the graph of f at (x_0, y_0). The two functions have the same derivative (their graphs have the same slope) at x_0, hence they have the same differential there: if x changes from x_0 to $x_0 + dx$, both functions change by $f'(x_0)dx$. In the case of the linear function this can be demonstrated by direct calculation,

$$[f'(x_0)(x_0 + dx - x_0) + y_0] - [f'(x_0)(x_0 - x_0) + y_0] = f'(x_0)\, dx;$$

for the original function f, it amounts to the identity $df = \dfrac{df}{dx}\, dx$ at x_0. With additional conditions on f the agreement of infinitesimals is supplemented by an approximate equality of finite increments (§§3.8.7–8).

Now consider a function $z = f(x, y)$ in the vicinity of values $x = x_0$ and $y = y_0$, and suppose that f has continuous partial derivatives. It can be shown that the surface which is the graph of f has a non-vertical tangent *plane* at $P(x_0, y_0, z_0 = f(x_0, y_0))$ (Fig. 4-29). Then f is comparable to the function whose graph that plane is. From analytic geometry we know that the equation of the plane has the form

$$z = a(x - x_0) + b(y - y_0) + z_0:$$

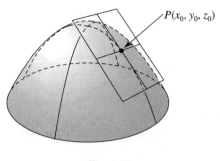

Fig. 4-29

z is a linear function of two variables (cf. §4.5.6.Q8 & Q9). The partial derivatives of this function are $\partial z/\partial x = a$, the slope of the plane in the x-direction,
and $\partial z/\partial y = b$, its slope in the y-direction. The graph of f has these same
slopes at P; in other words, the partial derivatives $f_x(x_0, y_0)$ and $f_y(x_0, y_0)$
are the same as the corresponding partial derivatives of the linear function
(cf. §4.5.6.Q6). Consequently the function that defines the tangent plane is

$$z = f_x(x_0, y_0)(x - x_0) + f_y(x_0, y_0)(y - y_0) + z_0 \, .$$

As in the one-variable case, if x changes from x_0 to $x_0 + dx$ and y changes
from y_0 to $y_0 + dy$, the change in this linear function is the same as the change
in $f(x, y)$. Direct calculation shows that for the linear function

$$dz = f_x(x_0, y_0) \, dx + f_y(x_0, y_0) \, dy \, ;$$

this is then also the differential of f,

$$df = f_x(x_0, y_0) \, dx + f_y(x_0, y_0) \, dy$$

or

(1)
$$df = \frac{\partial f}{\partial x} \, dx + \frac{\partial f}{\partial y} \, dy \, ,$$

where the partial derivatives are evaluated at (x_0, y_0).[27]

---

[27] If dx and dy are taken to be finite increments this equation can be used to define a finite
quantity df; cf. the comment at the end of §3.8.8.

Corresponding to this equation of infinitesimals is an approximate equation for finite changes,

$$\Delta f \approx \frac{\partial f}{\partial x}\, \Delta x + \frac{\partial f}{\partial y}\, \Delta y\,,$$

valid when f satisfies certain conditions near (x_0, y_0); it can be used for practical purposes, much as in §3.8.8.

The differential of a function of more than two variables is analogous to a two-variable differential: for $f(x, y, z)$ we have

$$df = f_x\, dx + f_y\, dy + f_z\, dz\,.$$

4.5.8   QUESTIONS

Q1.   Calculate df for the following functions $f(x, y)$. (a) x (b) $ax + by$ (a, b constants) (c) xy (d) y/x (e) $\arctan(y/x)$

Q2.   At a certain place near a heat source the temperature changes at the rate of $+5°$ per meter going east and $-2°$ per meter going north. How much does it change if you go 1 cm in the direction $30°$ east of north?

Q3.   Find the equation of the tangent plane to the paraboloid of Fig. 4-26 at any point (x_0, y_0, z_0).

Q4.   What linear function best approximates $f(x, y) = xy$ at $x = 1$, $y = 1$?

4.5.9   Certain composite functions and the chain rule. A. x and y dependent on t.

In a function $z = f(x, y)$ let us suppose that x and y both depend on a third variable t. As in §4.2.4 we can regard t as time, and think of $x(t)$ and $y(t)$ as the coordinates of a moving particle in the plane. Since they are related through their dependence on t, x and y are no longer independent of one another: as t varies, the particle at (x, y) is constrained to move along some curve in the plane whose equation expresses the connection between the two variables. If, for example, $x = 2\cos t$ and $y = 2\sin t$, as t increases the particle moves counterclockwise around the circle $x^2 + y^2 = 4$ of §4.2.4.

Through the function $f(x, y)$ we then define a composite function of the single variable t by

$$F(t) = f(x(t), y(t))\,.$$

The value of F at time t is the value of f at the point where the particle is located at that time. If, say, $f(x, y)$ is the temperature at (x, y), then $F(t)$ is the temperature the particle experiences at time t.

In order to differentiate F we evidently need an extension of the chain rule to the case where there are *two* intermediate variables (between t and z). An argument by infinitesimals will quickly give the correct formula, although its justification by limits would require some attention to detail. Let t change by dt, and x and y correspondingly by dx and dy. Even though x and y are not independent, we can say that the change in f caused by dx and dy is

$$df = \frac{\partial f}{\partial x}\, dx + \frac{\partial f}{\partial y}\, dy$$

(§4.5.7(1)), because this formula holds for any dx and dy. Now if the change in F caused by dt is dF we have $dF = df$, and dividing by dt gives

$$(2)\qquad \frac{dF}{dt} = \frac{\partial f}{\partial x}\frac{dx}{dt} + \frac{\partial f}{\partial y}\frac{dy}{dt}.$$

For an illustration let $f(x,y) = y/x$ ($x \neq 0$), where $x = 2\cos t$ and $y = 2\sin t$ as before. Then $f_x = -y/x^2$ and $f_y = 1/x$, while $dx/dt = -2\sin t$ and $dy/dt = 2\cos t$, so

$$\frac{dF}{dt} = \left(-\frac{y}{x^2}\right)(-2\sin t) + \left(\frac{1}{x}\right)(2\cos t),$$

or substituting the values of x and y

$$\frac{dF}{dt} = \left(-\frac{2\sin t}{4\cos^2 t}\right)(-2\sin t) + \left(\frac{1}{2\cos t}\right)(2\cos t) = \tan^2 t + 1 = \sec^2 t.$$

The answer is confirmed by noticing that

$$F(t) = f(x(t), y(t)) = \frac{2\sin t}{2\cos t} = \tan t,$$

the derivative of which we know to be $\sec^2 t$.

Abandoning now the interpretation of t as time, consider the special case $t = x$; that is, suppose that y is a function of x, so that (x, y) is confined to the graph of a function $y(x)$. Then a composite function of x is defined by $F(x) = f(x, y(x))$, and formula (2) becomes

$$(3)\qquad \frac{dF}{dx} = \frac{\partial f}{\partial x} + \frac{\partial f}{\partial y}\frac{dy}{dx}.$$

With $f(x, y) = y/x$ as before, if we define y as a function of x by $y = \sqrt{4 - x^2}$ (so that (x, y) is on the upper half of the same circle) we find

$$\frac{dF}{dx} = -\frac{y}{x^2} + \frac{1}{x}\left(-\frac{x}{\sqrt{4 - x^2}}\right) = -\frac{4}{x^2\sqrt{4 - x^2}}.$$

4.5.10 B. Implicit differentiation. Additional variables. These formulas shed new light on the technique of implicit differentiation introduced in §4.2.3. Given an equation $f(x, y) = 0$ which holds because either $y = y(x)$ or $x = x(t)$ and $y = y(t)$, we have $F(x) = 0$ or $F(t) = 0$ for all x or t, hence $F'(x) = 0$ or $F'(t) = 0$. Then in the former case (3) gives

$$\frac{\partial f}{\partial x} + \frac{\partial f}{\partial y}\frac{dy}{dx} = 0,$$

or

$$\frac{dy}{dx} = -\frac{\partial f}{\partial x}\bigg/\frac{\partial f}{\partial y},$$

or

$$y' = -\frac{f_x}{f_y};$$

this is actually the general form of the expression found for y' when $f(x, y) = 0$ is differentiated implicitly. For example, in §4.2.3 we (in effect) differentiated $f(x, y) = x^2 + y^2 - 4 = 0$ to obtain $y' = -x/y$; here we have $f_x = 2x$ and $f_y = 2y$, so that $-f_x/f_y = -x/y$.

Similarly, by (2) $F'(t) = 0$ is equivalent to

$$\frac{\partial f}{\partial x}\frac{dx}{dt} + \frac{\partial f}{\partial y}\frac{dy}{dt} = 0,$$

which is the relation between dx/dt and dy/dt obtained by implicit differentiation of $f(x, y) = 0$ with respect to t.

The form of the chain rule is similar when f is a function of more than two variables; e.g., the derivative of $F(t) = f(x(t), y(t), z(t))$ is

$$\frac{dF}{dt} = \frac{\partial f}{\partial x}\frac{dx}{dt} + \frac{\partial f}{\partial y}\frac{dy}{dt} + \frac{\partial f}{\partial z}\frac{dz}{dt}.$$

If x, y, etc., are functions of more than one variable, we must calculate *partial* derivatives of the composite function. The formulas will be sufficiently illustrated by the case of a function $f(x, y)$, where $x = x(u, v)$ and $y = y(u, v)$. Let $F(u, v) = f(x(u, v), y(u, v))$, then

$$\frac{\partial F}{\partial u} = \frac{\partial f}{\partial x}\frac{\partial x}{\partial u} + \frac{\partial f}{\partial y}\frac{\partial y}{\partial u}$$

and

$$\frac{\partial F}{\partial v} = \frac{\partial f}{\partial x}\frac{\partial x}{\partial v} + \frac{\partial f}{\partial y}\frac{\partial y}{\partial v}.$$

Each of these is deduced from (2) by holding one of the two variables u and v constant.

These formulas give us the means to derive partials with respect to polar coordinates r, θ from partials with respect to rectangular coordinates (Q4).

4.5.11   QUESTIONS

Q1.   Use the chain rule to find the derivative at $t = 1$ of $F(t) = f(x(t), y(t))$, where $f(x, y) = \sqrt{xy}$ and $x = e^{t-1}$, $y = t^3 + t$.

Q2.   A particle with coordinates $(x(t), y(t))$ moves in the plane so that at a certain instant t_0, $x'(t_0) = 1$ and $y'(t_0) = 2$. How fast is its distance from the origin changing at t_0? (Let $f(x, y) = \sqrt{x^2 + y^2}$.)

Q3.   A particle moves on a plane curve whose slope at any x is $y' = 2x$. If the temperature in the plane is given by $T(x, y) = x^2 + y^2$, what is the rate of change of temperature at the particle with respect to x?

Q4.   Express the partials of $F(r, \theta) = f(x(r, \theta), y(r, \theta))$, where $x = r\cos\theta$, $y = r\sin\theta$, in terms of the partials of $f(x, y)$.

4.6   METHODS OF INTEGRATION

4.6.1   Integration in terms of elementary functions.   Every continuous function has an integral function (primitive), but there are very simple combinations of elementary functions whose integrals cannot be expressed as such combinations—that is, as *formulas*, in the usual sense of the word. It can be proved, for example, that $\sin x^2$ and e^x/x do not have "elementary" integrals. Nevertheless, many common functions do have integrals of this kind, and many of those integrals can be obtained by the application of one or more of a number of powerful methods that go beyond the basic rules introduced in Art. 3.6. The most important method corresponds to the chain rule for differentiation; I will present it first and then briefly consider two others.

It should be mentioned that the results of integrating common types of functions by these and other methods are available in extensive printed tables and through computer programs. Such aids are of use for integrals that are neither so simple as to yield easily to a familiar method, nor so complex as not to have been tabulated.

Integrals that do not have formulas in terms of known functions can be used to define *new* functions; this mode of definition is illustrated in §4.7.4 and §6.2.1. They may also admit a representation by "infinite series": see §8.1.5.

4.6.2   Substitution, or change of variable.   It is possible to evaluate the integral

$$\int \sin x \cos x \, dx$$

by the following device. Let $u = \sin x$. Then

$$du = \frac{du}{dx} \, dx = \cos x \, dx \,,$$

hence

$$\int \sin x \cos x \, dx = \int u \, du = \frac{u^2}{2} + C = \frac{\sin^2 x}{2} + C \,.$$

The answer is correct, because

$$\frac{d}{dx} \left(\frac{\sin^2 x}{2} \right) = \frac{2 \sin x \cos x}{2} = \sin x \cos x \,.$$

The reason that the x-integral turned into such a simple u-integral is that du was already present in the former under the guise of $\cos x \, dx$. The integral

$$\int \cos 2x \, dx$$

does not at first appear to offer the same possibility of transformation. But if we let $u = 2x$ we find $du = 2 \, dx$, or $dx = \frac{1}{2} \, du$, and then

$$\int \cos 2x \, dx = \int (\cos u) \left(\frac{1}{2} \, du \right) = \frac{1}{2} \int \cos u \, du$$

$$= \frac{1}{2} \sin u + C = \frac{1}{2} \sin 2x + C \,.$$

This example is not essentially different from the preceding one, as we can see by writing the given integral as

$$\int \left[(\cos 2x) \cdot \frac{1}{2} \right] \cdot 2 \, dx \, .$$

Although the du was not obviously present in $\int \cos 2x \, dx$, dx differed from it only by a constant factor.

Both integrations are instances of the method of **substitution**, or **change of variable**, in which one variable is replaced by another for the purpose of transforming a given integral into a simpler one. The first example generalizes to the formula

$$(1) \qquad \int f(u(x)) \frac{du}{dx} \, dx = \int f(u) \, du \, ,$$

where $u(x)$ is to be substituted for u in the integral on the right (after evaluation) to make it a function of x. If in this formula we let $f(u) = u$ and $u(x) = \sin x$ it becomes

$$\int \sin x \cos x \, dx = \int u \, du$$

as in the example.

Formula (1) is nothing but an integral version of the chain rule. For let

$$F(u) = \int f(u) \, du \, ,$$

and substitute $u(x)$ for u; then the right side of (1) is $F(u(x))$,[28] so the formula asserts that

$$\frac{d}{dx} F(u(x)) = f(u(x)) \frac{du}{dx} \, ,$$

which is the chain rule exactly.

The second example differed only in one respect, that instead of having $\frac{du}{dx} dx = du$ present in the integral as given we had to solve for dx in terms of du, $dx = \frac{dx}{du} du$, and substitute the result for dx. As we saw by rewriting the given integral, we were still in effect using (1). In general, the validity of the

---

[28] More precisely, $F(u(x)) + C$, since F is a particular primitive of f; see the remark at the end of §3.5.7.

modified procedure is deducible from (1) in the same way (Q3). Note that (1) requires that $u(x)$ be continuously differentiable (in order that the integrand on the left may be continuous); if $\dfrac{dx}{du} = 1 \Big/ \dfrac{du}{dx}$ is to be used we must also have $du/dx \neq 0$, in which case $u(x)$ will be monotonic, hence invertible (§§3.9.1, 4.3.2). These requirements do not ordinarily cause any difficulty.

4.6.3　Substitution in a definite integral.　We can evaluate a *definite* integral by change of variable without performing the final step of substituting $u(x)$ to return to a function of x. Instead, we *transform the limits of integration* to their corresponding u-values. For example, with $u = \sin x$,

$$\int_0^{\pi/2} \sin x \cos x \, dx = \int_{u(0)}^{u(\pi/2)} u \, du = \int_{\sin 0}^{\sin(\pi/2)} u \, du$$

$$= \int_0^1 u \, du = \frac{u^2}{2} \bigg|_0^1 = \frac{1}{2}.$$

Formula (1) becomes

(2)
$$\int_a^b f(u(x)) \frac{du}{dx} \, dx = \int_{u(a)}^{u(b)} f(u) \, du,$$

as follows from the Fundamental Theorem. For we know that if $F(u)$ is an integral of $f(u)$ then $F(u(x))$ is an integral of $f(u(x)) \dfrac{du}{dx}$; hence both sides of (2) are equal to $F(u(b)) - F(u(a))$.

The meaning of this rule is best understood by regarding the integrals as sums. For simplicity let us assume $a < b$ and $u(x)$ increasing. Then as x increases from a to b, u correspondingly increases from $u(a)$ to $u(b)$. A u-increment du does not, however, necessarily equal its corresponding x-increment dx; rather, the change of variable produces a *change of scale* measured by the "scale factor" $\dfrac{du}{dx}$, for $du = \dfrac{du}{dx} dx$. But the value of the function f, considered as a function of u, is the same at u as the value of the same function considered as a function of x—that is, the value of the composite function $f \circ u$—at the corresponding x; i.e., if $u = u(x)$ then $f(u) = f(u(x))$. Consequently

$$f(u(x)) \cdot \frac{du}{dx} \, dx = f(u) \cdot du,$$

from which (2) follows by summing over the corresponding intervals of values of the two variables, namely $[a, b]$ on the left and $[u(a), u(b)]$ on the right.

4.6.4 Examples of substitution. There may be several choices of $u(x)$ by which a given integral can be transformed, and one choice may be preferable to another. Of course, there is no guarantee that a particular choice, or any choice at all, will simplify the integral. The fact is that the "method" of substitution has a strong admixture of art.

A number of examples follow.

EXAMPLE 1. $\int \dfrac{x+1}{2}\,dx.$

Let $u = x + 1$, then $du = dx$ and

$$\int \frac{x+1}{2}\,dx = \int \frac{u}{2}\,du = \frac{u^2}{4} + C = \frac{(x+1)^2}{4} + C.$$

Or let $u = \dfrac{x+1}{2}$, then $du = \frac{1}{2}\,dx$, $dx = 2du$, and

$$\int \frac{x+1}{2}\,dx = \int u \cdot 2\,du = u^2 + C = \left(\frac{x+1}{2}\right)^2 + C.$$

The two solutions agree, because $\dfrac{(x+1)^2}{4} = \left(\dfrac{x+1}{2}\right)^2.$

EXAMPLE 2. $\int xe^{x^2}\,dx.$

Let $u = x^2$, then $du = 2x\,dx$, $x\,dx = \frac{1}{2}\,du$, and

$$\int xe^{x^2}\,dx = \int e^u \cdot \frac{1}{2}\,du = \frac{1}{2}e^u + C = \frac{1}{2}e^{x^2} + C.$$

We did not need $\dfrac{dx}{du}$, since $\dfrac{1}{2}\dfrac{du}{dx}$ was present in the integrand; so it was of no concern that $u(x)$ is not monotone.

EXAMPLE 3. $\int \tan x\,dx.$

This does not look like a fit subject for substitution, but remember that $\tan x = \sin x / \cos x$. Let $u = \cos x$, then $du = -\sin x\,dx$ and

$$\int \tan x\,dx = \int \frac{\sin x}{\cos x}\,dx = \int \frac{-du}{u} = -\int \frac{1}{u}\,du$$

$$= -\log|u| + C \quad (\S 3.6.2)$$

$$= -\log|\cos x| + C.$$

For $-\pi/2 < x < \pi/2$, where $\cos x > 0$, this is $-\log \cos x + C$. The integral, like the integrand, does not exist when x is an odd multiple of $\pi/2$. Since $\sec x = 1/\cos x$ the answer can also be written $\log|\sec x| + C$ (§3.3.13.Q3(a)).

EXAMPLE 4. $\displaystyle\int_0^{\pi/4} \tan x\, dx$.

From the preceding example this equals

$$-\log \cos x \Big|_0^{\pi/4} = -\log \frac{1}{\sqrt{2}} - (-\log 1)$$

$$= -\log \frac{1}{\sqrt{2}} \quad (\text{since } \log 1 = 0)$$

$$= \log \sqrt{2}, \quad \text{or} \quad \frac{1}{2}\log 2$$

(§3.3.13.Q3). We can also evaluate the definite integral directly, using the same substitution as before:

$$\int_0^{\pi/4} \tan x\, dx = \int_0^{\pi/4} \frac{\sin x}{\cos x}\, dx = \int_1^{1/\sqrt{2}} \frac{-du}{u},$$

because $u(0) = \cos 0 = 1$ and $u(\pi/4) = \cos(\pi/4) = 1/\sqrt{2}$. Continuing,

$$-\int_1^{1/\sqrt{2}} \frac{du}{u} = -\log u \Big|_1^{1/\sqrt{2}} = -\log \frac{1}{\sqrt{2}} - (-\log 1), \quad \text{etc}.$$

EXAMPLE 5. $\displaystyle\int \sqrt{3x-1}\, dx$.

Let $u = 3x - 1$, then $du = 3dx$, $dx = \frac{1}{3}\, du$, and

$$\int \sqrt{3x-1}\, dx = \int u^{1/2} \cdot \frac{1}{3}\, du = \frac{2}{9} u^{3/2} + C = \frac{2}{9}(\sqrt{3x-1})^3 + C.$$

EXAMPLE 6. $\displaystyle\int x^5 \sqrt{1-x^2}\, dx$.

Let $u = 1 - x^2$, then $du = -2x\, dx$, $-\frac{1}{2}\, du = x\, dx$, and

$$\int x^5 \sqrt{1-x^2}\, dx = \int x^4 u^{1/2} \left(-\frac{1}{2}\, du\right).$$

To deal with the x^4, observe that $x^2 = 1 - u$, therefore $x^4 = (1 - u)^2$. Then the integral is

$$-\frac{1}{2} \int (1 - u)^2 u^{1/2} \, du = -\frac{1}{2} \int (1 - 2u + u^2) \, u^{1/2} \, du \,\cdot$$

$$= -\frac{1}{2} \int (u^{1/2} - 2u^{3/2} + u^{5/2}) \, du$$

$$= -\frac{1}{2} \int u^{1/2} \, du + \int u^{3/2} \, du - \frac{1}{2} \int u^{5/2} \, du$$

$$= -\frac{1}{3} u^{3/2} + \frac{2}{5} u^{5/2} - \frac{1}{7} u^{7/2} + C$$

$$= -\frac{1}{3} (1 - x^2)^{3/2} + \frac{2}{5} (1 - x^2)^{5/2} - \frac{1}{7} (1 - x^2)^{7/2} + C \,.$$

(We could factor $(1 - x^2)^{1/2}$ or $(1 - x^2)^{3/2}$ out of the non-constant part of the result, leaving a polynomial in x as the other factor.)

EXAMPLE 7. $\displaystyle \int \frac{1}{\sqrt{a^2 - x^2}} \, dx, \quad a > 0.$

Let $u = x/a$, then $du = dx/a$ and

$$\int \frac{dx}{\sqrt{a^2 - x^2}} = \int \frac{a \, du}{\sqrt{a^2 - a^2 u^2}} = \int \frac{du}{\sqrt{1 - u^2}}$$

$$= \arcsin u + C \quad (\S4.4.8)$$

$$= \arcsin \frac{x}{a} + C \,.$$

In a case like this it would be more natural to say, "Let $x = au$, then $dx = a \, du$," etc.: the idea is to replace x by an expression in u that brings the integral into recognizable form. The next example is a more impressive application of the same technique.

EXAMPLE 8. $\displaystyle \int \sqrt{1 - x^2} \, dx.$

With the identity $\sin^2 u + \cos^2 u = 1$ in mind, let $x = \sin u$, i.e. $u = \arcsin x$, $-\pi/2 \le u \le \pi/2$; then $dx = \cos u \, du$ and $\sqrt{1 - x^2} = \cos u$ (the cosine is non-negative on the prescribed u-interval), hence

$$\int \sqrt{1 - x^2} \, dx = \int (\cos u)(\cos u \, du) = \int \cos^2 u \, du \,.$$

By the half-angle formula $\cos^2 u = \frac{1}{2}(1 + \cos 2u)$ (§3.3.7) this is

$$\frac{1}{2}\left[\int du + \int \cos 2u \, du\right].$$

Using the value obtained in §4.6.2 for the second integral we find

$$\int \cos^2 u \, du = \frac{1}{2}\left(u + \frac{1}{2}\sin 2u\right) + C = \frac{1}{2}(u + \sin u \cos u) + C,$$

the second step by a double-angle formula. Since $u = \arcsin x$ this gives

$$\int \sqrt{1 - x^2}\, dx = \frac{1}{2}(\arcsin x + x\sqrt{1 - x^2}) + C.$$

The restriction of u to $[-\pi/2, \pi/2]$ in this argument corresponds to the restriction of x to $[-1, 1]$ necessitated by the requirement that $1 - x^2$ be non-negative.

EXAMPLE 9. $\displaystyle\int \sqrt{\frac{x}{1 - x}}\, dx, \quad 0 \le x < 1.$

Let $u = \sqrt{1 - x}$, then $du = -dx/2u$ and $x = 1 - u^2$, so

$$\int \sqrt{\frac{x}{1 - x}}\, dx = -2\int \sqrt{1 - u^2}\, du$$

$$= -(u\sqrt{1 - u^2} + \arcsin u) + C \quad \text{(by Example 8)}$$

$$= -(\sqrt{x(1 - x)} + \arcsin \sqrt{1 - x}) + C.$$

More generally, for $0 \le x/a < 1$,

$$(3) \qquad \int \sqrt{\frac{x}{a - x}}\, dx = -a\left[\sqrt{\frac{x}{a}\left(1 - \frac{x}{a}\right)} + \arcsin \sqrt{1 - \frac{x}{a}}\right] + C,$$

which is proved by using the substitution $u = x/a$. This integral comes up in gravitational theory (see §5.4.4).

A few more examples are proposed in the questions. By clever substitutions, and by substitutions combined in sequence, a surprisingly large number of apparently intractable integrals can be reduced to familiar form.

4.6.5   QUESTIONS

Q1.   Integrate. (a) $\int (2x - 1)^{10} dx$ (b) $\int_0^{\pi/6} \sin^2 x \cos x \, dx$ (as a definite integral)

(c) $\int \dfrac{1}{5t + 4} dt$ $(5t + 4 > 0)$ (d) $\int t^2 e^{t^3} dt$ (e) $\int \dfrac{1}{x^2 + 4} dx$ (cf. Example 7, §4.6.4)

(f) $\int \dfrac{1}{x^2 - 2x + 5} dx$ (complete the square, then see (e))

Q2.   Use the extraordinary substitution $u = \sec x + \tan x$ to find $\int \sec x \, dx$.

Q3 (§4.6.2).   To justify substituting $\dfrac{dx}{du} du$ for dx, prove that §4.6.2(1) implies

$$\int g(u(x)) \, dx = \int g(u) \frac{dx}{du} \, du \,,$$

where $u(x)$ is to be substituted on the right after evaluation.

Q4.   Use substitution to prove that the function $F(x) = \int_0^x f(x) dx$ is even if f is odd, and odd if f is even.

Q5.   (a) Find $\int \sqrt{x^2 - 1} \, dx$, $x \geq 1$. (Cf. Example 8. Let $x = \cosh u$ and apply §4.4.9.Q10 and §4.1.5.Q7.) (b) Find $\int \sqrt{x^2 + 1} \, dx$.

Q6.   (a) Find $\int \sqrt{\dfrac{x}{1 + x}} \, dx$, $x \geq 0$. (Cf. Example 9. Apply Q5(a).) (b) Find $\int \sqrt{\dfrac{x}{a + x}} \, dx$, $x/a \geq 0$.

4.6.6   Integration by parts.   The product formula

$$(fg)' = f'g + fg'$$

states that fg is an integral of $f'g + fg'$, i.e.

$$f(x)g(x) = \int f'(x)g(x) \, dx + \int f(x)g'(x) \, dx \,.$$

Rearrangement of this equation as

$$\int f'(x)g(x) \, dx = f(x)g(x) - \int f(x)g'(x) \, dx$$

gives us a formula by which the integral of a product of functions is expressed in terms of another integral.

We can rewrite the formula this way:

$$\int h(x)k(x)\,dx = H(x)k(x) - \int H(x)k'(x)\,dx\,,$$

where $H(x) = \int h(x)\,dx$ ($H(x)$ is any primitive of $h(x)$). In order to apply it we must as preliminary steps integrate one of the functions from which the given product is formed (from h find H) and differentiate the other (from k find k'). The object in choosing h and k is usually to make $\int H(x)k'(x)\,dx$ simpler than the original integral (but see Example 3).

For definite integration the rule takes the form

$$\int_a^b h(x)k(x)\,dx = \Big[H(x)k(x)\Big]_a^b - \int_a^b H(x)k'(x)\,dx\,.$$

Here are a few examples.

4.6.7   Examples of integration by parts.

EXAMPLE 1. $\int xe^x\,dx.$

We integrate e^x, which does not make it more complicated, and differentiate x, which makes it simpler. That is, we let $h(x) = e^x$, $k(x) = x$; then we can take $H(x) = e^x$, while $k'(x) = 1$, so

$$\int xe^x\,dx = xe^x - \int 1 \cdot e^x\,dx = xe^x - e^x + C\,.$$

EXAMPLE 2. $\int \log x\,dx.$

Regarding the integral as the product $1{\cdot}\log x$, we integrate 1 and differentiate $\log x$:

$$\int 1 \cdot \log x\,dx = x\log x - \int x \cdot \frac{1}{x}\,dx$$

$$= x\log x - \int 1\,dx$$

$$= x\log x - x + C\,.$$

EXAMPLE 3. $\int e^x \sin x\,dx.$

If we integrate e^x and differentiate $\sin x$ we find

$$\int e^x \sin x \, dx = e^x \sin x - \int e^x \cos x \, dx \,,$$

so we seem to have done no more than replace the given integral with another equally unknown. But let us treat the new integral in the same way:

$$\int e^x \cos x \, dx = e^x \cos x - \int e^x (-\sin x) \, dx$$

$$= e^x \cos x + \int e^x \sin x \, dx \,.$$

Now substituting this result into the first equation gives

$$\int e^x \sin x \, dx = e^x \sin x - e^x \cos x - \int e^x \sin x \, dx \,,$$

from which follows

$$\int e^x \sin x \, dx = \frac{e^x}{2} (\sin x - \cos x) + C \,.$$

EXAMPLE 4. $\displaystyle\int \cos^n x \, dx \quad (n > 1).$

We write $\cos^n x = \cos x \cdot \cos^{n-1} x$ and put $h(x) = \cos x$, $k(x) = \cos^{n-1} x$. Then

$$\int \cos^n x \, dx = \sin x \cos^{n-1} x - \int \sin x \, [(n-1) \cos^{n-2} x](-\sin x) \, dx$$

$$= \sin x \cos^{n-1} x + (n-1) \int \sin^2 x \cos^{n-2} x \, dx$$

$$= \sin x \cos^{n-1} x + (n-1) \int (1 - \cos^2 x) \cos^{n-2} x \, dx$$

$$= \sin x \cos^{n-1} x + (n-1) \int \cos^{n-2} x \, dx - (n-1) \int \cos^n x \, dx \,,$$

which implies

$$\int \cos^n x \, dx = \frac{1}{n} \sin x \cos^{n-1} x + \frac{n-1}{n} \int \cos^{n-2} x \, dx \,.$$

By repeated application of this formula the given integral is reduced to the known integral $\int 1 \, dx$ (if n is even) or $\int \cos x \, dx$ (if n is odd).

4.6.8   QUESTIONS

Q1. Integrate. (a) $\int x \sin x \, dx$ (b) $\int x^2 e^x \, dx$ (c) $\int e^x \cos x \, dx$

Q2. (a) Verify that the case $n = 2$ of Example 4, §4.6.7, gives the same result as that obtained in the course of Example 8, §4.6.4. (b) Use the formula of Example 4 to find $\int \cos^6 x \, dx$.

Q3. Find $\int \arcsin x \, dx$. (Let $h(x) = 1$, and evaluate the resulting integral by substitution.)

Q4. (a) Find $\int \sec^3 x \, dx$. (Let $h(x) = \sec^2 x$, so that $H(x) = \tan x$, and use $1 + \tan^2 x = \sec^2 x$ and §4.6.5.Q2.) (b) Hence find $\int \sqrt{x^2 + 1} \, dx$. (Let $x = \tan u$. Compare the result with that of §4.6.5.Q5(b).)

4.6.9   Partial fractions.

Any **rational function**, or quotient of polynomials $p(x)/q(x)$, can be integrated by expressing it as a sum of rational functions of certain simple types, once its denominator $q(x)$ has been factored. A pair of examples will illustrate this method.

EXAMPLE 1. $\displaystyle \int \frac{1}{x^2 - 3x + 2} \, dx.$

We have

$$\frac{1}{x^2 - 3x + 2} = \frac{1}{(x-2)(x-1)} = \frac{1}{x-2} - \frac{1}{x-1} :$$

this is the decomposition of the given rational function into its **partial fractions**. Then

$$\int \frac{1}{x^2 - 3x + 2} \, dx = \int \frac{1}{x-2} \, dx - \int \frac{1}{x-1} \, dx$$

$$= \log |x - 2| - \log |x - 1| + C$$

$$= \log \left| \frac{x-2}{x-1} \right| + C.$$

EXAMPLE 2. $\displaystyle \int \frac{x+2}{x^4 + x^2} \, dx.$

The denominator factors as $x^2(x^2 + 1)$, and we have

$$\frac{x+2}{x^2(x^2+1)} = \frac{1}{x} + \frac{2}{x^2} - \frac{x+2}{x^2+1}$$

(a way of discovering this decomposition is shown in Q2). Then

$$\int \frac{x+2}{x^2(x^2+1)}\,dx = \int \frac{1}{x}\,dx + \int \frac{2}{x^2}\,dx - \int \frac{x}{x^2+1}\,dx - \int \frac{2}{x^2+1}\,dx.$$

Each of these integrals can readily be evaluated, and the result is

$$\int \frac{x+2}{x^4+x^2}\,dx = \log|x| - \frac{2}{x} - \frac{1}{2}\log(x^2+1) - 2\arctan x + C$$

$$= \log\frac{|x|}{\sqrt{x^2+1}} - \frac{2}{x} - 2\arctan x + C.$$

In general, assuming that $q(x)$ is of higher degree than $p(x)$ the decomposition of $p(x)/q(x)$ into partial fractions involves a *constant* numerator for each power of a *linear* factor occurring in $q(x)$ (in Example 2, 1 for x and 2 for x^2), and a *linear* numerator for each power of a non-factorable *quadratic* factor ($x+2$ for x^2+1).[29]

4.6.10   QUESTIONS

Q1. Find $\int \dfrac{1}{x^2+x-6}\,dx$.

Q2 (§4.6.9). Let

$$\frac{x+2}{x^2(x^2+1)} = \frac{A}{x} + \frac{B}{x^2} + \frac{Cx+D}{x^2+1}$$

and determine the constants A, B, C, and D by first multiplying through by $x^2(x^2+1)$ and then equating coefficients of equal powers of x.

Q3. Find $\int \dfrac{x^3}{x^4+x^2-2}\,dx$. (Factor the denominator and apply the method of Q2, heeding the remark at the end of §4.6.9.)

4.7   IMPROPER INTEGRALS

4.7.1   Improper integrals.   A. Integrals over infinite intervals.
The definition of the integral can be stretched so as to accommodate, under certain circumstances, integrals over infinite intervals and integrals of

---

[29] If the assumption is not fulfilled we can divide $p(x)$ by $q(x)$ to obtain a polynomial plus a fraction having a numerator of lower degree than $q(x)$.

discontinuous functions. A few examples will show how these so-called **improper** integrals are defined.

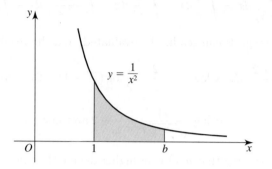

Fig. 4-30

EXAMPLE 1. The area under $y = 1/x^2$ from 1 to $b > 1$ increases as b does, but more and more slowly (Fig. 4-30); in fact, it approaches the limiting value 1 as $b \to \infty$, for

$$\int_1^b \frac{1}{x^2}\, dx = -\frac{1}{x}\bigg|_1^b = -\frac{1}{b} + 1 \to 1\,.$$

We regard this value as the area under the curve "from 1 to ∞," and define the integral of $1/x^2$ from 1 to ∞ accordingly:

$$\int_1^\infty \frac{1}{x^2}\, dx = \lim_{b\to\infty} \int_1^b \frac{1}{x^2}\, dx = 1\,.$$

This definition is quite independent of the geometrical interpretation of the integral as an area of infinite extent.

In like manner we define and calculate $\displaystyle\int_a^\infty \frac{1}{x^\alpha}\, dx$ for any $a > 0$ and $\alpha > 1$ (Q2). But for $\alpha \leq 1$ these integrals cannot be defined, as the requisite limits are not finite (Q3).

EXAMPLE 2. Consider $y = 1/(1 + x^2)$, which is defined for all x (Fig. 4-31). We have by definition

$$\int_0^\infty \frac{1}{1 + x^2}\, dx = \lim_{b\to\infty} \int_0^b \frac{1}{1 + x^2}\, dx = \lim_{b\to\infty} \arctan b = \frac{\pi}{2}\,.$$

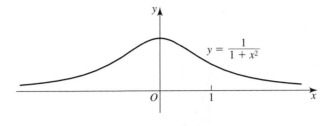

Fig. 4-31

Similarly, we can let the interval of integration be infinite in the opposite direction:

$$\int_{-\infty}^{0} \frac{1}{1+x^2}\, dx = \lim_{a\to-\infty} \int_{a}^{0} \frac{1}{1+x^2}\, dx = \lim_{a\to-\infty} (-\arctan a) = \frac{\pi}{2},$$

because as $a \to -\infty$, $\arctan a \to -\pi/2$. Of course, in view of the symmetry about the y-axis it was to be expected that the latter integral would agree with the former one (§4.6.5.Q4).

Combining the two integrals we can write a single one over the entire x-axis:

$$\int_{-\infty}^{\infty} \frac{1}{1+x^2}\, dx = \int_{-\infty}^{0} \frac{1}{1+x^2}\, dx + \int_{0}^{\infty} \frac{1}{1+x^2}\, dx = \frac{\pi}{2} + \frac{\pi}{2},$$

or

$$\int_{-\infty}^{\infty} \frac{1}{1+x^2}\, dx = \pi.$$

This expression for π, which is independent of the circle and indeed of geometry altogether, will find an application in §4.7.4.

> One of the uses of these integrals over infinite intervals is to provide simplification in physical applications where it is possible to regard a far-off point as infinitely distant. We will have an instance of this in §5.4.5.Q6.

4.7.2 B. Integrals of discontinuous functions. EXAMPLE 3. The area under $y = 1/\sqrt{x}$ from a to 1, where $0 < a < 1$, increases as a decreases, but so slowly that it approaches a finite limit as $a \to 0$ (Fig. 4-32):

$$\int_{a}^{1} \frac{1}{\sqrt{x}}\, dx = 2\sqrt{x}\,\Big|_{a}^{1} = 2 - 2\sqrt{a} \to 2.$$

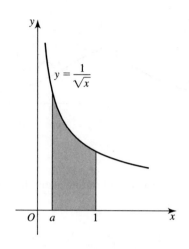

Fig. 4-32

Thus although $1/\sqrt{x}$ is not continuous at 0, it can be said to have an integral over the interval $[0, 1]$, with the definition

$$\int_0^1 \frac{1}{\sqrt{x}} \, dx = \lim_{a \to 0} \int_a^1 \frac{1}{\sqrt{x}} \, dx.$$

In the same way we define and calculate $\int_0^b \frac{1}{x^\alpha} \, dx$ for any $b > 0$ and $\alpha < 1$ (Q4). For $\alpha \geq 1$ the integrals fail to exist (Q5).

An improper integral of this kind can sometimes be defined even if the discontinuity of the integrand occurs in the middle of the interval of integration (Q6(a)).

EXAMPLE 4.   The function $1/\sqrt{1 - x^2}$ is defined only for $-1 < x < 1$ (Fig. 4-33), but we write

$$\int_0^1 \frac{1}{\sqrt{1 - x^2}} \, dx = \lim_{b \to 1} \int_0^b \frac{1}{\sqrt{1 - x^2}} \, dx = \lim_{b \to 1} \arcsin b = \frac{\pi}{2}.$$

Similarly $\int_{-1}^0 \frac{1}{\sqrt{1 - x^2}} \, dx = \frac{\pi}{2}$, and the integral over the whole interval $[-1, 1]$ is

$$\int_{-1}^1 \frac{1}{\sqrt{1 - x^2}} \, dx = \pi.$$

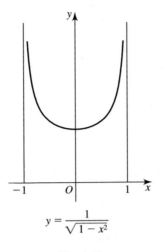

$$y = \frac{1}{\sqrt{1 - x^2}}$$

Fig. 4-33

4.7.3 QUESTIONS

Q1.   Which of the following improper integrals exist? Calculate the ones that do.

(a) $\displaystyle\int_1^\infty \frac{1}{x^3}\, dx$ (b) $\displaystyle\int_0^4 \frac{1}{x^3}\, dx$ (c) $\displaystyle\int_0^8 t^{-1/3}\, dt$

Q2 (§4.7.1, Example 1).   Give the definition and carry out the evaluation.

Q3 (§4.7.1, Example 1).   Verify this.

Q4 (§4.7.2, Example 3).   Give the definition and carry out the evaluation.

Q5 (§4.7.2, Example 3).   Verify this.

Q6.   The same as Q1, for the following integrals.   (a) $\displaystyle\int_{-1}^2 x^{-1/3}\, dx$ (cf. §3.6.2)

(b) $\displaystyle\int_0^\infty x e^{-x^2}\, dx$ (c) $\displaystyle\int_{-1}^1 \frac{1}{1 - x^2}\, dx$ (d) $\displaystyle\int_{-\infty}^\infty \cos x\, dx$

Q7.   Use §4.6.4(3) to calculate $\displaystyle\int_a^b \sqrt{\frac{x}{a - x}}\, dx,\ 0 \le b/a < 1.$

Q8.   Answer Q4(b) of §4.4.9.

4.7.4 Definition of the trigonometric functions by means of an integral.[30]   The expression of the inverse trigonometric functions as integrals (§4.4.8) opens up the prospect of founding the trigonometric functions upon

---

[30]This section can be omitted.

purely analytical considerations, without reference to geometry. Let us have a look at how the theory begins.

Assuming ignorance of the trigonometric functions and their inverses, we *define* a function "arctan" by

$$\arctan x = \int_0^x \frac{1}{1+x^2}\,dx\,.$$

By the Fundamental Theorem arctan is differentiable and its derivative is the positive function $1/(1+x^2)$. Therefore it is increasing. It is also an odd function, because the integrand is even (§4.6.5.Q4). Furthermore, the limit of arctan x as $x \to \infty$ is finite, i.e. $\int_0^\infty \frac{1}{1+x^2}\,dx$ exists. For the substitution $u = 1/x$ transforms $\int_1^b \frac{1}{1+x^2}\,dx$ into $-\int_1^{1/b} \frac{1}{1+u^2}\,du = \int_{1/b}^1 \frac{1}{1+u^2}\,du$, which as $b \to \infty$ approaches the finite value $\int_0^1 \frac{1}{1+u^2}\,du$; thus $\int_1^\infty \frac{1}{1+x^2}\,dx$ exists and equals $\int_0^1 \frac{1}{1+x^2}\,dx$, so $\int_0^\infty \frac{1}{1+x^2}\,dx$ exists too (and equals $2\int_0^1 \frac{1}{1+x^2}\,dx$). If therefore we *define* a number "π" by

$$\frac{\pi}{2} = \int_0^\infty \frac{1}{1+x^2}\,dx\,,$$

or equivalently

$$\pi = \int_{-\infty}^\infty \frac{1}{1+x^2}\,dx\,,$$

we can say that $y = \arctan x$ takes all values $-\pi/2 < y < \pi/2$.

The next step is to define a function "tan" to be the inverse of arctan. Then $\tan x$ is continuous and increasing, and defined for $-\pi/2 < x < \pi/2$. Functions "cos" and "sin" are then defined over the same interval in terms of tan:

$$\cos x = \frac{1}{\sqrt{1+\tan^2 x}}\,, \quad \sin x = \frac{\tan x}{\sqrt{1+\tan^2 x}}$$

(these are recognizable as trigonometric identities valid where $\cos x > 0$, e.g. for $-\pi/2 < x < \pi/2$). We further define $\cos(\pi/2) = 0$ and $\sin(\pi/2) = 1$. Since $\tan x \to \infty$ as $x \to \pi/2$ from the left, $\cos x \to 0$ and $\sin x \to 1$ (for the latter see §4.1.2.Q4(c)). Thus we have defined $\cos x$ and $\sin x$ as continuous functions for $-\pi/2 < x \leq \pi/2$.

We now extend $\cos x$ and $\sin x$ to all x by repeated application of the conditions

$$\cos(x + \pi) = -\cos x, \quad \sin(x + \pi) = -\sin x.$$

That is: any x is located in some interval $-\dfrac{\pi}{2} + n\pi < x \leq \dfrac{\pi}{2} + n\pi$, so that $-\pi/2 < x - n\pi \leq \pi/2$; by applying the first condition $|n|$ times we can determine $\cos x = \cos\left[(x - n\pi) + n\pi\right]$ in terms of $\cos(x - n\pi)$, and similarly for $\sin x$.[31] The extended functions are continuous (Q1). Having extended sin and cos we can extend tan to all x except odd multiples of $\pi/2$ by $\tan x = \sin x / \cos x$ (or equivalently by $\tan(x + \pi) = \tan x$), and similarly define the other trigonometric functions.

This completes the geometry-independent definition of the trigonometric functions. It remains to show that the functions so defined possess the properties expected of them; the essential step is the proof of the addition formulas.[32] I will not carry out this development, which involves some difficulty. However, the idea of obtaining a function as the inverse of an integral will be of use in the next two chapters (§§5.3.5–6, 5.4.4; §6.2.1).

4.7.5   QUESTION

Q1 (§4.7.4).   Establish the continuity of $\sin x$ as defined.

4.8   INTEGRATION IN THE PLANE AND IN SPACE

4.8.1   Integrals over geometrical regions.   The length L of an x-interval $[a, b]$ can be expressed as a definite integral, the sum of infinitesimal increments dx:

$$\int_a^b dx = b - a = L([a, b]).$$

More generally, according to the Fundamental Theorem the total quantity F associated with any (continuous) function f defined on $[a, b]$ is given by a

---

[31]Cf. §3.3.9.Q7(b) & Q14(b).

[32]See G. H. Hardy, *A Course of Pure Mathematics*, 10th edition (1952 and reprints), §225.

definite integral, the sum of infinitesimals $dF = f(x)\,dx$:

$$\int_a^b f(x)\,dx = F(b) - F(a) = F([a, b]).^{33}$$

For example, the mass of a segment of wire of line density $\lambda(x)$ is the sum of differentials of mass $dm = \lambda(x)\,dx$:

$$M([a, b]) = \int_a^b \lambda(x)\,dx\,.$$

Higher-dimensional analogues of these quantities are (i) the *area of a region of the plane*, and a total quantity associated with that region, such as the *mass of a metal sheet* that occupies it; and (ii) the *volume of a region of space*, and a total quantity associated with that region, such as the *mass of a metal solid* that occupies it. The definite integral can be generalized to express such quantities as well, and make possible their calculation. The full generalization requires a careful study of functions of several variables (Art. 4.5), which we will not undertake; but the one-variable theory of integration we have already developed is enough to deal with many cases of interest. The present article will consider some of these.

For the sake of brevity I will be relying entirely on the language of infinitesimals, although (as you may want to show) justification by limit arguments is not far to seek.

4.8.2 Integrals over regions of the plane. A. Area. Strip elements.

If $f(x)$ is a non-negative function, the integral $\int_a^b f(x)\,dx$ expresses the area under the graph of f as a sum of infinitesimal areas $f(x)\,dx$ of parallel rectangular strips (§2.4.1). The area of a plane region which is not "under a curve" can similarly be expressed as an integral, by dividing the region into parallel strips or other infinitesimal parts and summing by integration. If dA denotes the area of a typical infinitesimal part, or element,[34] of a region R, we

---

[33] Here F is being used to denote both a primitive of f and the total quantity associated with f.

[34] The distinction between the infinitesimal part and its area is not usually maintained: both can be called elements of area and denoted dA.

write symbolically

$$\text{area } A \text{ of } R = \int_R dA$$

("integral over R of dA"); to calculate such an integral we represent it somehow as an integral in the ordinary sense, the integral of a function over an interval.

Suppose to begin with that R is divided into strips perpendicular to the x-axis, as in the familiar case of a region under a graph. Let $L(x)$ denote the length of the strip at location x along the axis; then $dA = L(x)\,dx$ and the area of R is

$$A = \int_R dA = \int_a^b L(x)\,dx\,,$$

where a and b are the x-values that mark the extremities of the region.

As an example let us calculate the area between a curved graph and a slanted line.

EXAMPLE 1. The region R between the parabola $y = x^2 - 2$ and the line $y = x$ is made up of vertical strips dA (Fig. 4-34). The length, or height, of a strip is the y-coordinate of its upper end minus the y-coordinate of its

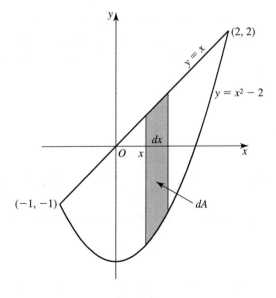

Fig. 4-34

lower end:

$$L(x) = x - (x^2 - 2) = -x^2 + x + 2$$

(we neglect the variation in height that occurs across the infinitesimally thin strip). Hence the area of the strip is

$$dA = L(x)\,dx = (-x^2 + x + 2)\,dx,$$

and the area of the region is

$$A = \int_R dA = \int_R (-x^2 + x + 2)\,dx.$$

The limits between which x varies are the x-values at which the parabola and the line intersect; i.e., where $x^2 - 2 = x$. Solution of this equation gives $x = -1$ and $x = 2$. The area of R is then

$$A = \int_{-1}^{2} (-x^2 + x + 2)\,dx = \left[-\frac{x^3}{3} + \frac{x^2}{2} + 2x \right]_{-1}^{2} = \frac{9}{2}.$$

The next examples are set up without a formal coordinate system.

EXAMPLE 2. (a) It is proved in elementary plane geometry that parallelograms with equal bases and equal heights are equal in area. The method of strips makes this obvious at a glance (Fig. 4-35): parallelogram II is made of the same strips dA as parallelogram I, displaced horizontally. A consequence is that the area of a parallelogram is given by the formula for the area of a rectangle, base × height.

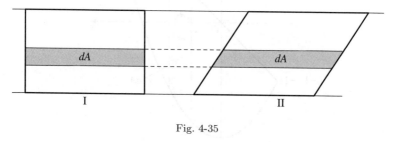

Fig. 4-35

(b) Since any triangle is half a parallelogram, the area of a triangle is $\frac{1}{2}$ base × height. But let us also prove this directly by integration. In Fig. 4-36

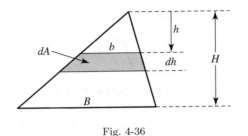

Fig. 4-36

we have $dA = b\,dh$, and h (measured downward from the top[35]) increases from 0 to H; therefore the area is $\int_0^H b\,dh$. In order to integrate we express b in terms of h by similar triangles: $b/h = B/H$, so the integral is

$$A = \int_0^H \frac{B}{H} h\,dh = \frac{B}{H}\frac{h^2}{2}\bigg|_0^H = \frac{1}{2}\,BH\,.$$

4.8.3   B. Integrals of surface density. Ring elements.   A total quantity other than area can be found in the same way, provided that the corresponding specific quantity can be considered constant on each element of area. To find the mass of a region R of a plane sheet of metal, for example, we integrate its surface density σ (§4.5.1) over R. Assuming that σ is constant over dA, the mass of the element (the element of mass) is

$$dm \text{ (kg)} = \sigma \text{ (kg/m}^2) \times dA \text{ (m}^2) = \sigma\,dA \text{ (kg)}\,,$$

so that

$$\text{mass } M \text{ of } R = \int_R dm = \int_R \sigma\,dA\,.$$

EXAMPLE 3.   Suppose that the surface density of a metal triangle as in Fig. 4-36 is constant along horizontal lines $h = $ constant, but may vary from the top of the triangle to the base. This means that σ is a function of h (its values are determined by h). We treat σ as constant on the thin strip from h to $h + dh$; then the mass of that strip is

$$dm = \sigma\,dA = \sigma b\,dh = \sigma\frac{B}{H}h\,dh\,,$$

---

[35] Thus h here is "depth" rather than height. The formulas are neater that way.

and the mass of the whole triangle is

$$M = \frac{B}{H} \int_0^H \sigma(h) h \, dh$$

(here $\sigma(h)$ denotes the value of the function σ at h).

To be specific, suppose that σ increases uniformly from 2 kg/m^2 at the top to 5 kg/m^2 at the base. Then the increase in σ from 2 to its value at level h bears the same ratio to its total increase as h does to H:

$$\frac{\sigma - 2}{3} = \frac{h}{H}, \quad \text{or } \sigma = 2 + \frac{3}{H} h.$$

The total mass is

$$M = \frac{B}{H} \int_0^H \left(2 + \frac{3}{H} h\right) h \, dh = \frac{B}{H} \left[h^2 + \frac{h^3}{H}\right]_0^H = 2BH.$$

It may be appropriate to divide R into elements that are not straight strips, as the next example shows.

EXAMPLE 4. The surface density of a disk of radius ρ has its greatest value σ_0 at the center, and falls off uniformly in all directions with increasing radius r to the value 0 at $r = \rho$. To find the mass of the disk we use ring-shaped elements, on which σ is constant (Fig. 4-37). The area element dA can be calculated from the formula for the area of a circle of radius r: differentiation of $A(r) = \pi r^2$ shows that the change in area corresponding to dr is $dA = 2\pi r \, dr$.

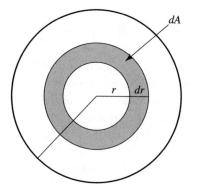

Fig. 4-37

The corresponding element of mass is

$$dm = \sigma \, dA = 2\pi r \sigma(r) \, dr \,,$$

where $\sigma(r)$ is the function

$$\sigma(r) = \sigma_0 \left(1 - \frac{r}{\rho} \right).$$

Then the total mass is

$$M = \int dm = \int_0^\rho 2\pi r \sigma_0 \left(1 - \frac{r}{\rho} \right) dr$$

$$= 2\pi\sigma_0 \left[\frac{r^2}{2} - \frac{r^3}{3\rho} \right]_0^\rho = \frac{\pi \rho^2 \sigma_0}{3} \,.$$

Straight strips would have been useless for this problem, because σ would not have been constant over them, so we could not have said $dm = \sigma \, dA$. Yet the equation of infinitesimals $dA = 2\pi r \, dr$ amounts to regarding the ring element as a rectangular strip (Fig. 4-38); for the difference between the inner and outer circumferences of the ring is $2\pi \, dr$, which is negligible in comparison to $2\pi r$ (§2.4.8).

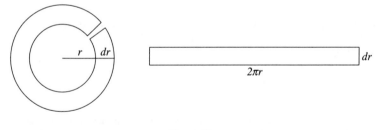

Fig. 4-38

4.8.4   C. Elements $dA = dx \, dy$.   The method of the preceding examples depends on decomposing R into elements of finite length and infinitesimal width, in such a way that (i) the areas of all elements can be expressed in terms of a single variable, and (ii) the surface density σ or other specific quantity associated with the points of R can be regarded as constant on each element. When this kind of decomposition is unavailable we employ elements infinitesimal in *diameter* (greatest dimension), such as little rectangles (Fig. 4-39). Then σ has to be considered constant only in the immediate vicinity of each

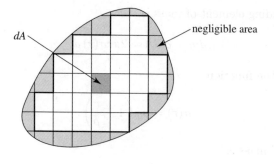

Fig. 4-39

point, which is merely a matter of continuity. On the other hand, it is necessary to use *two* variables to locate an element and specify its area, and σ must therefore be treated as a function of two variables, $\sigma = \sigma(x, y)$ (§4.5.1).

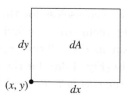

Fig. 4-40

If we take a little rectangle at (x, y) with sides parallel to the axes (Fig. 4-40), its area is $dA = dx\,dy$, and its mass is $\sigma dA = \sigma(x, y)dx\,dy$. The total mass is then

$$\int_R \sigma\,dA = \int\!\!\int_R \sigma(x, y)\,dx\,dy,$$

where the two integral signs indicate that two variables are involved. The theory of such integrals belongs to the study of functions of several variables (Q7).

Calculation of area in polar coordinates is addressed in Q9.

4.8.5   QUESTIONS

Q1.   Sketch and calculate the area between the parabola $y = -x^2 + 1$ and the line $2y = -x - 1$.

Q2.   Find the area of one of the infinitely many congruent figures contained between the graphs of the sine and cosine functions.

Q3.   Write and explain a formula or formulas for the area between two continuous graphs $y = f(x)$ and $y = g(x)$.

Q4.   To each point $P(x, y)$ of the circle $x^2 + y^2 = b^2$ is assigned the point $P'(x', y')$ defined by $x' = (a/b)x$, $y' = y$, where $a > b > 0$ (Fig. 4-41). (a) Show that the points P' constitute the ellipse $\dfrac{x^2}{a^2} + \dfrac{y^2}{b^2} = 1$. (b) Determine the area of this ellipse by means of strips, arguing in terms of integrals but not performing an integration.

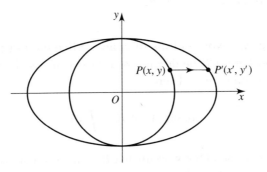

Fig. 4-41

Q5.   The water **pressure**, or force per area (newton/m^2), on each point of a certain flat vertical parabolic dam (Fig. 4-42) is proportional to the depth of the water behind it at that point. Represent the underwater portion of the face of the dam as the area between $y = x^2 - 1$ and the x-axis, and find an expression for the total force of water on the dam by integrating with respect to y.

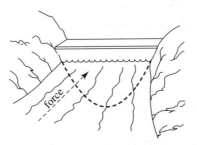

Fig. 4-42

Q6.   The rate at which gas passes through each point of a circular filter of radius 0.2 m, in kg/m^2 per second, is proportional to the distance from the center of the filter. Find an expression for the total rate of flow through the filter, in kg/sec.

Q7 (§4.8.4).   If this theory is developed, what in it will be the analogue to the "area under a curve" that stands upon an interval of the line?

Q8. In plane geometry it is proved that the areas of similar rectilinear figures are to one another as the squares of their corresponding linear dimensions. Extend this proposition to similar curved figures (curved figures of the same shape but drawn on a different scale).[36]

Q9. Let a curve have equation $r = f(\theta)$ in polar coordinates. (a) Show that the infinitesimal sector contained by the curve and the radius vectors at θ and $\theta + d\theta$ has area $dA = \frac{1}{2}r^2\, d\theta$. (Use §3.3.6.Q7.) (b) Find the area of a sector of the logarithmic spiral $r = r_0 e^{\alpha\theta}$ (§3.3.12) from θ_1 to θ_2.

4.8.6   Integrals over regions of space. Slice elements.

Let a region of space (a solid figure) R be decomposed into elements of volume, i.e. solids of infinitesimal volumes dV. Then

$$\text{volume } V \text{ of } R = \int_R dV .$$

To find the mass of a solid body occupying R we integrate its **volume density**, or simply **density**, δ (kg/m^3):

$$\text{mass } M \text{ of } R = \int_R dm = \int_R \delta\, dV ,$$

δ being regarded as constant over (or throughout) each volume element. Other total quantities associated with a region of space are calculated from their specific quantities in the same way.

There are a variety of useful shapes for volume elements, but I will consider only a spatial analogue of the plane strip, the **slice**, and that very briefly. (One other is called for in Q4.) A slice of a solid is a flat cross-section of infinitesimal thickness. Let R be divided (like a loaf of bread) into parallel slices perpendicular to the x-axis (say), and let the area of a face of the slice at location x be $A(x)$ (Fig. 4-43); then

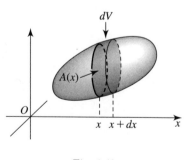

Fig. 4-43

---

[36]Cf. Newton's *Principia*, Book 1, Lemma 5.

$dV = A(x)\,dx$ (Q6(b)), and the volume of R is therefore

$$V = \int_R dV = \int_a^b A(x)\,dx\,,$$

where a and b are the x-coordinates of the extremities of R.

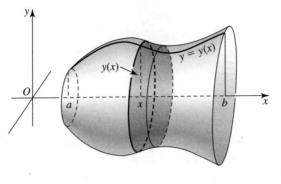

Fig. 4-44

Integration by slices can be applied to calculate the volume of a **solid of revolution** obtained by revolving the region under a graph $y = y(x) \geq 0$ about the x-axis (Fig. 4-44). Here the slice at x is a disk of radius $y(x)$, so $A(x) = \pi y(x)^2$ and

$$V = \int_a^b \pi y(x)^2\,dx\,.$$

EXAMPLE 1. A sphere of radius ρ can be obtained by revolving $y = \sqrt{\rho^2 - x^2}$ about the x-axis (Fig. 4-45). Its volume is then

$$\int_{-\rho}^{\rho} \pi(\rho^2 - x^2)\,dx$$

$$= \pi\left[\rho^2 x - \frac{x^3}{3}\right]_{-\rho}^{\rho} = \frac{4}{3}\pi\rho^3\,.$$

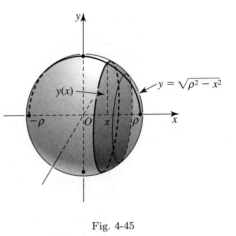

Fig. 4-45

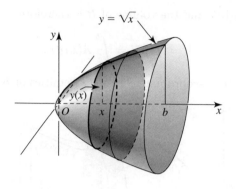

Fig. 4-46

EXAMPLE 2. A paraboloid is obtained by revolving $y = \sqrt{x}$ about the x-axis (Fig. 4-46). Its volume up to $x = b$ is

$$\int_0^b \pi x \, dx = \frac{1}{2} \pi b^2 \, .$$

If the density δ of the solid of revolution varies with x, its mass is

$$M = \int_a^b \pi y(x)^2 \delta(x) \, dx \, .$$

A different type of application is to a solid **cone**, in the generalized sense—a solid circular cone, or pyramid, or other figure formed by joining the points of a plane region enclosed by a curve to a point P not in the plane (Fig. 4-47). Let H be the altitude of the vertex P above the base plane, and measure h downward from P (cf. §4.8.2, Example 2(b)); let B be the area of the base and $B(h)$ the area of the slice at h. Then $B(h)/B = h^2/H^2$, because the face of the slice and the base, being similar figures (Q7), are in area as the squares of their linear dimensions (§4.8.5.Q8). Consequently the volume of the cone is

$$V = \int_0^H B(h) \, dh = \int_0^H \frac{B}{H^2} h^2 \, dh = \frac{1}{3} BH \, .$$

The remarks in §4.8.4 concerning integration of arbitrary quantities over arbitrary plane regions apply, *mutatis mutandis*, to integration over regions of space.

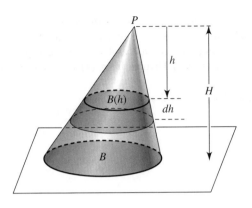

Fig. 4-47

4.8.7 QUESTIONS

Q1. (a) Calculate the volume of a right circular cone of height H by treating it as the solid of revolution of the straight line $y = \alpha x$, $x \geq 0$. Compare with the formula $\frac{1}{3} BH$. (b) Let the density of the (solid) cone be $\delta(x) = 1 + 2x$. Find its mass.

Q2. (a) The sphere of §4.8.6, Example 1, is filled with electric charge (§2.10.5). If the charge density is $\delta(x) = x - 1$ coulomb/m^3, find the total (i.e. net) charge. (b) Explain why the answer is familiar.

Q3. (a) Find the volume of the horn-shaped surface obtained by revolving $y = 1/x$, $1 \leq x \leq b$, about the x-axis. (b) What happens when $b \to \infty$?

Q4. The density of a sphere of radius ρ is δ_0 at the center and falls off uniformly in all directions with increasing radius r to the value 0 at $r = \rho$. Find its mass. (Follow §4.8.3, Example 4.)

Q5. Find the volume of the **ellipsoid of revolution** obtained by revolving the ellipse $\dfrac{x^2}{a^2} + \dfrac{y^2}{b^2} = 1$ about the x-axis.

Q6. A **right cylinder**, in the generalized sense, is a solid as in Fig. 4-48: the curved surface is generated by moving a straight line, held perpendicular to a plane p, around a closed curve of arbitrary shape in p; the congruent parallel flat faces

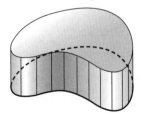

Fig. 4-48

are the region of p enclosed by that curve and the region cut out by the surface from another plane parallel to p. (a) Show that the volume of a right cylinder is given by base × height. (Decompose the base and use the corresponding formula for the volume of a prism.) (b) (§4.8.6) Deduce from (a) that a slice has the volume it is claimed to have.

Q7 (§4.8.6).   Establish the similarity.

CHAPTER 5

Differential Equations of Rectilinear Motion

INTRODUCTION

Let a particle of mass m move on the x-axis under the sole influence of a force f directed along that axis.[1] The location coordinate of the particle is a function of time $x(t)$, the first and second derivatives of which are respectively the velocity and acceleration of the particle: $v(t) = dx/dt$, $a(t) = dv/dt = d^2x/dt^2$. The motion is governed by Newton's law $f = ma$, which in terms of $x(t)$ is

$$m\frac{d^2x}{dt^2} = f.$$

This is a **differential equation** for $x(t)$—an equation that involves a derivative of the function. Every function $x(t)$ that describes a possible motion of the particle satisfies the equation; conversely, every solution of the equation describes a possible motion.

If the force is known but the motion is not, we can undertake to solve the equation for $x(t)$ by integration. The feasibility of this project depends on the form of f. In this chapter, $x(t)$ will be determined more or less completely in three cases of great physical interest: uniform gravitation (Art. 5.1); the kind of oscillatory motion a spring imparts (Arts. 5.2, 5.3); and gravitation under the inverse-square law that holds in nature (Art. 5.4). Direct integration of the Newtonian equation will be possible only for the first of these; for the second and especially the third we will use a different approach, based on

---

[1]On force see Art. 1.4. Articles 5.1 and 5.2 depend on §§1.4.1–4, and Arts. 5.3 and 5.4 on §§1.4.6–8.

considerations of *energy* to be introduced in Art. 5.3, including the concept of potential energy and the principle of conservation of energy.

The methods applied to rectilinear motion in this chapter can be extended to problems of *curvilinear* motion, such as the determination of planetary orbits. Although we will not pursue the general subject, in Chapter 7 (§7.5.5) there will be a glimpse of the way in which motion along plane curves is related to rectilinear motion.

5.1 UNIFORM GRAVITATION

5.1.1 The equation of motion and its solution. Near the surface of the earth a body of mass m experiences constant downward gravitational force of magnitude mg, where g is the acceleration of gravity, approximately 9.8 m/sec^2 (§1.3.5).

If a finite body moves it also feels air resistance (Q9), which is ignored here; I am in effect imagining gravitation in a terrestrial vacuum. In addition, g varies a little with altitude, as we will see in §5.4.1, and also with location on the earth.

The term "body" can be understood here as equivalent to "particle," but in fact our theory applies well to real bodies. Part of the reason is that a body as a whole is attracted by the earth as though all of its mass were concentrated at a single point (its center of mass).

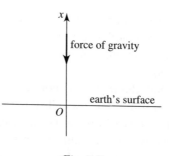

Fig. 5-1

Let the x-axis be drawn vertically upward from a given point on the earth, so that the zero coordinate is at that point (Fig. 5-1). Then the differential equation of motion takes the form

$$m\frac{d^2x}{dt^2} = -mg$$

—the force is negative because it acts in the negative direction (§2.10.7)—or

$$\frac{d^2x}{dt^2} = -g.$$
(1)

Since m has dropped out, the motion of a particle is independent of its mass.
Integrating once with respect to t gives

$$v(t) = \frac{dx}{dt} = -gt + c_1.$$

The constant of integration can be determined by setting $t = 0$ (cf. §§3.6.4, 3.8.2):

$$v(0) = -g \cdot 0 + c_1 = c_1.$$

If as usual the motion is considered to begin at $t = 0$, this constant is called the **initial velocity** of the particle. Denoting it by v_0, we have

$$v(t) = \frac{dx}{dt} = -gt + v_0.$$
(2)

A second integration yields

$$x(t) = -\frac{1}{2}gt^2 + v_0 t + c_2.$$

Here we find $c_2 = x(0)$, the **initial position** (or location) x_0 of the particle, so that finally

$$x(t) = -\frac{1}{2}gt^2 + v_0 t + x_0.$$
(3)

This is the solution of the differential equation (1) with **initial conditions** $x(0) = x_0$, $v(0) = v_0$.

What we have proved is that any solution $x(t)$ of (1) has the form (3), where $x_0 = x(0)$ and $v_0 = x'(0)$. Conversely, differentiation shows that any function $x(t)$ of the form (3), where x_0 and v_0 are constants, satisfies (1), and that $x(0) = x_0$, $x'(0) = v_0$ (Q1). Thus the numbers x_0 and v_0 can be chosen arbitrarily, and once chosen they determine $x(t)$ completely. In other words, the particle can have any initial position and velocity at all, and its motion is completely determined by them. It was indeed obvious that the motion depends upon initial position and velocity; calculus assures us that in consequence of Newton's law it depends on nothing else, and specifies the exact nature of the dependence.

5.1.2   Consequences of the solution.   Let us consider the motion for $t \geq 0$. If the particle is released from *rest* (dropped) at $t = 0$, $v_0 = 0$ and the formulas (2) and (3) are

$$v(t) = -gt,$$

$$x(t) = -\frac{1}{2}gt^2 + x_0.$$

The velocity is downward and *proportional to the time*. The change of position is

$$x - x_0 = -\frac{1}{2}gt^2,$$

downward and *proportional to the square of the time* (Fig. 5-2).

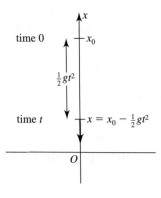

Fig. 5-2

Assuming that the particle is released above the surface of the earth, i.e. $x_0 > 0$, it reaches the ground at a time t_1 when $x = 0$, which is the same as $x_0 = \frac{1}{2}gt_1^2$ or

$$t_1 = \sqrt{\frac{2x_0}{g}}.$$

At that moment its speed is

$$|v| = gt_1 = \sqrt{2gx_0}.$$

A particle dropped from a height of 3.0 meters (about ten feet) will strike the earth after about $\sqrt{\dfrac{2 \times 3.0}{9.8}} = 0.78$ sec with speed $9.8 \times 0.78 = 7.6$ m/sec, or 17 mph.

Suppose now that the particle is projected upward at $t = 0$ rather than dropped, so that $v_0 > 0$. Formula (2),

$$v = -gt + v_0 \, ,$$

shows that the velocity decreases steadily from its initial positive value v_0, reaching zero at

$$t = \frac{v_0}{g}$$

and then becoming negative. At time v_0/g the particle ceases to rise and begins to fall; it is at its highest point, which by (3) is

$$x\left(\frac{v_0}{g}\right) = -\frac{1}{2} g \cdot \left[\frac{v_0}{g}\right]^2 + v_0 \cdot \frac{v_0}{g} + x_0 = \frac{v_0^2}{2g} + x_0 \, .$$

In the language of the theory of maxima and minima (Art. 3.9), we have found the maximum value of the function

$$x(t) = -\frac{1}{2} gt^2 + v_0 t + x_0 \, ,$$

which occurs at the critical point $t = v_0/g$ (where its derivative $v(t)$ vanishes); the second derivative $a(t) = -g$ is negative, as expected. The graph of $x(t)$ is of course a portion of a parabola (Fig. 5-3). It is not hard to calculate the time t_1 at which the particle reaches the ground (Q6).

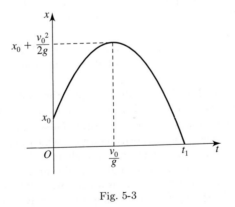

Fig. 5-3

Reviewing the preceding treatment of gravitational motion, we see that calculus was essential for

(i) the formulation of the problem as a differential equation, and

(ii) the solution of the equation.

It could also be invoked for

(iii) the interpretation of the solutions found.

Had the solutions been more complicated, calculus would have been indispensable for (iii) as well as (i) and (ii).

> Galileo achieved a description of motion equivalent to (3) without calculus—so not, of course, as the solution of a differential equation. See his *Two New Sciences* (Third and Fourth Days), and also §7.5.8.Q10(a).

5.1.3   QUESTIONS

Q1 (§5.1.1).   Verify this, starting with a given function $x(t) = -\frac{1}{2}gt^2 + v_0 t + x_0$, where x_0 and v_0 are constants.

Q2.   A body is dropped from a height of 20 m. Find $x(t)$, $v(t)$, the time when the body reaches the ground, and the speed with which it strikes.

Q3.   If a body falls p m in the first q sec, how far does it fall in the next q sec?

Q4.   (a) From what height must a body be dropped to strike the ground at 12 mph? (b) What height will be reached by a body projected upward from the ground at 12 mph?

Q5.   (a) Explain the significance of each of the three terms that make up the expression (3) for $x(t)$. (b) If g were zero, what would the motion be?

Q6 (§5.1.2).   Calculate t_1 in two ways: (a) by solving $x(t) = 0$; (b) by regarding the body as falling from rest at its highest point (where $v = 0$) and using the formula for time of fall.

Q7.   Show that a body projected upward from ground level returns in twice the time it takes to reach its highest point, and at the same speed with which it left the earth.

Q8.   The parabola of Fig. 5-3 is only a graph: the body itself remains on the x-axis. Can the parabolic path nevertheless be seen by a suitable observer?

Q9.   In air a moving body experiences a resisting force approximately proportional to speed, $f_r = -\rho v$, where ρ is a positive constant which depends on the size and shape of the body as well as other factors. If a body falls through the air ($v < 0$) the total force on it is the sum of the force of gravity (downward, < 0) and the resisting force (upward, > 0). (a) Show that there is a **terminal velocity** v_T such that $|v_T|$ is the least number the speed of a body falling from rest cannot exceed, and find it for a body of mass m. (It is the velocity at which acceleration would vanish.) In §6.2.10.Q17 we will determine $v(t)$ under this hypothesis of air resistance. (b) For a typical skydiver ρ is about 13 kg/sec. Estimate $|v_T|$ in m/sec and mph.

5.2   SIMPLE HARMONIC MOTION

5.2.1   The equation of spring-driven oscillation, and a solution.

A coil spring under tension or compression, provided that it is not stretched or squeezed too far, exerts a force proportional to the departure from its unstressed length (Hooke's law). Let a particle of mass m be attached to one end of a spring, the other end of which is fixed to a point of the x-axis so located that the particle is at rest at $x = 0$ when the spring is not under stress (Fig. 5-4). If the particle is displaced from 0 in either direction the spring will tend to restore its position. Let us suppose that a displacement occurs, and consider the resulting motion. I assume that the particle can only move on the x-axis, and that after the initial disturbance it is subject to no other force that affects its motion along that axis; we might imagine a block sliding on frictionless horizontal rails (cf. §§1.4.1 and 2.10.2).

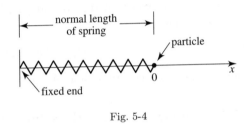

Fig. 5-4

According to the stated law, with this arrangement force is proportional to displacement, and opposed to it. That is, there is a positive constant c such that the force on the particle at any location x is

$$(1) \qquad f(x) = -cx$$

(Fig. 5-5). (We can regard c, the **spring constant**, as a measure of the stiffness or power of a particular spring.) The differential equation of motion is then

$$m \frac{d^2 x}{dt^2} = -cx \,,$$

or

$$(2) \qquad \frac{d^2 x}{dt^2} = -\omega^2 x \,,$$

where I have put $\omega = \sqrt{c/m}$, a quantity that will turn out to be significant.

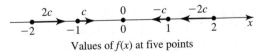

Values of $f(x)$ at five points

Fig. 5-5

Unlike the equation of uniform gravitation (§5.1.1(1)), this one is not ready to be integrated, because the right side involves the very function we seek to determine. Nevertheless, we can easily discover solutions.

Consider first the special case $\omega = 1$, when the equation is

$$\frac{d^2x}{dt^2} = -x:$$

we want a function $x(t)$ which when differentiated twice yields the negative of itself. One such is $\sin t$, and another is $\cos t$. These immediately suggest the solutions $x(t) = \sin \omega t$ and $x(t) = \cos \omega t$ to equation (2). A step further leads to solutions $\alpha \sin(\omega t + \beta)$ and $\alpha \cos(\omega t + \beta)$, where α and β are constants; differentiation verifies that they do satisfy (2). Our purposes will be served by solutions of the form

$$(3) \qquad\qquad x(t) = A \cos(\omega t + \delta),$$

where A and δ are constants and $A \geq 0$.

5.2.2   Interpretation of the solution: simple harmonic motion.

If $A = 0$ formula (3) yields the trivial solution $x(t) = 0$, representing a particle at rest at the origin. Let us assume $A > 0$ and begin by examining the solution in the case $\delta = 0$, which is

$$x = A \cos \omega t.$$

This expression tells the location of the particle as it varies with time. By differentiation we find the velocity of the particle,

$$v = -A\omega \sin \omega t,$$

and its acceleration,

$$a = -A\omega^2 \cos \omega t.$$

Each of the three functions is periodic (§3.3.7), passing through a complete cycle of values as ωt increases by 2π; i.e., as t increases by $\tau = 2\pi/\omega$. The latter quantity is the **period**, or **periodic time**, of the motion—the *number of seconds per cycle*. Since the range of values of the sine and cosine is $[-1, 1]$, the range of values of x is $[-A, A]$, of v $[-A\omega, A\omega]$, and of a $[-A\omega^2, A\omega^2]$.

At $t = 0$ the particle is at $x = A$; it is momentarily at rest, $v = 0$, but accelerating leftward, $a = -A\omega^2$. This and subsequent points of the cycle

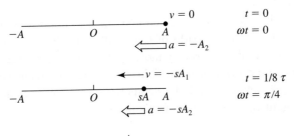

$v = 0$ $t = 0$

$-A$ O A $\omega t = 0$

$a = -A_2$

$v = -sA_1$ $t = 1/8\ \tau$

$-A$ O sA A $\omega t = \pi/4$

$a = -sA_2$

$v = -A_1$ $t = 2/8\ \tau$

$-A$ O A $\omega t = \pi/2$

$a = 0$

$v = -sA_1$ $t = 3/8\ \tau$

$-A$ $-sA$ O A $\omega t = 3\pi/4$

$a = sA_2$

$v = 0$ $t = 4/8\ \tau$

$-A$ O A $\omega t = \pi$

$a = A_2$

$v = sA_1$ $t = 5/8\ \tau$

$-A$ $-sA$ O A $\omega t = 5\pi/4$

$a = sA_2$

$v = A_1$ $t = 6/8\ \tau$

$-A$ O A $\omega t = 3\pi/2$

$a = 0$

$v = sA_1$ $t = 7/8\ \tau$

$-A$ O sA A $\omega t = 7\pi/4$

$a = -sA_2$

$v = 0$ $t = \tau$

$-A$ O A $\omega t = 2\pi$

$a = -A_2$

Fig. 5-6

from $t = 0$ to $t = \tau$ are shown in Fig. 5-6, where I have put $A_1 = A\omega$, $A_2 = A\omega^2$, and $s = 1/\sqrt{2}$.

The motion depicted is oscillation about the zero point to a distance on each side equal to A, the **amplitude** of the oscillation. Because it is very nearly the same as the simplest vibratory motion of a point on a sounding string, it is called **simple harmonic motion**. The connection between the trigonometric functions which define it and the geometry of the circle makes possible the following elegant representation of this motion.

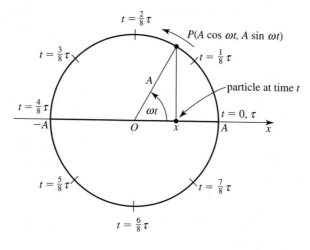

Fig. 5-7

With the x-axis embedded in the xy-plane in the usual way, let point P travel at constant speed counterclockwise around the circle of radius A centered at the origin, so as to pass the point $(A, 0)$ at $t = 0$ (Fig. 5-7). The **angular velocity** of P, by which I mean the rate at which radius OP turns, is assumed to be ω radians per second. Then at time t OP has turned through angle ωt from the positive x-axis; consequently the perpendicular from P meets that axis at $x = A\cos\omega t$, the location of the particle. As P circles the origin x shuttles back and forth between A and $-A$.

By study of this representation the simultaneous variation of x, v, and a can be thoroughly understood (Q3–Q5). To observe the motion, set a small object near the rim of a constant-speed turntable and view it from the side as the turntable rotates (Fig. 5-8).

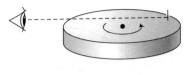

Fig. 5-8

A continuous picture of the variation is obtained by graphing x, v, and a together. Fig. 5-9 has $A = 3$ and $\omega = \pi/6 \approx 0.52$, hence $\tau = 12$ and

$$x = 3\cos(\pi/6)t\,,$$

$$v = -(\pi/2)\sin(\pi/6)t \approx -1.57\sin(\pi/6)t\,,$$

$$a = -(\pi^2/12)\cos(\pi/6)t \approx -0.82\cos(\pi/6)t\,.$$

(Since the units of the three quantities are different, it is useless to compare the heights of their graphs.) This much will suffice for an account of the case $\delta = 0$ of formula (3).

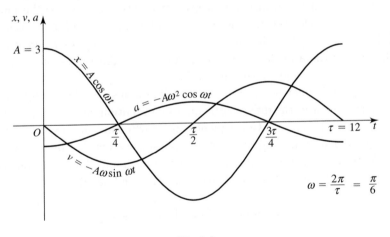

Fig. 5-9

The new element in the general formula $x = A\cos(\omega t + \delta)$ is a possible displacement of the motion in time. At $t = 0$ the particle is at $x = A\cos\delta$, which is different from $x = A$ unless $\delta = 2\pi n$ for some integer n. In the circular representation of Fig. 5-7, $\angle AOP$ is $\omega t + \delta$ rather than ωt, so that the times are shifted around the circle by the angle δ (Fig. 5-10). Apart from this adjustment in timekeeping the motion proceeds just as before.

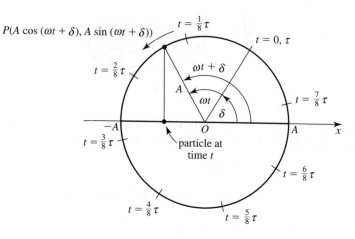

Fig. 5-10

5.2.3 Specifying a particular solution.

We call $\omega t + \delta$ the **phase** of the motion, and δ the **phase constant**.[2] Since the period τ is the number of seconds per cycle, the **frequency** of oscillation, or *number of cycles per second*, is $1/\tau = \omega/2\pi$; therefore ω is the *number of cycles in 2π seconds*, which is called the **angular frequency**. In the circular representation, as we know, it equals the angular velocity of P.

The angular frequency ω, as a number determined by the spring and the particle, is present in the differential equation of motion, and so is the same in all solutions $A \cos(\omega t + \delta)$. Consequently any particular solution of this type is entirely specified by its amplitude A and phase constant δ. Conversely, given a solution of this form, the amplitude is determined and the phase constant is determined up to an additive multiple of 2π. In symbols, if

$$A' \cos(\omega t + \delta') = A \cos(\omega t + \delta)$$

for all t, where A and A' are > 0, then $A' = A$ and $\delta' = \delta + 2\pi n$ for some integer n. (For the proof, see Q6.) Thus a motion of the type $x = A \cos(\omega t + \delta)$ corresponds to an amplitude and a setting of the clock.

Instead of identifying a motion by A and δ, we can refer to the initial position x_0 and velocity v_0 of the particle, as we did with uniform gravitation. These

---

[2]The terms **phase displacement** and **epoch** are also used for δ.

are determined by A and δ as

$$x_0 = x(0) = A \cos \delta$$

and

$$v_0 = v(0) = -A\omega \sin \delta \,.$$

Conversely, given any x_0 and v_0 not both zero, a corresponding $A > 0$ and δ are determined, the former uniquely and the latter up to a multiple of 2π (Q7(a)).

In the next article (§5.3.7) it will be proved that *all* solutions of our differential equation

$$\frac{d^2 x}{dt^2} = -\omega^2 x$$

can be put in the form we have been considering,

$$x(t) = A \cos(\omega t + \delta) \,, \quad A \geq 0 \,.$$

5.2.4   Constancy of the frequency. The pendulum. We have already noticed that the angular frequency ω is determined by the spring constant and the mass of the particle. This has the interesting consequence that the frequency $\omega/2\pi$ is independent of the quantities A and δ, x_0 and v_0: all oscillations of a given spring-particle system, however they begin and whatever their amplitude, exhibit the same number of cycles per second.

The same phenomenon is to be seen in an ordinary pendulum, which, provided that the amplitude (half the breadth) of swing is short in comparison to the pendulum's length, will maintain a very nearly constant period as friction causes the amplitude to decay. One might conjecture that pendulum motion is approximately simple harmonic motion, and this is in fact so. A demonstration follows.

Let a particle of mass m be suspended by a massless cord or (better) massless inflexible rod of fixed length l free to pivot about the point of suspension (Fig. 5-11). If initially displaced from its point of rest the particle will swing back and forth in a vertical plane, its motion driven by its weight mg. The circular arc it follows will be almost straight if the displacement was much less than the radius l.

Not all of the particle's weight acts to impel it along the arc; most of it is occupied with pulling on the cord. In the figure I have represented the weight by arrow XW. A fundamental principle of physics implies that if the length of XW represents the magnitude mg of this downward force, the magnitude of the force it produces *in the direction of motion* is proportionately represented by the length of arrow XY, obtained by projecting XW perpendicularly onto the straight line in that direction (the tangent to the circular arc at X). If the angle of displacement OSX is θ, a little geometry shows that $\angle XWY$ is also θ, and therefore the length of XY is $mg\sin\theta$. This is the magnitude of the effective force on the particle tending to restore it to position O.

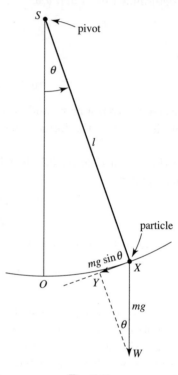

Fig. 5-11

In the situation pictured θ is positive, because the displacement of SX is counterclockwise from SO; the restoring force, which acts in the opposite direction, should then be assigned the negative value $-mg\sin\theta$. When X is on the other side of O, $\theta < 0$ and $-mg\sin\theta > 0$.

The signed length of arc OX is $l\theta$. Hence the ratio of restoring force to displacement is

$$\frac{-mg\sin\theta}{l\theta} = -\frac{mg}{l}\frac{\sin\theta}{\theta}.$$

This is not a constant $-c$ as required by the spring law (§5.2.1(1)); however, as $\theta \to 0$, $(\sin\theta)/\theta \to 1$ (§4.4.2(1)), so the ratio approaches $-mg/l$. For small values of θ, then, the motion approximates that of a particle driven by a spring with constant $c = mg/l$.

Such a motion has frequency

$$\frac{\omega}{2\pi} = \frac{1}{2\pi}\sqrt{\frac{c}{m}} = \frac{1}{2\pi}\sqrt{\frac{g}{l}}.$$

This quantity is independent not only of the amplitude of swing, as anticipated, but also of the mass of the particle. The remarkable fact is that the frequency is completely determined by l, the length of the cord. (See, however, Q11.)

5.2.5   QUESTIONS

Q1.   (a) What is the period of the simple harmonic motion $x = 3\cos(2t - 1)$ m? (b) What is the maximum speed it attains?

Q2.   A pendulum swings from one end of its arc to the other in 2 sec. Assuming conditions as in §5.2.4, how long is the cord?

Q3.   The quantity ω is on the one hand a number of radians/sec (Fig. 5-7) and on the other a number of cycles in 2π sec. Reconcile these interpretations.

Q4.   At what speed does P in Fig. 5-7 move? What relation does its speed bear to that of the particle?

Q5.   (a) Explain carefully how the signs of v and a vary for an oscillator. (b) Taking $A = \omega = 1$ in Fig. 5-7, find line segments that represent v and a as Ox represents x.

Q6 (§5.2.3).   Prove this. (To show $A = A'$, consider maximum values of the functions. For the second part see §3.3.9.Q7(b).)

Q7.   (a) (§5.2.3) Find an explicit expression for A in terms of x_0 and v_0 and thereby show that A and δ are determined. (b) What restrictions on x_0 and v_0 are imposed by an actual spring that cannot be compressed or extended more than 0.3 m without its elasticity being affected?

Q8.   (a) Show that $\alpha\cos\omega t + \beta\sin\omega t$, where α and β are arbitrary constants, is a solution of equation (2), §5.2.1. (b) Is the solution $A\cos(\omega t + \delta)$ of this type? (c) Can every solution of the type in (a) be put in the form $A\cos(\omega t + \delta)$?

Q9.   Verify by direct differentiation that the force on the oscillator $x = A\cos\omega t$ at any time t is the time derivative of its momentum $p = mv$ (§2.8.4).

Q10.   Find the mean acceleration of the oscillator $x = A\cos\omega t$ over the quarter-cycle starting at $t = 0$. (See §3.7.16. Here the independent variable is time.)

Q11.   (a) Take a pendulum to the moon; how will its behavior change? (b) What "system" really constitutes a pendulum, as the system of movable mass + fixed spring constitutes the spring-driven oscillator?

Q12.   (a) Design and carry out a pendulum experiment to measure g. With very crude equipment you can get it within a few percent. (b) How should the period vary with the length? Test this.

Q13.   Reinvent the pendulum clock.

5.3   ENERGY

5.3.1   Motion as a result of difference of height.   Why does a body fall? The discussion of falling in Art. 5.1 was based on the answer, "Because

it is heavy"—invoking *weight*, a kind of force. Description of the fall followed from the relation $f = ma$, which states in general the exact connection between variable force and motion.

The question has another reasonable answer, to wit: "Because it is high up"; for things tend to move from a higher place to a lower one. It is natural then to try to generalize *height*, as weight has been generalized to force, in such a way that motion can be accounted for and accurately described by use of the new generalization. That is what I am about to do. We may anticipate that advantages will come with a new point of view, but should not expect different results, since the phenomena are the same as before.

To guide our investigation, we will begin as before with a particle subject to a given force. There is one limitation to observe: we will have to assume that *the force on the particle depends only on its location.* This is trivially true for constant gravitational force (Art. 5.1), by hypothesis the same everywhere and always—it depends on nothing but the particle's mass—and is true as well for the force exerted by a fixed spring (Art. 5.2). It will also hold for the variable gravitational force to be considered in the next article. But a frictional force such as air resistance, which depends on velocity (§5.1.3.Q9), does not satisfy the condition, nor does a time-varying force such as that of an electromagnet powered by a changing current. Since our interest at present is in how relative location (e.g. height) determines motion, it is permissible to exclude forces like these from consideration.

> The assumption made here is strong enough for the present case of rectilinear motion. For more generality see the remark at the end of §7.5.3.

5.3.2   Potential energy.

Suppose, then, that a particle of mass m is subject to force $f(x)$, a function of x alone. Following the idea of height we seek a way of assigning a number to each location on the x-axis in such a fashion that the force is directed from higher to lower numbers. Besides the *direction* of the force, the assignment of numbers should take account of its *magnitude*. A body that falls on the moon, in order to attain the same velocity it would reach by falling on the earth, must fall farther; if our generalization of height is to explain velocity, the numbers must change more slowly with vertical distance on the moon than on the earth. In general, the numbers should change rapidly with x where $f(x)$ is large in magnitude, slowly where it is small—as the height of a hill falls off quickly where the downward slope is steep, slowly where it is gentle—so as to represent by their pattern the whole variation of the force.

Formulating the problem in this way suggests the following definition. Let $\phi(x)$ be any integral of $-f(x)$,

$$\phi(x) = \int -f(x)\,dx\,;$$

then ϕ assigns a number to every x, and

$$\phi'(x) = -f(x)\,,$$

or

$$f(x) = -\phi'(x)$$

—the force is represented as the *negative of the rate of change* of numbers $\phi(x)$ along the axis. The negative sign makes the direction of force be the direction of decreasing $\phi(x)$, from higher to lower numbers.

As a test of this solution let us apply it to the case of constant gravity, $f(x) = -mg$. An integral of $-f(x)$ is $\phi(x) = mgx$, which is proportional to the height x above the earth's surface (the coordinate introduced in §5.1.1). Height is exactly what we intended to generalize; as for the constant of proportionality, which is the magnitude of the force, besides fitting the given mass it takes care of that difference between earth and moon alluded to a moment ago. The moon's gravity is one sixth as strong as the earth's, i.e. $g_{moon} = \frac{1}{6}\,g_{earth}$, so the scales of numbers $\phi(x)$ compare as shown in Fig. 5-12.

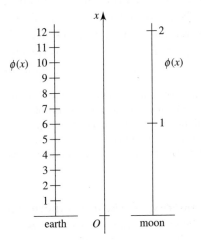

Fig. 5-12

What if we choose a different integral of mg, one which is not zero at $x = 0$? Since integrals of the same function differ by a constant, the only effect of the change is to shift the scale of numbers $\phi(x)$ by a constant. But what explains falling is *difference* of height, not height itself—the location of the zero of coordinates is of no importance when distance of fall is to be calculated. So the revised scale is as good for our purpose as the other.

To return to the general situation, we call $\phi(x) = -\int f(x)\,dx$ the **potential energy** of the particle—more correctly, of the system composed of the particle and the source of the force $f(x)$ (the earth, a spring, or something else). The name will be explained in the next section. As in the special case of gravity, potential energy depends on the choice of integral: a different choice means that $\phi(x)$ is changed at every x by the same additive constant. This indeterminacy makes it clear that the value of the potential energy at a single point cannot by itself be a meaningful quantity. On the other hand the *change* in potential energy from one point x_1 to another point x_2 is independent of the choice of integral, for by the Fundamental Theorem

$$(1) \qquad \phi(x_2) - \phi(x_1) = -\int_{x_1}^{x_2} f(x)\,dx\,,$$

the same for every ϕ. In the gravitational example this is $mg(x_2 - x_1)$, proportional to the difference in height of the two locations.

It should be apparent therefore that we are really concerned with a *total* quantity, $\phi(x_2) - \phi(x_1) = \Phi([x_1, x_2])$ (for $x_1 < x_2$). Whatever potential energy is, it assigns numbers to *intervals* primarily; its definition as a function of points $\phi(x)$ is only for the sake of convenience, and involves an arbitrary choice. To take account of the direction of motion let us write $\Phi_{12} = \phi(x_2) - \phi(x_1)$ for any x_1, x_2 (i.e. whether or not $x_1 < x_2$): Φ_{12} is the change in ϕ from x_1 to x_2. A similar notation will be used for other total quantities. (Cf. §2.10.3.)

> If the gravitational potential energy of a particle at height x is mgx, the potential energy *per unit mass* is gx, an integral of acceleration g rather than force mg. Being independent of the mass, this quantity is more strictly analogous to height than is the potential energy. In the theory of inverse-square gravitation (Art. 5.4) the corresponding quantity is called the gravitational **potential**.

5.3.3   Potential energy, work, and kinetic energy.

Let a potential energy function $\phi(x)$ be chosen, and suppose at first that $\phi(x)$ decreases from $x = x_1$ to $x = x_2$. If the particle is initially at rest at x_1 it will be driven by the force "down" to x_2, and in the course of its motion will have work

$$W_{12} = \int_{x_1}^{x_2} f(x) \, dx \text{ done upon it.}[3] \text{ Then by (1) we have}$$

(2) $\phi(x_1) - \phi(x_2) = W_{12}$

—the amount by which the potential energy of the particle at x_1 exceeds that at x_2 is the amount of work done on the particle as it moves from x_1 to x_2.

The work W_{12} is equal to the increase in the particle's kinetic energy, namely

$$\frac{1}{2} m v_2^2 - \frac{1}{2} m v_1^2 = k(x_2) - k(x_1) = K_{12} \, ,$$

where v_1 and v_2 are the velocities at the respective points. Consequently

(3) $\phi(x_1) - \phi(x_2) = K_{12}$

—the amount by which the potential energy of the particle at x_1 exceeds that at x_2 is the amount of kinetic energy gained by the particle in moving from x_1 to x_2.

These arguments do not need the hypothesis that ϕ is decreasing, nor the assumption that the particle starts from rest. Whenever the particle moves from a point x_1 to a point x_2, W is the integral of f and the increment of k, and therefore (2) and (3) are valid. Nevertheless there is reason to distinguish two cases.

(i) If $\phi(x_1) > \phi(x_2)$ (as was the case under the initial hypothesis), the earth, or spring, or whatever else produces the force $f(x)$, does positive work on the particle equal to the decrease in potential energy, and the kinetic energy of the particle increases by the same amount. This happens when a body falls from x_1 to x_2, for example. Since the loss of potential energy equals the gain in kinetic energy, we interpret the event as an *exchange* of potential for kinetic energy, or a *conversion* of one into the other. The falling body swaps height, the sign of potential energy, for speed, which confers kinetic energy.

(ii) If $\phi(x_1) < \phi(x_2)$, the earth or spring does negative work on the particle equal in magnitude to the *increase* in potential energy, and the kinetic energy of the particle *decreases* by the same absolute amount. This is the case when a body *rises* from x_1 to x_2. The cost of the gain in potential energy is the equal loss of kinetic energy: kinetic energy is exchanged for, or converted into, potential energy. Thus a body projected upward slows as it rises.

---

[3]For work and kinetic energy see §§1.4.7–8, and also §§2.3.4, 2.10.2, 2.10.7.

In the light of this discussion I can say something about the meanings of our terms. "Energy" may be understood in terms of force as the *capacity to do work*. A particle acquires a quantity of kinetic energy as an equal quantity of work is done upon it, and in giving up that kinetic energy it can do the same amount of work on another particle (§2.10.9.Q12). When a particle moves from a location x_1 where its potential energy is greater to a location x_2 where it is less, work equal to the potential energy difference $\phi(x_1) - \phi(x_2)$ is done upon it to accelerate it, and that quantity of potential energy is thereby used up. Both kinds of energy therefore deserve the name. There is however a difference in the ways in which they are associated with bodies. While kinetic energy is thought of as localized in the moving body (particle), potential energy belongs to a *system* of bodies—particle and earth, particle and spring, etc.—of which the *configuration* determines the capacity to do work on its members, as e.g. greater separation of particle and earth means higher potential energy. As such this energy can be regarded as latent in the system, hence "potential."

Although localized in the separate moving parts of a system, kinetic energy can be transferred between them. When a spring previously at rest is being compressed by a sliding block its parts are moving, and so have non-zero kinetic energy, which must have come from the block; thus some of the block's kinetic energy is turned into kinetic rather than potential energy of the spring. Only if the mass of the spring is negligible in comparison to that of the block can we regard all of the spring's energy as potential. The assumption of negligible spring mass, which we can adopt in our discussion here, is cousin to the tacit understanding that the gravitational attraction exerted by a projectile does not appreciably move the earth—the earth has negligible velocity, hence negligible kinetic energy.

5.3.4   Conservation of energy.   In terms of the total quantities K and Φ equation (3) can be written

$$-\Phi_{12} = K_{12}$$

—the increments of kinetic and potential energy associated with a change of place are negatives of each other. They always sum to zero:

$$K_{12} + \Phi_{12} = 0 \,.$$

Another way of putting this is to rewrite (3) as

$$k(x_1) + \phi(x_1) = k(x_2) + \phi(x_2) \,,$$

which implies that $k(x) + \phi(x)$ is the same for all values of x, or

$$(4) \qquad\qquad\qquad k(x) + \phi(x) = E$$

for some constant E. *The sum of the kinetic energy and the potential energy is constant.* This is the principle of **conservation of energy**. The constant E is called the **total energy** of the particle.[4] As the particle moves, its total energy is conserved (does not change), although its kinetic and potential energies swap about. Of course, the value of E depends on the choice of ϕ; as a matter of fact, it also depends on something we have taken for granted, the choice of a "stationary" background relative to which v, which determines k, is measured. But once these choices are made, that value is fixed. Hence if we know both k and ϕ at *one* point we can find E, and then can use E to determine k from ϕ or ϕ from k at any *other* point. The power of this idea will soon be demonstrated.

5.3.5   Uniform gravitation.   In order to apply the considerations of the last two sections to uniform gravitation (Art. 5.1), we will use the potential energy function $\phi(x) = mgx$ introduced in §5.3.2.

A particle at rest at height x_0 has zero kinetic energy, but potential energy mgx_0; its total energy is then mgx_0. Let it fall to the earth ($x = 0$). When it reaches it, the potential energy has fallen to zero, because $\phi(0) = 0$, while the kinetic energy has risen to $\frac{1}{2}mv^2$, where v is the velocity with which the particle strikes the earth. Since the total energy is constant, we have

$$\frac{1}{2}mv^2 = mgx_0\,,$$

from which we can determine $|v|$ in terms of x_0:

$$|v| = \sqrt{2gx_0}\,,$$

in agreement with §5.1.2. In fact we know that $v = -\sqrt{2gx_0}$, because the particle moves from higher to lower potential energy, i.e. downward.

Suppose now that the particle falls from rest at x_0 to an arbitrary height x. The same argument leads to the expression

$$v = -\sqrt{2g(x_0 - x)}$$

---

[4]More accurately, E is the total *mechanical* energy, and the principle is of conservation of mechanical energy (in the limited circumstances of our discussion). A more general conservation principle applies in the presence of frictional and other forces.

for the velocity at x. Assuming that the particle leaves x_0 at $t = 0$ we can go on to find the relation between x and t, in the following way. We have

$$\frac{dx}{dt} = v = -\sqrt{2g(x_0 - x)},$$

therefore

$$\frac{dt}{dx} = -\frac{1}{\sqrt{2g(x_0 - x)}}$$

for $x < x_0$ (Q7(a)). The latter equation can be integrated:

$$t = \int \frac{dt}{dx}\, dx = \int -\frac{1}{\sqrt{2g(x_0 - x)}}\, dx = -\frac{1}{\sqrt{2g}} \int \frac{1}{\sqrt{x_0 - x}}\, dx$$

$$= -\frac{1}{\sqrt{2g}} \left(-2\sqrt{x_0 - x}\right) + C = \sqrt{\frac{2(x_0 - x)}{g}} + C.$$

As $t \to 0$, $x \to x_0$ and $\sqrt{\dfrac{2(x_0 - x)}{g}} \to 0$, so $C = 0$; then

$$t = \sqrt{\frac{2(x_0 - x)}{g}}, \quad x \leq x_0.$$

This expresses t as a function of x. The inverse function is found by solving for x:

$$x = -\frac{1}{2} gt^2 + x_0,$$

which again agrees with §5.1.2.

With respect to the preceding integration it should be noticed that we could have expressed t as a determinate function of x in the first place by writing

$$t = \int_{x_0}^x \frac{dt}{dx}\, dx = \int_{x_0}^x -\frac{1}{\sqrt{2g(x_0 - x)}}\, dx.$$

This is an *improper* integral (§§4.7.1–2) which upon evaluation yields the same expression for t (Q7(b)).

The general case, in which the particle leaves x_0 with velocity v_0, can be handled in the same way (Q8). In contrast to §5.1.2, force does not enter the argument. We obtain a differential equation for $x(t)$ by reasoning not about force but about integrals of force, namely kinetic and potential energy.

5.3.6   A differential equation of motion resulting from the principle of conservation of energy.   The method we have applied to uniform gravitation leads to a solution, in principle, to the general problem of determining motion from potential energy.

From the principle of conservation of energy

$$\frac{1}{2} mv^2 + \phi(x) = E \tag{5}$$

we obtain, by solving for $v = dx/dt$,

$$\frac{dx}{dt} = \pm\sqrt{\frac{2}{m}\left(E - \phi(x)\right)}\,, \tag{6}$$

where the sign depends on the direction of motion. Let us assume that the particle moves in one direction only, so that the $\pm$ represents one definite sign (if this is not the case, we may have to consider separately the intervals where v has constant sign). Then we have a determinate equation

$$\frac{dt}{dx} = \frac{1}{\pm\sqrt{\dfrac{2}{m}\left(E - \phi(x)\right)}}\,.$$

By integration we can express t as a function of x:

$$t = \int_{x_0}^{x} \frac{1}{\pm\sqrt{\dfrac{2}{m}\left(E - \phi(x)\right)}}\, dx\,, \tag{7}$$

where as usual x_0 is the location of the particle at time $t = 0$ (the integral vanishes at $x = x_0$). The function $x(t)$ is then the inverse of that function. If as in §5.3.5 the integration can be performed and equation (7) solved for x, we will have an explicit expression for $x(t)$.

Even when $x(t)$ cannot be found from (7), knowledge of $\phi(x)$ makes it possible to obtain information about $x(t)$ from (5) or (6), inasmuch as the derivative of a function indicates essential features of its behavior (Art. 3.9). Here, of course, the derivative of x is known as a function of x, not of t, but the information can be valuable nonetheless. A simple application of this idea will be made in the next section, and a more complex one in Art. 5.4, where (7) will also be used again.

5.3.7   Simple harmonic motion.   Recall the equation of simple harmonic motion,

$$(8) \qquad \frac{d^2x}{dt^2} = -\omega^2 x$$

(§5.2.1(2)). By applying $f = ma$ we convert it into the expression for the force,

$$f(x) = -cx \,,$$

where $c = m\omega^2$, with which we began in §5.2.1. The simplest choice for the potential energy function is then

$$\phi(x) = \frac{1}{2}\, cx^2 \,.$$

With that choice the total energy is

$$E = \frac{1}{2}\, mv^2 + \frac{1}{2}\, cx^2 \,,$$

a constant sum of two non-negative quantities.

This representation of E has an immediate consequence. Suppose that the particle is initially at rest at the center, $x_0 = v_0 = 0$. Then $E = 0$. Since the summands of E are non-negative, both must be zero at all times, so that $x = v = 0$ always.[5] The meaning of this is that *a particle initially at rest at the center will remain so*—a sufficiently obvious conclusion for spring-driven motion, here deduced from the differential equation by means of the principle of conservation of energy.

We are now able to prove in general that *a particle's motion is completely determined by its initial position and velocity.* Let functions $x_1(t)$ and $x_2(t)$ both satisfy equation (8), and suppose in addition that $x_1(0) = x_2(0)$ and $x_1'(0) = x_2'(0)$. I claim that $x_1(t)$ and $x_2(t)$ are the same function, i.e. $x_1(t) = x_2(t)$ for all t. In other words, two motions with the same initial location and velocity must agree. To put it still another way: a solution to the differential equation is uniquely determined by given initial conditions $x(0) = x_0$, $v(0) = v_0$, just as we found to be the case for the differential equation of uniform gravitation in §5.1.1.

The proof is as follows. Define a new function by $x(t) = x_1(t) - x_2(t)$. It is easily verified that $x(t)$ satisfies the differential equation and $x(0) = x'(0) = 0$.

---

[5]This conclusion can be reached by a similar argument if a different $\phi(x)$ is used.

By the result just established, $x(t) = 0$ for all t, and this is equivalent to $x_1(t) = x_2(t)$ for all t.

A corollary is the proposition asserted at the end of §5.2.3, that *every so-lution of the differential equation can be expressed in the form* $A\cos(\omega t + \delta)$, *where* $A \geq 0$. For let $x(t)$ be any solution, with initial location x_0 and velocity v_0. As we know from §5.2.3, there is a solution $A\cos(\omega t + \delta)$ which satisfies the same initial conditions. But this implies that it is the same solution as $x(t)$, so $x(t)$ admits the desired form.

The expressions for the energies as functions of time in terms of A and δ have considerable interest. Let

$$x(t) = A\cos(\omega t + \delta),$$

then

$$v(t) = -A\omega\sin(\omega t + \delta)$$

and we have

$$k(x(t)) = \frac{1}{2}mv^2 = \frac{1}{2}mA^2\omega^2\sin^2(\omega t + \delta) = \frac{1}{2}cA^2\sin^2(\omega t + \delta)$$

and

$$\phi(x(t)) = \frac{1}{2}cx^2 = \frac{1}{2}cA^2\cos^2(\omega t + \delta).$$

The total energy is then

$$E = k(x(t)) + \phi(x(t)) = \frac{1}{2}cA^2(\sin^2(\omega t + \delta) + \cos^2(\omega t + \delta)) = \frac{1}{2}cA^2.$$

Here we must remember that we *chose* $\phi(x)$ to be $\frac{1}{2}cx^2$, the only integral of $-f(x)$ that vanishes at $x = 0$. Taking this choice into account, we can say that *if the potential energy is zero at $x = 0$, the total energy is proportional to the square of the amplitude*. The kinetic and potential energies vary in time periodically between 0 and the total energy $\frac{1}{2}cA^2$.

5.3.8 QUESTIONS

Q1. As in §5.1.3.Q2, a body is dropped from a height of 20 m. Reasoning directly from the principle of conservation of energy (rather than relying on formulas we have derived from it), find v when $x = 10$ m.

Q2. A certain spring obeys the force law $f(x) = -\frac{4}{3}x$ within its elastic limit, 1.1 m of displacement. If a particle of mass 0.4 kg, attached to the spring, is propelled from the center at 2 m/sec, will the spring be stretched beyond the elastic limit? (Calculate the total energy, and compare with the potential energy at $x = 1.1$.)

Q3. On a pair of coordinate axes, of which the horizontal axis is the x-axis and the vertical axis is energy, graph the kinetic, potential, and total energies of a 1 kg particle governed by the equation of motion $d^2x/dt^2 = -2x$ and oscillating with amplitude 2 m.

Q4. An electrical device called a parallel-plate capacitor (Fig. 5-13) produces a constant force of magnitude α newtons per coulomb perpendicular to the plates in the space between them, tending to drive a positively charged particle from plate A to plate B and a negatively charged one in the opposite direction. What is the potential-energy difference between $x = a$ and $x = b$ for a particle of charge q coulomb? (Electric charge was introduced in §2.10.5.)

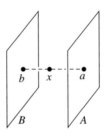

Fig. 5-13

Q5. How do (a) the total energy and (b) the maximum speed of a pendulum vary with its length and amplitude of swing?

Q6. (a) Deduce Newton's law $f = dp/dt$, where f is force and p is momentum, directly from the principle of conservation of energy. (Refer to §4.2.8.) (b) Conversely, obtain the conservation equation directly from Newton's law.

Q7. (a) (§5.3.5) Justify this step. How do we know $x(t)$ is invertible? (See §§4.3.2, 4.3.4.) (b) (§5.3.5) Relate the improper integral to Example 3 of §4.7.2, and evaluate it.

Q8 (§5.3.5). Work out the general case, starting from the principle of conservation of energy.

Q9. A particle subject to the equation of simple harmonic motion §5.3.7(8) is released at time $t = 0$ from rest at $x = A > 0$. (a) Show without using the calculation at the end of §5.3.7 that if the potential energy is zero at $x = 0$, the total energy is $\frac{1}{2}cA^2$. (b) Apply §5.3.6(7) to show that $x(t) = A\cos\omega t$.

Q10. (a) The potential energy of a particle at any x in some particular situation of rectilinear motion is a given function $\phi(x)$. The particle has total energy E. With no further information, can anything be said about where on the x-axis it is located

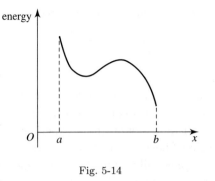

Fig. 5-14

at a given moment? (b) Let a portion of the graph of $\phi(x)$ have the form shown in Fig. 5-14. The particle, which has mass m, is released at some point x_0 between a and b with initial velocity $v_0 \geq 0$. Will it reach b?

5.4 INVERSE-SQUARE GRAVITATION

5.4.1 The inverse-square law and the equation of motion. The law of uniform gravitation near the earth's surface is only a local approximation to a more general law, according to which the magnitude of the force of gravity diminishes with increasing height. Let the positive x-axis be drawn outward from the center of the earth (not from the surface, as it was before). A particle of mass m located on or outside the earth at distance x from its center is attracted to the center by a force

$$f(x) = -\frac{GMm}{x^2}, \quad x \geq R,$$

where R is the radius and M the mass of the earth and G is a number called the **gravitational constant**. Approximate values of these quantities are:[6]

$$R = 6.4 \times 10^6 \text{ m} \quad (4000 \text{ miles}),$$

$$M = 6.0 \times 10^{24} \text{ kg} \quad (6.6 \text{ sextillion tons}),$$

$$G = 6.7 \times 10^{-11} \text{ newton m}^2/\text{kg}^2.$$

---

[6]The meaning of the unit in which G is expressed, if it has a meaning, is of no importance to us.

At the surface, where x is equal to R, the force is

$$f(R) = -\frac{GM}{R^2}\, m\,,$$

and calculation shows that $GM/R^2 = 9.8$ m/sec^2, the value of the acceleration of gravity g. Even at a height of two miles above the earth GM/x^2 is only about one tenth of one percent less than g (Q1); in consequence $f(x) = -gm$ is quite accurate over distances of ordinary experience.

The inverse-square law I have stated for $x \geq R$ results from the application to the earth (regarded as spherically symmetrical in respect of mass[7]) of Newton's law of universal gravitation, which tells us, among other things, that the earth's attraction for external bodies is the same as it would be if all the mass M were concentrated at the center. Inside the earth the attraction lessens with x, falling to zero at $x = 0$. Since we are not going to need this internal law, we can pretend that the inverse-square attraction holds at all points other than the center:

$$(1) \qquad\qquad f(x) = -\frac{GMm}{x^2}\,, \quad x > 0\,.$$

As the force law of a fictitious earth located at a single point this represents the point of view of celestial mechanics, for whose purposes the radius of the earth can be neglected, being insignificant in comparison to astronomical distances. Allowing x to approach zero is useful also in the study of electrical and other phenomena in which exactly similar inverse-square attractions (and repulsions) are produced by particles of the very smallest size (cf. Example 4, §7.5.3). In the vicinity of $x = 0$ the force is large in magnitude, and as $x \to 0$, $f(x) \to -\infty$.

By substituting the force law into $f = ma$—as usual, we neglect air resistance and other non-gravitational forces[8]—we obtain the differential equation of motion

$$(2) \qquad\qquad \frac{d^2x}{dt^2} = -\frac{GM}{x^2}\,, \quad x > 0\,,$$

in which the mass does not appear. As with the equation of simple harmonic motion (§5.2.1(2)) the x on the right prevents direct integration. Moreover, while there is again a solution that can be found easily (Q12), in this case

---

[7]That is, the earth is considered to be a body the density of which at any point P depends only on the distance of P from the center, as is the case for the sphere in §4.8.7.Q4.

[8]We also ignore the gravitational attraction of bodies other than the earth.

it is not a general one. Nevertheless, by means of energy arguments we can investigate the nature of the motion and even determine, if not explicit formulas for all the possible solutions $x(t)$ $(t \geq 0)$, at least formulas for their inverse functions.

5.4.2 Motion of a particle subject to inverse-square gravitation.
A. The energy equation. The simplest choice for a potential energy function corresponding to the force law (1) of the preceding section is

$$\phi(x) = -\frac{GMm}{x}, \quad x > 0.$$

This is a negative increasing function which approaches zero as $x \to \infty$ (Fig. 5-15)—we say that the potential energy is "zero at infinity."

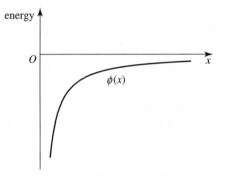

Fig. 5-15

With this choice the total energy of a particle is

$$E = \frac{1}{2}mv^2 - \frac{GMm}{x}.$$

Let the particle have initial location x_0 and velocity v_0. The total energy at any time equals the total energy at $t = 0$:

$$\frac{1}{2}mv^2 - \frac{GMm}{x} = \frac{1}{2}mv_0^2 - \frac{GMm}{x_0}.$$

Hence

(3)
$$v^2 = v_0^2 + 2GM\left(\frac{1}{x} - \frac{1}{x_0}\right).$$

Thus we have a fairly simple expression for v as a function of x (for graphs see Q10). By the method of §5.3.6 we can determine the inverse of $x(t)$ from this formula. Before undertaking any integration let us examine certain features of two special cases.

5.4.3 B. Fall and launch. Escape velocity.

The first case is that of a body falling from rest to the surface of the earth. We have $v_0 = 0$ and upon impact $x = R$, therefore

$$v^2 = 2GM \left(\frac{1}{R} - \frac{1}{x_0} \right),$$

where $x_0 \geq R$. A body falling from the height of one earth radius ($x_0 = 2R$), for example, would reach the ground with speed

$$|v| = \sqrt{2GM \left(\frac{1}{R} - \frac{1}{2R} \right)} = \sqrt{\frac{GM}{R}} = 7.9 \times 10^3 \text{ m/sec} = 18,000 \text{ mph}.$$

This is in the absence of air resistance, which would greatly slow the descent (§5.1.3.Q9). But since the existing atmosphere is almost entirely concentrated within a few tens of miles of the earth's surface (§6.2.10.Q20), the speed we have calculated would in fact be nearly attained in the region of the lower artificial satellites (Q2).

Now imagine dropping the body from greater and greater heights; i.e., let $x_0 \to \infty$. The velocity with which it reaches earth does not approach infinity, as one might expect, but instead we clearly have

$$v^2 \to \frac{2GM}{R},$$

so that

$$|v| \to \sqrt{\frac{2GM}{R}} = 1.1 \times 10^4 \text{ m/sec} = 25,000 \text{ mph}.$$

This limiting speed will be denoted by v_R.

The magnitude of the limiting speed can be derived by reflecting that the body acquires only so much kinetic energy as is supplied by the potential energy difference through which it falls. As $x_0 \to \infty$ that difference, which is $\phi(x_0) - \phi(R)$, approaches $-\phi(R) = GMm/R$; therefore $\frac{1}{2}mv^2 \to GMm/R$, which implies $|v| \to v_R$.

The second special case is of a projectile launched outward from the earth at velocity v_0. (A launch from above the atmosphere of the real earth would avoid the air resistance disregarded here.) From the symmetry of this situation with the case of a falling body we can conclude that the projectile will return to earth only if $v_0 < v_R$. For if it falls back after reaching some height, it will strike the earth with the same kinetic energy as it had when it left, so at the speed v_0; but we have just seen that when a body falls from any finite height its final speed is less than v_R.

The same conclusion can be arrived at more directly by asking whether the initial kinetic energy k_0 of the projectile is sufficient to do the work required to carry it off "to infinity," i.e. beyond all bounds. The work needed to attain height x is the potential energy difference $\phi(x) - \phi(R)$; if the projectile is to go beyond *every* finite height, then, its initial kinetic energy has to exceed $\phi(x) - \phi(R)$ for every x. Since $\phi(x)$ is an increasing function with limit 0, the smallest possible value for the kinetic energy is $-\phi(R) = GMm/R$, the potential energy difference "from R to ∞." Therefore the condition is

$$k_0 \geq -\phi(R) \,,$$

or

$$\frac{1}{2} m v_0^2 \geq GMm/R \,,$$

which is equivalent to $v_0 \geq v_R$ (Q6).

When this condition is fulfilled, the projectile has non-zero velocity at any height x it attains, because the amount of kinetic energy it uses up in reaching x is less than its initial kinetic energy:

$$\phi(x) - \phi(R) < -\phi(R) \leq k_0 \,.$$

And it does attain *every* height x, i.e.

$$\lim_{t \to \infty} x(t) = \infty \,,$$

as the following argument demonstrates.

> Suppose on the contrary that it does not reach some height x_1, but $x(t) < x_1$ for all t. Then the kinetic energy at any x is
>
> $$k(x) = k_0 - [\phi(x) - \phi(R)] > k_0 - [\phi(x_1) - \phi(R)] > 0 \,.$$
>
> Let $k_1 = k_0 - [\phi(x_1) - \phi(R)]$, and let $v_1 = \sqrt{2k_1/m}$. Then $\frac{1}{2} m v^2 = k(x) > k_1 = \frac{1}{2} m v_1^2$, so $v > v_1 > 0$ at all x: the velocity is always greater than v_1.

But this is impossible, since it implies that in less time than $(x_1 - R)/v_1$ the projectile will go distance $x_1 - R$ and so reach x_1.

Because v_R is the least launch speed with which a projectile can escape the earth's gravity it is called the **escape velocity**. To accelerate a mass of one gram from rest to this velocity we must do work upon it equal to the kinetic energy

$$\frac{1}{2} m v_R^2 = \frac{1}{2} (1 \times 10^{-3})(2GM/R) = 6.3 \times 10^4 \text{ joule}.$$

This is approximately the energy needed to lift a placid African bull elephant to a height of one meter (Q5).

The preceding analysis in no way depended on the value of R. Consequently there is an "escape velocity" for any initial position (location of launch) x_0, namely

$$v_{x_0} = \sqrt{\frac{2GM}{x_0}}$$

In terms of this quantity equation (3) can be put in the forms

(4) $$v^2 = v_0^2 - v_{x_0}^2 + \frac{2GM}{x}$$

and

(5) $$v^2 = v_0^2 + v_{x_0}^2 \left(\frac{x_0 - x}{x} \right).$$

5.4.4   C. Solutions for launch at escape velocity and fall. I will now apply integration to the cases $v_0 = v_{x_0}$ and $v_0 = 0$. The other possibilities are treated in Q13.

1. If $v_0 = v_{x_0}$ equation (4) becomes

$$v^2 = \frac{2GM}{x}.$$

Then

$$\frac{dx}{dt} = \sqrt{\frac{2GM}{x}},$$

so that

$$\frac{dt}{dx} = \sqrt{\frac{x}{2GM}}$$

and

(6) $$t = \frac{1}{\sqrt{2GM}} \int_{x_0}^{x} \sqrt{x}\, dx = \sqrt{\frac{2}{9GM}}\left(x^{3/2} - x_0^{3/2}\right).$$

If we put $\alpha = (9GM/2)^{1/3}$, this can be written

$$x^{3/2} = \alpha^{3/2}t + x_0^{3/2}$$

—the $\frac{3}{2}$ power of x is a linear function of t. By raising both sides to the power $\frac{2}{3}$ we obtain x explicitly:

$$x(t) = \left(\alpha^{3/2}t + x_0^{3/2}\right)^{2/3}.$$

As $t \to \infty$, $x(t)$ comes to resemble $\alpha t^{2/3}$, in the sense that

$$\frac{x(t)}{\alpha t^{2/3}} \to 1$$

(Q11). Thus for large t a projectile launched at escape velocity has approximately the motion of a particle whose distance from $x = 0$ is proportional to the $\frac{2}{3}$ power of time.

This confirms the earlier result that $\lim_{t\to\infty} x(t) = \infty$ when $v_0 \geq v_{x_0}$, since if the limit is ∞ for $v_0 = v_{x_0}$ it is surely the same when $v_0 > v_{x_0}$.

2. If $v_0 = 0$ equation (5) becomes

$$v^2 = v_{x_0}^2 \left(\frac{x_0 - x}{x}\right),$$

so that

$$\frac{dx}{dt} = -v_{x_0}\sqrt{\frac{x_0 - x}{x}},$$

the sign being negative because the particle is falling. Then

$$\frac{dt}{dx} = -\frac{1}{v_{x_0}}\sqrt{\frac{x}{x_0 - x}},$$

so

$$t = -\frac{1}{v_{x_0}} \int_{x_0}^{x} \sqrt{\frac{x}{x_0 - x}}\, dx.$$

By §4.7.3.Q7 the improper integral has the value

$$-x_0 \left[\sqrt{\frac{x}{x_0} \left(1 - \frac{x}{x_0}\right)} + \arcsin\sqrt{1 - \frac{x}{x_0}} \right].$$

Then

$$t = \frac{x_0}{v_{x_0}} \left[\sqrt{\frac{x}{x_0} \left(1 - \frac{x}{x_0}\right)} + \arcsin\sqrt{1 - \frac{x}{x_0}} \right],$$

or equivalently

(7) $$t = \frac{x_0^{3/2}}{\sqrt{2GM}} \left[\sqrt{\frac{x}{x_0} \left(1 - \frac{x}{x_0}\right)} + \arcsin\sqrt{1 - \frac{x}{x_0}} \right].$$

The inverse of this function is $x(t)$.

5.4.5 QUESTIONS

Q1 (§5.4.1). Verify this.

Q2 (§5.4.3). What would be the speed at a height of 100 miles?

Q3.   (a) How long would it take a body to fall to earth from rest at a height of one earth radius? (Apply formula (7), §5.4.4.) (b) Find how long it would take a body to fall to the center ($x = 0$) from rest at a distance of one earth radius from the center. Then show that the body would flout relativity by exceeding the speed of light $c = 3.0 \times 10^8$ m/sec, and determine where the transgression would occur.

Q4.   How long would a projectile launched from the earth at escape velocity take to reach the orbit of the moon (mean radius 3.8×10^8 m) in the absence of gravitational attraction by the moon?

Q5 (§5.4.3). Verify this.

Q6 (§5.4.3). Obtain this result by expressing the work required to carry the projectile off to infinity as an improper integral (§4.7.1) and evaluating the integral directly.

Q7.   (a) What is the total energy E of a projectile launched at escape velocity? Comment on its kinetic and potential energies. (b) Substitute E into §5.3.6(7) and show that the result is the same as the integral used to find $t(x)$ in §5.4.4 (case 1).

Q8.   Is there any way to escape earth's gravity without attaining escape velocity?

Q9.   Let a projectile be launched outward from $x = x_0$. (a) If $v_0 < v_{x_0}$, find the maximum height $\bar{x}$ the projectile attains. (b) If $v_0 \geq v_{x_0}$, find its limiting velocity.

Q10.   Sketch the graph of $v(x)$ in each of the three cases (a) $v_0^2 = v_{x_0}^2$, (b) $v_0^2 > v_{x_0}^2$, and (c) $v_0^2 < v_{x_0}^2$. Indicate the direction of motion along each graph. (Equation (4), §5.4.3, defines x as a function of v; first graph $x(v)$.)

Q11 (§5.4.4, case 1).   Prove this.

Q12.   (a) Find a solution to equation (2), §5.4.1, without integration, by trying a power of t: substitute $x = \alpha t^\beta$ into the equation and thereby determine the constants α and β. (b) For what values of t is the solution valid? Interpret it. (c) Obtain the solution of case 1, §5.4.4, by substituting $x = (\gamma t + \delta)^\beta$.

Q13 (§5.4.4).   Determine $t(x)$ when v_0 is not v_{x_0} or 0. (Treat separately the cases $v_0^2 < v_{x_0}^2$ and $v_0^2 > v_{x_0}^2$.)

The Differential Equation of Intrinsic Growth: the Exponential and Logarithmic Functions

INTRODUCTION

The exponential functions are prominent in mathematical science, partly because they describe a certain natural pattern of growth or decay. Analysis of that pattern will lead in this chapter to a definition of the functions. The first article is devoted to giving the concept of organic growth the form of a differential equation and investigating the nature of the solutions of that equation, by which growth must be represented. In the second article the equation is solved and the desired functions are defined together with the logarithms. A number of applications of these ubiquitous functions then follow.

I will usually allow "growth," "increase," etc., to stand for "growth (or decay)," "increase (or decrease)," etc. The other case should be mentally supplied, as can be done by a straightforward analogy.

6.1 INTRINSIC GROWTH

6.1.1 Extrinsic and intrinsic uniform growth. Let a homogeneous substance increase continuously in quantity—mass, volume, or some other measure. The growth is *uniform*, as we understood the term in Art. 1.2, if it is as described in §3.6.5; in particular, if either of the two following equivalent conditions ((iii) and (iv) from that section) is satisfied.

(1) In any equal periods of time the substance is increased by addition of the same quantity.

(2) The rate of increase (increase per time) is constant.

Denoting the rate of increase by a and the quantity present at time t by $u(t)$ we can express the second of these by the differential equation[1]

$$\frac{du}{dt} = a.$$

Integration shows at once that its solutions are the *linear* functions of the form

$$u(t) = u_0 + at,$$

where $u_0 = u(0)$, and a process so described is often called **linear** growth.

As an example of this kind of growth I have used the increase of water in a vessel being filled by a steady stream. The source of the uniform growth is outside the growing substance, and the growth may be called **extrinsic** to it, because the quantity added in any interval of time depends only on the length of that interval and not on the amount of substance present. Equivalently, the rate of growth, being constant, is unaffected by the growing substance. This is the idea of uniform growth we have accepted up to now.

There is another interpretation of the notion, which can likewise be summed up in two conditions.

(3)   In any equal periods of time the substance is increased in the same ratio—that is, by multiplication by the same number.

(4)   The rate of proportional increase (relative or percentage increase per time) is constant.

Setting aside (4) for the moment, let us consider the meaning of (3). If a substance growing in the way prescribed doubles in one hour it will double again in the next, and in any other one-hour period; similarly if it triples in an hour; so in general, if in any time interval of length T such a growing substance increases in the ratio $1:p$ it will increase in that same ratio $1:p$ in any other time interval of length T. It does not matter when the intervals occur: if the quantity increases in the ratio $1:2.7$ between $t = 3.0$ and $t = 4.2$, it increases in that same ratio not only over the succeeding t-interval of equal length, namely $[4.2, 5.4]$, but also over the interval from $t = 3.5$ to $t = 4.7$.

Growth of this type occurs when every part of a substance independently grows in the same way as every other part. Imagine for the sake of illustration

---

[1] Recall from the introduction to Chapter 5 that this term means an equation involving a derivative.

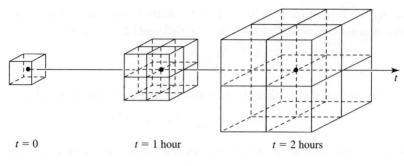

$t = 0$ $t = 1$ hour $t = 2$ hours

Fig. 6-1

a large body of tiny identical individual living cells, each of which divides after a short period of growth—an ideal colony of bacteria, say, amply provided with nutrients and undisturbed by accumulation of waste or debris. Suppose that in one hour it increases eightfold in volume (or mass). Each eighth of the enlarged colony is the same in every respect as the original colony, hence experiences the same increase in the next hour; the result is another eightfold increase in the whole, as Fig. 6-1 illustrates.[2] In general, growth of the colony by a factor p in time T is due to growth of each part by the same factor, and will therefore occur in any other time interval of equal length. In contrast to extrinsic uniform growth, the increase here arises out of the substance, so we may call this **intrinsic** uniform growth, or just intrinsic growth for short.

A colony of finite cells will satisfy (3) only approximately, one reason being that we cannot take parts smaller than a certain size without noticing deviation from identical behavior in given time. Nevertheless the approximation will be good if the number of cells is so large, and the time between cell divisions so short, that at any time cells in all stages of their life cycles are distributed densely throughout the colony.

6.1.2 The differential equation of intrinsic growth, and a solution.

In order to understand (4) we must be clear about the notion of proportional increase, which was introduced in §3.8.8 under the name of relative error. Let $q(t)$ denote quantity of substance, and suppose that at a given time $q(t) = 20$, while a bit later we find $q(t + \Delta t) = 25$. Then the increase in time Δt is

---

[2]Measurement will verify that the portions of a succeeding cube each equal the preceding cube, although an optical illusion makes them appear smaller.

$\Delta q = 5$, while the proportional or relative increase is

$$\frac{\Delta q}{q(t)} = \frac{5}{20} = 0.25 \,.$$

The corresponding percentage increase is 100 times that, or 25%. If instead $q(t) = 100$ and $q(t + \Delta t) = 105$ the increase Δq is still 5, but the proportional increase is only

$$\frac{\Delta q}{q(t)} = \frac{5}{100} = 0.05 \,,$$

and the percentage increase is 5%.

Now these are proportional increases in a given time Δt. To find the *rate* of proportional increase we divide by Δt and pass to the limit as $\Delta t \to 0$. We have

$$\frac{1}{\Delta t}\frac{\Delta q}{q} = \frac{1}{q}\frac{\Delta q}{\Delta t} \to \frac{1}{q}\frac{dq}{dt} \,;$$

this is the desired rate. In the language of infinitesimals: the proportional increase in time dt is dq/q, so the rate of that increase is

$$\frac{dq/q}{dt} = \frac{1}{q}\frac{dq}{dt} \,.$$

If we apply this idea to the linear function $q(t) = u(t) = u_0 + at$ of the last section, we find that while the rate of increase a is constant, the rate of proportional increase $\dfrac{1}{u}\dfrac{du}{dt} = \dfrac{a}{u_0 + at}$ is always falling. The reason is that the constant quantity added each second is added to a larger and larger existing quantity.

In contrast, what (4) describes is growth at a rate increasing just fast enough that the proportional increase per time is constant:

$$\frac{1}{q}\frac{dq}{dt} = \alpha$$

for some positive constant α, called the **growth constant**. Rewriting this as

(5)
$$\frac{dq}{dt} = \alpha q$$

we obtain an alternate form of (4):

The rate of increase is proportional to the quantity present.

This is exactly what we would expect in the case of a colony of cells.

Since (3) and (4) both describe the intrinsic growth of the colony, they would appear to be equivalent, like (1) and (2), but the fact still requires demonstration. Besides this, the analogy to extrinsic growth calls for an expression for $q(t)$, an explicit solution of the differential equation (5).

With knowledge of the exponential function we can supply the second of these missing elements by noting the obvious solution

$$q(t) = ce^{\alpha t},$$

where c is any constant; it is not hard to show that this solution is general (Q9). For our present purpose c will be positive, and substituting $t = 0$ shows that $c = q(0) = q_0$, so the solution is

$$q(t) = q_0 e^{\alpha t}.$$

However, we have yet to define the exponential function, so we cannot legitimately claim to know what the expression we have written for $q(t)$ means. The problem before us is then to show that our differential equation (5) *has* a suitable solution, that intrinsic uniform growth admits mathematical description by a differentiable function.[3]

6.1.3 Growth functions. A. Uniqueness.
Guided by the idea that intrinsic growth proceeds from the substance itself, we expect a solution of equation (5) to be a *positive, increasing function defined for all t*, $-\infty < t < \infty$. For substance arises only from previously existing substance, so if the quantity present is non-zero at some time t it can never have been zero *before* t; and existing substance always grows. Thus true intrinsic growth is eternal, unlike the temporary imitation of it by a physical colony of bacteria, the beginning and end of which belong to processes other than division of cells. We also understand the function to be *differentiable*, seeing that we have a differential equation for it.

If two such solutions have the same value at $t = 0$ they are identical. For let $q_1(t)$ and $q_2(t)$ be two such functions. They need not in fact be assumed positive or increasing, but one of them, say q_2, must be nowhere zero. Consider their quotient

$$f(t) = \frac{q_1(t)}{q_2(t)}.$$

---

[3]The reader interested only in applications should here turn to §6.2.5.

This is a constant function, because

$$f'(t) = \frac{q_1'(t)q_2(t) - q_1(t)q_2'(t)}{q_2(t)^2}$$

$$= \frac{\alpha q_1(t)q_2(t) - q_1(t)\alpha q_2(t)}{q_2(t)^2} = 0.$$

If $q_1(0) = q_2(0)$, the constant value of $f(t)$ is

$$f(0) = \frac{q_1(0)}{q_2(0)} = 1.$$

Consequently $q_1(t) = q_2(t)$ for all t: the two functions are the same.

6.1.4   B. Equivalence of defining conditions. Now let $q(t)$ be any positive, increasing, differentiable function defined for all t. Regarding $q(t)$ as quantity of substance, let us prove that *if q satisfies (3) it satisfies (4), and conversely.* To this end we may assume that *the quantity present at $t = 0$ is one unit, $q_0 = 1$.* For if it is not, we divide the given function by q_0; the resulting function $q(t)/q_0$ does have the value 1 at $t = 0$, and it is easy to see that the new function satisfies (3) or (4) if and only if the given one does. To set $q_0 = 1$ is after all no more than to choose a scale, or unit of quantity.

As a preliminary to the proof, observe that (3) can be expressed by a proportion: for all u and v,

$$q(u) : q(u + v) = q(0) : q(v).$$

Here if $v > 0$ both sides represent the ratio of increase in time v, while if $v < 0$ the inverse proportion

$$q(u + v) : q(u) = q(v) : q(0)$$

has a similar meaning. (The case $v = 0$ is trivial.) Now since we are assuming $q(0) = 1$, an equivalent expression is

(6) $$q(u + v) = q(u)q(v).$$

This is then a new formulation of (3): the function *transforms sums into products.*

The equivalence of (3) and (4) can now be demonstrated by proving (6) equivalent to (5).

Suppose first that (6) holds. The derivative of $q(t)$ is the limit as $\Delta t \to 0$ of

$$\frac{q(t + \Delta t) - q(t)}{\Delta t} = \frac{q(t)q(\Delta t) - q(t)}{\Delta t} \qquad \text{(by (6))}$$

$$= \left[\frac{q(\Delta t) - 1}{\Delta t} \right] q(t) = \left[\frac{q(\Delta t) - q(0)}{\Delta t} \right] q(t),$$

and that limit is $q'(0)q(t)$. This proves (5), with $q'(0)$ as the constant α.

Conversely, given that q is a solution of (5), let s be any fixed number and consider the functions $q_1(t) = q(s + t)$ and $q_2(t) = q(s)q(t)$. They satisfy (5), are nowhere zero, and have the same value at $t = 0$; hence by §6.1.3 they are the same function. Since s was arbitrary, we have $q(s + t) = q(s)q(t)$ for all s and t, which is the same as (6).

Now that it is established that conditions (3) and (4) equally describe intrinsic uniform growth, let us define a **growth function** to be any positive, increasing, differentiable function, defined for all t, that satisfies these conditions. Then given positive numbers α and q_0 we know that there is at most one growth function $q(t)$ such that

$$q'(t) = \alpha q(t)$$

and

$$q(0) = q_0 .$$

Furthermore, if $q_0 = 1$ then (6) holds for all u and v.

6.1.5   A special growth function. Assume for the moment that a growth function exists in the special case $\alpha = q_0 = 1$, and let it be denoted by $p(t)$. Then

$$p'(t) = p(t)$$

and

$$p(0) = 1 .$$

Let a number e be *defined* by

$$e = p(1) .$$

I am going to prove that

$$p(t) = e^t,$$

if t is a *rational* number—the only kind whose use as an exponent has been justified (§3.3.2). In fact, for later use I will prove the more general proposition

(7) $$p(t\alpha) = p(\alpha)^t,$$

where α may be any number; it reduces to the former one when $\alpha = 1$.

The proof is in two parts.

(i) First we show that for any integer n,

$$p(n\alpha) = p(\alpha)^n.$$

This is obvious if $n = 0$ or 1. For $n > 1$ it holds because (by (6))

$$p(2\alpha) = p(\alpha + \alpha) = p(\alpha)^2,$$

$$p(3\alpha) = p(2\alpha + \alpha) = p(2\alpha)p(\alpha) = p(\alpha)^3,$$

and so forth. Finally, if $n < 0$ then $-n$ is a positive integer, so that by what has just been proved

$$p(-n\alpha) = p(\alpha)^{-n} = \frac{1}{p(\alpha)^n};$$

but

$$1 = p(0) = p(n\alpha + (-n\alpha)) = p(n\alpha)p(-n\alpha),$$

so

$$p(n\alpha) = \frac{1}{p(-n\alpha)} = p(\alpha)^n.$$

(ii) Now let $t = m/n$, where m and n are integers, $n > 0$. Then

$$p(t\alpha)^n = p(nt\alpha) \quad \text{by part (i), with } t\alpha \text{ in place of } \alpha$$

$$= p(m\alpha)$$

$$= p(\alpha)^m \quad \text{by part (i)}.$$

Taking nth roots we obtain

$$p(t\alpha) = p(\alpha)^{m/n} = p(\alpha)^t,$$

q.e.d.

With this result in hand, once $p(t)$ has been shown to exist we will be justified in making the *definition*

$$e^t = p(t)$$

for all irrational numbers t. Indeed, we will know that $p(t)$ is the only possible continuous extension of the function e^t to irrational values of t (§3.1.6.Q5).

Another consequence of (7) is that $p(t)$ assumes all positive numbers as values. Thus besides being eternal, this intrinsic growth produces every possible quantity of substance. The proof is indicated in Q11; I omit it here, because the result will be gotten by other means in §§6.2.1–2.

The existence of $p(t)$ will be established in the next article. When this is done we will have the general growth function as well, for if $\alpha > 0$ the function defined by

$$q(t) = q_0 p(\alpha t)$$

will clearly be positive and increasing everywhere and satisfy $q'(t) = \alpha q(t)$ and $q(0) = q_0$.

6.2 THE EXPONENTIAL AND LOGARITHMIC FUNCTIONS AND THEIR APPLICATIONS

6.2.1 Solution of the special growth equation. The natural logarithm and the exponential function.

In the first article we were led to look for a positive, increasing function $p(t)$, defined for all t, such that $p' = p$ and $p(0) = 1$. To solve the differential equation for p we can use the device of *inversion* already employed in Chapter 5 (§§5.3.5, 5.3.6, 5.4.4). From

$$\frac{dp}{dt} = p(t),$$

where $-\infty < t < \infty$ and $p > 0$, we obtain

$$\frac{dt}{dp} = \frac{1}{p},$$

and then by integration

$$t(p) = \int_1^p \frac{1}{p}\, dp.$$

The lower limit has been chosen to make $t(1) = 0$, corresponding to $p(0) = 1$. The desired function $p(t)$ is the inverse of this function.

The expression for t as a function of p rests on the assumption that $p(t)$ exists as described. But that is only a hypothesis; we have not verified the existence of a growth function. Let us therefore adopt the integral formula as the *definition* of a new function $t(p)$. We can of course recognize this function as the natural logarithm, now defined for the first time. Writing it as a function of t, we have by definition

$$\log t = \int_1^t \frac{1}{t}\, dt, \quad t > 0.$$

The derivative of $\log t$ is the positive function $1/t$. Consequently the logarithm is increasing, and so invertible. Let $p(t)$ denote its inverse. Does $p(t)$ have the expected properties? It is positive because its inverse $\log t$ is defined only for $t > 0$, and increasing because $\log t$ is. Clearly $p(0) = 1$, and $p' = p$ follows by reversing the steps that led us to the definition of the logarithm. In detail: if $p = \log^{-1} t$ (log inverse of t) we have $t = \log p$, so $dt/dp = 1/p$, and therefore $dp/dt = p$. Then the only property still to verify is that $p(t)$ is defined for all t. This is equivalent to the statement that the logarithm takes all numbers as values; and since the logarithm is increasing, that is equivalent to the statements

(1) $\lim_{t\to\infty} \log t = \infty \quad \text{and} \quad \lim_{t\to 0} \log t = -\infty$.[4]

(The latter equivalence follows from the intermediate value property (Q10).) These will be proved in the next section.

Given (1), we have shown that $p(t)$ is the special growth function we were seeking, and the definitions

$$e = p(1)$$

and

$$e^t = p(t)$$

proposed in §6.1.5 are justified. A way of calculating e^t is given in §6.2.4, and another one in §8.2.7. The graphs of e^t and $\log t$ appear in Fig. 4-16(a)

---

[4]The second limit is as $t \to 0$ from the right.

(§4.3.1). Since $\log t$ is defined for all positive t, the range of values of e^t is all the positive numbers, as was asserted in §6.1.5.

Note that once p is known to be defined everywhere, the law

(2) $$\log uv = \log u + \log v$$

follows from §6.1.4(6) for p, because $p(\log u + \log v)$ is defined and equal to $p(\log u)p(\log v) = uv$ (cf. §3.3.11).

6.2.2   Limits of $\log t$.

The first limit of (1), as $t \to \infty$, can be obtained directly from the expression for $\log t$ as an integral. Since $\log t$ is increasing, it is enough to find a sequence of t-values t_n such that $\log t_n \to \infty$. We let $t_n = 2^n$ and argue as follows.

The value $\log t_n$ is the integral of $1/t$ over the interval from 1 to t_n. Partition that interval for $n > 1$ by the points of division $t_1, t_2, \ldots, t_{n-1}$; then $\log t_n$ is the sum of the integrals of $1/t$ over the subintervals $[1, t_1], [t_1, t_2], \ldots, [t_{n-1}, t_n]$. I will show that each of these integrals is $\geq 1/2$. Since there are n of them, it will follow that $\log t_n \geq n/2$, which $\to \infty$ with n.

The integral estimates rely on §3.7.2(1). On the first subinterval $[1, t_1]$ or $[1, 2]$, which has length 1, we have $t \leq 2$, hence $1/t \geq 1/2$ and

$$\int_1^{t_1} \frac{1}{t}\, dt \geq \int_1^2 \frac{1}{2}\, dt = \frac{1}{2} \cdot 1 = \frac{1}{2}.$$

On the second subinterval $[t_1, t_2]$ or $[2, 4]$, which has length 2, we have $t \leq 4$, hence $1/t \geq 1/4$ and

$$\int_{t_1}^{t_2} \frac{1}{t}\, dt \geq \int_2^4 \frac{1}{4}\, dt = \frac{1}{4} \cdot 2 = \frac{1}{2}$$

(Fig. 6-2: each of the two shaded blocks has area $1/2$). And in general, on the ith subinterval $[t_{i-1}, t_i]$ or $[2^{i-1}, 2^i]$, which has length 2^{i-1}, we have $t \leq 2^i$, hence $1/t \geq 1/2^i$ and

$$\int_{t_{i-1}}^{t_i} \frac{1}{t}\, dt \geq \int_{2^{i-1}}^{2^i} \frac{1}{2^i}\, dt = \frac{1}{2^i} \cdot 2^{i-1} = \frac{1}{2}.$$

The second limit of (1) is reduced to the first by a transformation of the integral. In $\log t = \int_1^t \frac{1}{u}\, du$ (introducing a new letter for the sake of clarity) substitute $v = 1/u$ and transform the limits of integration (§4.6.3). We have

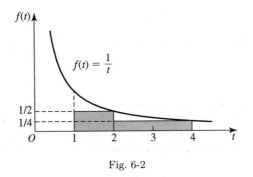

Fig. 6-2

$dv = -(1/u^2)\,du = -v^2\,du$, so the integral becomes $-\displaystyle\int_1^{1/t} \frac{1}{v}\,dv$, which is to say that $\log t = -\log(1/t)$. Then as $t \to 0$ through positive values, $1/t \to \infty$ and therefore $\log t \to -\infty$.

A different proof of these results is suggested in Q13(b).

6.2.3   The number e as the limit of sequences. A. Sequences and limit. The number e is the limit of the *increasing* sequence $(1+(1/n))^n$ and also of the *decreasing* sequence $(1-(1/n))^{-n}$. That is, for any $n > 1$

$$\left(1+\frac{1}{n}\right)^n < e < \left(1-\frac{1}{n}\right)^{-n},$$

and both sides approach e monotonically. Calculation yields the following inequalities. (The values for $n > 2$ are not exact: those on the left are rounded down, those on the right rounded up.)

$n = 2$	$2.25 < e < 4$
$n = 10$	$2.59 < e < 2.87$
$n = 100$	$2.70 < e < 2.74$
$n = 1000$	$2.716 < e < 2.720$
$n = 10000$	$2.7181 < e < 2.7185$

To prove that the sequences approach e, write the derivative of $\log(1+x)$ at $x = 0$, whose value is 1, as the limit of a difference quotient, with $h = \Delta x$:

$$\lim_{h \to 0} \frac{\log(1+h)}{h} = 1.$$

Put $h = 1/k$; then $h \to 0$ as $k \to \infty$ or $k \to -\infty$. Thus in either case $k\log(1+(1/k)) \to 1$. Since e^x is continuous, $e^{k\log(1+(1/k))} \to e^1 = e$.

If we restrict k to rational values, §6.1.5(7) implies that $e^{k \log (1+(1/k))} = (1 + (1/k))^k$, so for such k, $(1 + (1/k))^k \to e$. Now if we restrict k first to positive and then to negative integral values, we find that as $n \to \infty$, $(1 + (1/n))^n \to e$ and $(1 - (1/n))^{-n} \to e$.

6.2.4 B. Monotonicity. Theorem of the means. Generalization to e^t.

It remains to show that the sequences are monotonic. This is a consequence of the notable theorem of the arithmetic and geometric means. The *arithmetic mean* of n non-negative numbers $a_1, a_2, \ldots, a_n$ is (as we know)

$$A = \frac{a_1 + a_2 + \cdots + a_n}{n} \, ;$$

the **geometric mean** of the numbers is

$$G = (a_1 a_2 \cdots a_n)^{1/n} \, .$$

Theorem. *Given $n > 1$ non-negative numbers, not all equal, their geometric mean is less than their arithmetic mean.*

I will prove this in a moment. Assuming its truth, let a be any non-negative number not equal to 1, and apply the theorem (with $n + 1$ in place of n) to the set of $n + 1$ numbers consisting of 1 and n a's. It yields

$$(a^n)^{1/(n+1)} < \frac{na + 1}{n + 1} \, ,$$

so that

$$a^n < \left(\frac{na + 1}{n + 1} \right)^{n+1} \, .$$

Now let $a = 1 + (1/n)$:

$$\left(1 + \frac{1}{n} \right)^n < \left[\frac{n \left(1 + \frac{1}{n} \right) + 1}{n + 1} \right]^{n+1} = \left(1 + \frac{1}{n + 1} \right)^{n+1} \, ,$$

which means that the lower sequence increases. Again, let $a = 1 - (1/n)$:

$$\left(1 - \frac{1}{n} \right)^n < \left[\frac{n \left(1 - \frac{1}{n} \right) + 1}{n + 1} \right]^{n+1} = \left(1 - \frac{1}{n + 1} \right)^{n+1} \, ,$$

and taking reciprocals

$$\left(1 - \frac{1}{n}\right)^{-n} > \left(1 - \frac{1}{n+1}\right)^{-(n+1)},$$

so that the upper sequence decreases.

A slight generalization of these arguments shows that for any $t > 0$, e^t is the limit of the increasing sequence $(1 + (t/n))^n$ and of the decreasing sequence $(1-(t/n))^{-n}$; the same then follows for $t < 0$ by way of the identity $e^{-t} = 1/e^t$ (Q23).

The proof of the theorem uses the obvious facts that (i) at least one of the a_i is $< A$ and at least one is $> A$ and (ii) the geometric mean of n *equal* numbers a is equal to a.

Suppose (by (i)) that $a_1 < A$ and $a_2 > A$, and in the set of numbers a_1, $a_2, \ldots, a_n$ replace a_1 by $a_1' = A$ and a_2 by $a_2' = a_1 + a_2 - A$. Since $a_1' + a_2' = a_1 + a_2$, the arithmetic mean is unchanged by this replacement. But the geometric mean is increased, because in view of the following inequality the product of the n numbers is now larger:

$$a_1' a_2' - a_1 a_2 = A(a_1 + a_2 - A) - a_1 a_2 = (A - a_1)(a_2 - A) > 0.$$

By repeating this replacement process, if necessary, we arrive after not more than $n-1$ steps at a set of n numbers all equal to A, having a geometric mean larger than G. But (by (ii)) their geometric mean is A; thus $G < A$, q.e.d.

6.2.5   The general growth function and the exponential and logarithmic functions.   We said at the end of Art. 6.1 that from $p(t) = e^t$ the general growth function is obtained as

$$q(t) = q_0 p(\alpha t) = q_0 e^{\alpha t}.$$

Here α, the growth constant or rate of proportional increase (§6.1.2), is positive. But there is no reason why in the expression for $q(t)$ we cannot take $\alpha = 0$ or $\alpha < 0$. In the former case $e^{\alpha t} = e^0 = 1$, so $q(t) = q_0$: instead of growing the quantity remains constant. If $\alpha < 0$ there is negative increase, which means decrease; and as $t \to \infty$, $\alpha t \to -\infty$, so $q(t) \to 0$. The quantity of substance diminishes beyond any limit of smallness that can be named, without ever becoming zero. In this case α is often called the **decay constant** instead of the growth constant.

Let us now generalize the exponential function to an arbitrary base. If a is any positive number, there is a unique number α such that

$$a = e^\alpha,$$

namely

$$\alpha = \log a.$$

Clearly α is positive, negative, or zero according as $a > 1$, $a < 1$, or $a = 1$. We *define*

$$a^t = e^{\alpha t}$$

for all t. By §6.1.5(7) this definition agrees with the elementary one for t rational, since the left side is $p(\alpha)^t$ and the right side is $p(\alpha t)$. If α is replaced by its expression in terms of a, the definition becomes

$$a^t = e^{t \log a}.$$

The function $\log_a t$ is now defined as the inverse of this function. Then

$$t = a^{\log_a t} = e^{(\log a) \log_a t},$$

and taking natural logarithms shows that

$$\log_a t = \frac{\log t}{\log a}.$$

The laws of exponents and the other expected properties of these functions are derivable from what has already been established (Q4, Q14, Q15).

By means of the generalized exponential function the growth formula

$$q(t) = q_0 e^{\alpha t}$$

can be written more concisely as

$$q(t) = q_0 a^t,$$

where $a = e^{\alpha}$. Substituting $t = 1$ shows that $a = q(1)/q(0)$, the ratio of the quantity present after one unit of time to the initial quantity. If, for example, a growing substance doubles each second, $a = 2$ and the growth function is $q(t) = q_0 2^t$; if a shrinking substance is halved each second, $a = \frac{1}{2}$ and $q(t) = q_0(\frac{1}{2})^t = q_0 2^{-t}$.

We have already noticed that the constant q_0 can be eliminated by adjusting the unit of quantity (§6.1.4). The same end can be attained by resetting the clock (altering the time-measurement by an additive constant), for the function $\bar{q}$ defined by $\bar{q}(t) = q(t - \log_a q_0)$ has the formula

$$\bar{q}(t) = a^t.$$

On account of these formulas, what we have called intrinsic uniform growth is known as **exponential** growth. Note that when a quantity grows exponentially, $q(t) = q_0 e^{\alpha t}$, its logarithm grows *linearly*:

$$\log q(t) = \log q_0 + \alpha t \,,$$

a linear function of time. This has the practical consequence that observed growth can be recognized as exponential in nature, and its constant α determined, by plotting the logarithm of quantity against time: a straight-line graph indicates exponential growth, and its slope is α (Q6).

6.2.6   The power functions. The general exponential function furnishes a definition for the *power functions* x^α that were introduced in §3.3.2. If x is positive and α is arbitrary, x^α is defined as the base x raised to the power α, i.e.

$$x^\alpha = e^{\alpha \log x} \,.$$

The laws of exponents show that this definition agrees with the elementary one for rational α, and the formula for the derivative of a power, which we have long taken for granted, is verified by the chain rule:

$$\frac{d}{dx}\left(x^\alpha\right) = e^{\alpha \log x} \cdot \alpha \cdot \frac{1}{x} = x^\alpha \cdot \frac{\alpha}{x} = \alpha x^{\alpha-1} \,.$$

6.2.7   Applications.   A. Bacterial growth and radioactive decay. Exponential growth and decay are found throughout nature, wherever a relation $q' = \alpha q$ arises. I will here illustrate this by a few diverse examples, which will bring the preceding discussion down to earth. Additional examples are presented in the questions.

EXAMPLE 1. The ideal colony of bacteria described in §6.1.1 increases eightfold in each hour. From this fact alone we can determine its volume or mass $q(t)$ at any time t in terms of the initial quantity q_0. For we know that

$$q(t) = q_0 a^t$$

for some positive a, and if time is measured in hours $a = q(1)/q_0 = 8$. Then

$$q(t) = q_0 8^t = q_0 2^{3t} \,.$$

Thus, for example, after 20 minutes the quantity present is $q(\frac{1}{3}) = 2q_0$: it has doubled.

The same problem can be approached by way of the formula

$$q(t) = q_0 e^{\alpha t}.$$

We have $e^\alpha = 8$, or $\alpha = \log 8 = 3 \log 2$. Then

$$q(t) = q_0 e^{(\log 8)t} = q_0 e^{(3 \log 2)t}.$$

This is of course equivalent to the earlier expression. A calculator gives $\log 8 \approx 2.08$, so

$$q(t) \approx q_0 e^{2.08t},$$

where t is measured in hours.

The time required for the colony of Example 1 to double was found to be $\frac{1}{3}$ hour. In general, the **doubling time** T for any exponential growth bears a simple relation to the growth constant α. From $t = 0$ to $t = T$ the quantity grows to $2q_0$, so

$$2q_0 = q_0 e^{\alpha T};$$

dividing by q_0 and taking logarithms gives

$$T = \frac{\log 2}{\alpha} \approx \frac{0.693}{\alpha}.$$

In the case of exponential *decay*, we speak analogously of the **half-life** τ of a substance, the time required for half of it to disappear. In this case $\alpha < 0$; let $\beta = -\alpha$, so that

$$q(t) = q_0 e^{-\beta t},$$

where $\beta > 0$. Then

$$\frac{q_0}{2} = q_0 e^{-\beta \tau}$$

and therefore

$$\tau = \frac{\log 2}{\beta} \approx \frac{0.693}{\beta}.$$

EXAMPLE 2. In the phenomenon of *radioactive decay* a substance loses mass as particles or rays or both are emitted from time to time by the disintegrating nuclei of its atoms. Whatever the cause of this process may be, it occurs in the same way in every part of the decaying matter independently; therefore the rate of decay is proportional to the quantity present,

$$q' = -\beta q\,,$$

where $\beta > 0$, and

$$q(t) = q_0 e^{-\beta t}\,.$$

The half-life of the radioactive substance is then

$$\tau = \frac{\log 2}{\beta}\,.$$

An instance of this is the gradual disappearance of carbon-14, an isotope of carbon which occurs as a certain definite proportion of the carbon present in any newly formed organic matter. As the carbon-14 is transmuted into nitrogen, the portion it makes of all the carbon diminishes, and by measurement the ratio q/q_0 can be determined at any time t not too distant from $t = 0$. In this way we can first determine β, or equivalently τ, experimentally, by using the formula $\log(q/q_0) = -\beta t$ with known t, and then conversely find the age t of a given organic specimen from the same formula with known β.

Experiment shows that for carbon-14 τ is about 5750 years. Suppose that 20% of the isotope is gone from an organic sample; how old is it? We have $q/q_0 = 0.8$, so

$$-\log 0.8 = \beta t = \frac{\log 2}{\tau} t \approx \frac{0.693}{5750} t$$

and calculation yields $t = 1850$ years.

In the two remaining examples the independent variable is not time.

6.2.8   B. Atmospheric pressure.   EXAMPLE 3. As one ascends from a depth under water, the **pressure**, or force exerted by the water on unit area of a surface in it (newton/m^2), decreases proportionately to distance of rise. The corresponding phenomenon in the case of the atmosphere is not analogous: air pressure falls off more rapidly than height increases, and in fact its decay is approximately exponential. This I will now demonstrate, under the rough hypothesis that the temperature and composition of the air are unaffected by altitude. The proof will show that it is the compressibility of the air that is responsible for the exponential law.

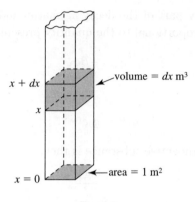

Fig. 6-3

Consider the column of air vertically above a square meter of earth at sea level. Let x be the height above the earth in the column, and let $\delta(x)$ (kg/m³) be the density of the air and $p(x)$ the pressure at height x (Fig. 6-3).[5] The volume of air from any x to $x + dx$ is equal to dx (m³), so its mass is $dm = \delta(x)\, dx$ and its weight is $dm \cdot g = g\delta(x)\, dx$, where g is the acceleration of gravity. This weight is the force it exerts on unit area below it; consequently as we rise from x to $x + dx$ that much pressure is lost, i.e.

$$dp = -g\delta(x)\, dx \,,$$

or

$$\frac{dp}{dx} = -g\delta(x)\,.$$

Now according to a form of Boyle's law, at constant temperature the pressure of a given gas is directly proportional to its mass and inversely proportional to its volume; that is, proportional to its density. Then

$$p(x) \propto \delta(x)$$

and it follows that

$$\frac{dp}{dx} = -\beta p(x)\,,$$

---

[5]It is assumed that x is so much smaller than the radius of the earth that the increase in column width with height can be neglected.

where β is some positive constant. Hence

$$p(x) = p_0 e^{-\beta x},$$

p_0 being the atmospheric pressure at sea level, which is one **atmosphere**, or about 1.0×10^5 newton/m^2 (15 lb/in^2). By measuring p at a known x we can determine β, and then if any p or x is specified, the other can be calculated.

Measurement shows that the atmospheric pressure falls to half its sea-level value at the height of about 3.4 miles (5.5×10^3 m). Then if distance is measured in miles we have $\beta = (\log 2)/3.4 \approx 0.20$. The pressure in atmospheres ($p_0 = 1$) at the top of Mt. Everest, 5.5 miles high, would be estimated at $e^{-\beta x} = e^{-(0.20)(5.5)} = 0.33$, a third of its sea-level pressure.

6.2.9   C. A self-similar shell.   EXAMPLE 4. The spiral shell of the living nautilus, of which the cross-section by its plane of symmetry is shown in Fig. 6-4, grows in *size* by addition to the outer end QR in such a way as to maintain the *shape* of the whole: it remains similar to itself (cf. §4.8.5.Q8). In order that the proportions may be preserved, it is necessary that a triangle OPQ, determined by any most recently added portion $PQRS$ of central angle $\Delta\theta$, be similar to every other triangle $OP'Q'$ corresponding to a portion $P'Q'R'S'$ with the same central angle, which formerly was the terminal portion. Then $OQ : OP = OQ' : OP'$, so the radius of the curve $P'Q'PQ$ is increased in the same ratio for any equal increments of the central angle. Applying characterization (3), §6.1.1, of the exponential function, we conclude

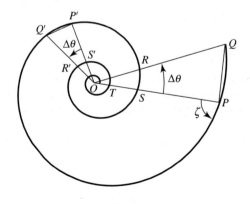

Fig. 6-4

that the equation of the curve in polar coordinates has the form

$$r = r_0 e^{\alpha\theta} \, ,$$

or

$$r = r_0 a^\theta \, ;$$

thus the curve is a *logarithmic* spiral (§3.3.12).

In this formula the radius r_0 is determined by the arbitrary choice of a radius vector to be assigned the angle $\theta = 0$. If it is eliminated from the expression by choosing that vector appropriately, the equation of the spiral becomes

$$r = e^{\alpha\theta} = a^\theta$$

(cf. the remarks in §6.2.5 on resetting the clock). The constant $\alpha = \log a$ fixes the shape of the curve; its value is $\cot \zeta$, where ζ is the constant angle between radius vector and curve (§3.3.13.Q9).

To determine α for a given shell, a good method is to measure the widths of successive whorls, as ST and PS; the direction to O will of course be a little uncertain, but can be identified with some accuracy even from a shell fragment (Q21). When θ increases by 2π, r is multiplied by $e^{2\pi\alpha}$; it follows that $PS/ST = e^{2\pi\alpha}$, and therefore

$$\alpha = \frac{1}{2\pi} \log \frac{PS}{ST} \, .$$

For *Nautilus pompilius* the ratio PS/ST is about 3, which yields $\alpha \approx 0.17$, or $\zeta \approx 80°$, as illustrated in Fig. 6-4.[6]

6.2.10   QUESTIONS

Q1.   Verify directly that the function $q(t) = q_0 e^{\alpha t}$ satisfies (3), §6.1.1.

Q2.   A colony of bacteria has mass 0.01 g at $t = 0$ and 0.03 g 20 seconds later. Assuming exponential growth, what will its mass be (a) at the end of 2 minutes? (b) at the end of 2.5 minutes?

Q3.   Find the half-life of radium, given that in 1 year a 1.00000 g sample decays to 0.99957 g.

---

[6]See D'Arcy W. Thompson, *On Growth and Form*, Revised Edition (1943), Chapter 11, from which this example is borrowed.

Q4. Let $a > 0$. Prove the formulas (i) $a^{b+c} = a^b a^c$, (ii) $\log a^b = b \log a$, and (iii) $(a^b)^c = a^{bc}$.

Q5. Show that if $q(t) = q_0 a^t$ the doubling time T equals $\log_a 2$.

Q6. Suppose that the x-axis of a sheet of ordinary graph paper (a grid of squares) is labeled with a scale 0, 1, 2, 3, etc., in the usual way, one number at each vertical line. How can you label the y-axis so that the graph of any function of the form $y = ba^{cx}$ ($b > 0$) is a straight line? Explain and illustrate. (This kind of paper is used for recognizing data as indicating exponential growth.)

Q7. (a) Show that the statements $2 < e < 3$ and $\log 2 < 1 < \log 3$ are equivalent. (b) Prove them by interpreting $\log 2$ and $\log 3$ as areas under the graph of $y = 1/x$. (To show $\log 3 > 1$ approximate the integral by a sum based on a fine enough subdivision of $[1, 3]$, as in Art. 2.1.)

Q8. What portion of the original carbon-14 would you expect to find in a first edition of Newton's *Principia* (1687)?

Q9 (§6.1.2). Given the exponential function, show that every solution of §6.1.2(5) has the form $ce^{\alpha t}$ for some constant c. (Let q be a solution and consider $qe^{-\alpha t}$.)

Q10 (§6.2.1). Give the details of this equivalence.

Q11 (§6.1.5). Give the direct proof. (As with §6.2.1(1), it must be shown that $p(t) \to \infty$ as $t \to \infty$ and $p(t) \to 0$ as $t \to -\infty$. For the former limit use §3.3.3.Q6(a).)

Q12. Let $s_n = 1 + \dfrac{1}{2} + \dfrac{1}{3} + \cdots + \dfrac{1}{n}$. Show that $s_n \to \infty$. (Compare s_n to $\log(n+1)$. This sequence defines the **harmonic series**: see §8.2.10.Q16.)

Q13. (a) Prove §6.2.1(2) directly from the definition of $\log t$ by temporarily fixing v and comparing the derivatives of $\log uv$ and $\log u$. (b) Deduce §6.2.1(1). (First show that $\log 2^n = n \log 2$.)

Q14. Prove that if $a_n \to a$ ($a_n, a > 0$) and $b_n \to b$ then $a_n^{b_n} \to a^b$.

Q15. Prove that for any $a > 1$ and any non-negative integer n, $\displaystyle\lim_{x \to \infty} \frac{a^x}{x^n} = \infty$. (First use §3.3.13.Q8 to show $a^x/x \to \infty$.)

Q16. Solve the differential equation $u' = \alpha u - c$, where α and c are non-zero constants, for $u(t)$. (Let $q(t) = u(t) - (c/\alpha)$.)

Q17. In §5.1.3.Q9 the equation $-mg - \rho v = ma$ was shown to describe fall against air resistance $-\rho v$. (a) Solve it for v, assuming that the falling body is dropped ($v_0 = 0$). (Apply Q16 with $u = v$.) (b) Verify from the solution that $v \to v_T$. (c) What if $v_0 \neq 0$?

Q18. The steady-state current I in an electric circuit is given by Ohm's law $V = RI$, where V is the electromotive force (emf) or voltage that drives the current and R is the resistance that opposes it. When the circuit is first closed, however, an opposing emf proportional to the rate of increase of current arises, namely $L(dI/dt)$, where L is the self-inductance of the circuit. The effective voltage is then reduced, and Ohm's law becomes $V - L(dI/dt) = RI$. Determine $I(t)$ and show that it approaches V/R as $t \to \infty$. (Use Q16.)

Q19. Suppose that each cell in a colony produces $p + 1$ progeny upon division, and that the time between divisions is T. Let the *number* of cells at time t be (approximately) represented by a differentiable function $N(t)$. Then $N(t) = N_0 e^{\alpha t}$ for some $\alpha > 0$. Express α in terms of p and T. (Calculate dN by determining the fraction of cells that divide in time dt.)

Q20. (a) Assuming that the result of §6.2.8, Example 3, applies well enough as long as x is not more than 1% of the earth's radius, find the air pressure at the height of 40 miles and compare it to that on Mt. Everest. (b) If the earth were flat, what part of the mass of its atmosphere would be contained within 40 miles of its surface? (Assume that gravity and the atmosphere are as usual. Refer to §6.2.8, Example 3, recalling that pressure and density are proportional.)

Q21 (§6.2.9). If given a fragment such as $PQRS$ in Fig. 6-3, how can we locate the center O?

Q22. Principal A_0 is invested at annual interest of $100\alpha\%$, compounded n times a year (that is, at the end of each nth part of a year interest of $100(\alpha/n)\%$ is added). (a) Show that at the end of t years, t being a whole number, the value of the investment is $A = A_0(1 + (\alpha/n))^{nt}$. (b) Let the frequency of compounding increase, and show that in the limit as $n \to \infty$, $A = A_0 e^{\alpha t}$. When A is determined by this formula the interest is said to be compounded *continuously*. (c) Calculate the advantage for a 10-year period in continuously compounded interest over interest compounded 12 times a year, if the rate is 5%. (d) Extend the formula in (b) to arbitrary t. (First extend the formula in (a) to $t = m/n$, then use Q14.)

Q23 (§6.2.4). Supply the arguments.

CHAPTER 7

Length and Curvature of Plane Curves

INTRODUCTION

Our geometrical investigations have so far neglected two essential features of plane curves, their *length* and *curvature*.[1] The present chapter will address these subjects with a mixture of geometry and analysis. After the basic concepts and formulas have been developed and applied in the first two articles, the relation between curvature and deviation from a tangent line is examined in Art. 7.3. Article 7.4 is concerned with the representation of a curve by "parametric equations," in which x and y play equal roles, and by an "intrinsic equation," in which arc length and curvature replace the usual coordinates. In the last article integration along a curve is turned to a variety of uses, and the length and curvature of a curve are related to the velocity and acceleration of a particle that moves along it, so that geometry and motion are brought together in a new way.

With one or two exceptions in examples and remarks, the curves considered in this chapter lie in the plane and are to be assumed continuous (in one piece free of breaks and gaps) and *smooth* in the sense of §2.9.6.[2] In addition, while we will allow closed curves such as the circle as well as arcs having two ends, to avoid complication we will exclude self-intersecting curves, such as a figure eight.

---

[1] The only exception was the length of circular arcs, considered in §§4.4.4–5.

[2] See also §2.9.9 and §3.1.7.

7.1   ARC LENGTH AND CURVATURE

7.1.1   Arc length. A. Curve and polygon, arc and chord.   Consider a curve with endpoints A and B, which we understand to coincide if the curve is closed. In order to discuss its length we have to generalize several terms of elementary geometry. An **arc** of the curve is any continuous portion (possibly the whole) with endpoints; if the endpoints are distinct, the straight line joining them is a **chord**, which **subtends** the arc. Let one of the two directions along the curve be designated as positive, the other as negative; then the curve is said to be **directed**, or **oriented**. Suppose that the positive direction is from A to B, and choose points $V_0 = A$, $V_1, V_2, \ldots, V_{n-1}$, $V_n = B$ along the curve in the order specified by the positive direction. (Here $n > 1$, or if the curve is closed, $n > 2$.) Join the successive points by straight lines V_0V_1, $V_1V_2, \ldots$, $V_{n-1}V_n$. The line $V_0V_1V_2\cdots V_n$ composed of these straight segments is a **polygon**, which is **inscribed** in the curve. Its **vertices** are V_0, V_1, etc., and its **sides** are V_0V_1, V_1V_2, etc., the chords that subtend successive arcs (Fig. 7-1).

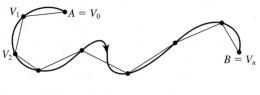

Fig. 7-1

The following two propositions are fundamental.[3]

> **Proposition 1.**   *The length of the curve is the limit of the lengths of its inscribed polygons, as the number of sides of a polygon is increased in such a way that the length of the longest side approaches zero.*

> **Proposition 2.**   *Let P be a fixed point on the curve. The ratio of the length of an arc PQ of the curve to the length of the chord that subtends it approaches unity as $Q \to P$, i.e. as the length of the chord approaches zero.*

Both are founded upon the idea that the length of a curved line can be approximately measured by laying a straight measuring-stick along it, and that

---

[3]Cf. Newton's *Principia*, Book 1, Lemma 3 (Cor. 4) and Lemma 7.

the approximation can be made arbitrarily accurate by using a short enough stick. In §4.4.5 the two propositions were proved equivalent to one another in the case of the circle, with the modification that in Proposition 1 the polygons were regular.

The intuitive notion of **length of arc** can be made precise by taking Proposition 1 as its *definition*, and that is what I will do here. Thus I declare that the curve has length L if the limit described in Proposition 1 exists, is unique, and has the value L. Here L is finite, a number; the limit may exist and be infinite, but we do not call infinity a length. Uniqueness of the limit means that the same number L is approached no matter how the polygons are chosen. (That they can be chosen with sides approaching zero is a preliminary requirement.) A curve that has length L under this definition is said to be **rectifiable** (capable of being straightened). Among the rectifiable curves are the graphs of smooth functions; this will be proved in §7.2.3, and in §7.4.3 a more comprehensive argument will be indicated. Examples show that an arbitrary continuous curve need not have the property (§7.2.6.Q13).

In the case of a graph, from Proposition 1 we can prove Proposition 2, and conversely if a quantity "length of arc" satisfies Proposition 2, Proposition 1 follows—the deductions will be carried out in §7.2.3. Another characterization of arc length, which does not involve a limit, is offered in §7.1.3.

In what follows the term "curve," unlike "arc," will often be applied (as in the past) to a line that has no specified endpoints, hence no definite length.

> We will not need a general definition of "curve," since graphs of smooth functions, and simple combinations of such graphs, suffice to exhibit the features of curves discussed in this chapter. A definition can be based on §7.4.1.

7.1.2   B. Signed arc length.   Length of arc is a kind of total quantity, associated with each arc or interval of a curve. It becomes a signed quantity, which I will denote by S, if the curve is oriented. Then to define a corresponding cumulative quantity s we choose a point P_0 on the curve and denote the signed length of arc from P_0 to any point P on the curve by $s(P)$. This makes s a *coordinate* on the curve, which indicates the location of any P relative to P_0 (Fig. 7-2(a)). If the curve is closed $s(P)$ admits multiple values (Fig. 7-2(b)), a complication which will cause us no difficulty.

As an example, let the curve be the x-axis with its customary positive direction, and let P_0 be the origin; then s is x, i.e. $s(P)$ is the x-coordinate of P. Just as in this case the signed length of the x-interval from x_1 to x_2 is given by the coordinate difference $\Delta x = x_2 - x_1$, so more generally the signed length

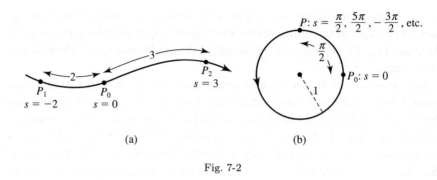

Fig. 7-2

of a curve from P_1 to P_2 is given by $S = \Delta s = s_2 - s_1$, where $s_1 = s(P_1)$ and $s_2 = s(P_2)$.

The introduction of s gives us the means to express any function of the points of a curve as a function of numbers. For example, the temperature at any point of a curve in space becomes a function $f(s)$, so that at P_1 in Fig. 7-2(a) the temperature is $f(-2)$ and at P_2, $f(3)$.

The next section outlines the alternative approach to arc length mentioned previously.

7.1.3   C. Arc length characterized without a limit.

Recall that a *concave* arc is one such that as you move along it from one end to the other your heading, or tangent direction arrow, turns in one direction only, either clockwise or counterclockwise (§§2.9.6, 3.9.6); straight segments are not allowed. Inequalities (4) of §4.4.4 can be generalized as follows.

> **Proposition 3.**   *Let a concave arc a have (distinct) endpoints A and B, and suppose that from A to B the tangent direction turns through an angle of magnitude less than π. Let T be the intersection of the (one-sided) tangents to a at A and B (Fig. 7-3(a)), and draw chord AB. Then of the three lines joining A and B the chord is the shortest and the path formed by the tangents the longest,*
>
> $$AB < a < ATB.$$

It is not hard to show that the three lines lie as pictured. The hypothesis of concavity of a keeps the arc from gaining length by meanders, as happens in Fig. 7-3(b), where $a > ATB$.

As I am about to demonstrate, this proposition is implied by Proposition 1, and in turn implies Proposition 2 for any curve that consists of finitely many concave arcs and straight segments (the hypothesis used in §2.9.6). For a

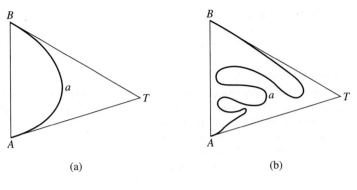

(a)   (b)

Fig. 7-3

graph of this type, then, it is equivalent to both Proposition 1 and Proposition 2, since they are equivalent to one another.

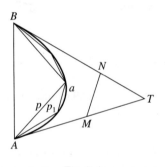

Fig. 7-4

Assuming Proposition 1, in the situation of Proposition 3 let p be any polygon inscribed in a, and let MN be any straight line from the interior of AT to the interior of BT that does not cross a (Fig. 7-4). Elementary geometry shows that

$$AB < p < AMNB < ATB.$$

Since (as Proposition 1 requires) the arcs between the vertices of p admit inscribed polygons having arbitrarily short sides, the length of a can be expressed as the limit of a sequence of inscribed polygons p_n such that the vertices of each polygon include among them the vertices of p. Clearly each p_n (other than p itself) is longer than p, so we have

$$AB < p < p_n < AMNB < ATB,$$

and hence for the limit a

$$AB < p \le a \le AMNB < ATB,$$

which proves Proposition 3.

Now suppose that Proposition 3 holds, and let P be a point on a curve made of finitely many concave arcs and straight segments. If in Proposition 2 arc PQ is straight there is nothing to prove; so assume it concave, and take Q so close to P that the tangent direction turns along PQ by less than $\pi/2$ (Fig. 7-5). Draw tangents at P and Q intersecting at T and drop QR perpendicular to tangent PR. Denoting the arc and its chord by a and c respectively and angle RPQ by θ, we have by Proposition 3

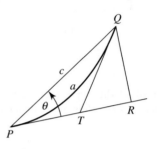

$$c < a < PTQ < PRQ,$$

Fig. 7-5

whence

$$1 < \frac{a}{c} < \frac{PR}{c} + \frac{RQ}{c} = \cos\theta + \sin\theta.$$

As $Q \to P$, i.e. $c \to 0$, also $\theta \to 0$ (§2.9.6), hence $\cos\theta + \sin\theta \to 1$, and this implies $a/c \to 1$, as Proposition 2 asserts.[4]

7.1.4 Curvature.

Any circle is uniformly curved, but a smaller circle is more sharply curved than a larger one, a fact that is brought home by the experience of driving at constant speed around different circular arcs. What exactly makes the difference between driving on this arc and on that? It is the angle you turn through in going a given distance. If your heading changes by $10°$ in 100 feet, the curve is gentle; if by $100°$, it is much sharper, and may reasonably be called ten times sharper, since the angle turned through is ten times as great. Generalizing, we arrive at the idea of *angle per distance* as the measure of the curvature of a circular arc. Of course, angles are to be measured in radians rather than degrees (and distances in meters, when a physical unit is needed).

---

[4]This proof of Proposition 2 resembles Newton's proof of his Lemma 7 (preceding footnote).

On a trip all the way around a circle the heading changes by 2π (radians), while the distance traveled is $2\pi r$, where r is the radius; then the angle per distance is $2\pi/2\pi r = 1/r$, and since the circle is everywhere the same this ratio holds for any arc of it. The conclusion is that *the curvature of a circle of radius r is $1/r$*. The other kind of uniform curve, the straight line, evidently has curvature zero.

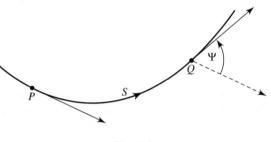

Fig. 7-6

Along a more general curve the curvature varies from point to point. Let P be a given point on an oriented curve and Q a nearby point, and draw tangent arrows in the positive direction at P and Q (Fig. 7-6). As we move a tangent arrow from P to Q it turns through a (signed) angle Ψ from the positive direction at P to the positive direction at Q, which is made visible by shifting the arrow at P parallel to itself over to Q (see the figure). We call Ψ the **total curvature** (or **integral curvature**) from P to Q. Let S be the (signed) arc length from P to Q. The ratio Ψ/S of total curvature to distance is regarded as the **mean curvature** of the arc from P to Q. Its limit as $Q \to P$ (assuming the limit to exist) is then the **curvature** at P, denoted by κ:

$$\kappa(P) = \lim_{Q \to P} \frac{\Psi}{S}\,.$$

So defined, the curvature is a *signed* quantity, which is generally *positive where the curve is concave towards the left* as we move along it in the positive direction, *negative where it is concave towards the right*; for in the former case Ψ and S have the same sign (regardless of which side of PQ is on), in the latter case opposite signs. Consequently the sign of κ will be changed if the direction called positive is reversed. Thus a circle oriented counterclockwise has $\kappa = 1/r$ at all points, but the same circle oriented clockwise has $\kappa = -1/r$ (Fig. 7-7).

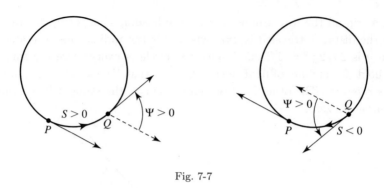

Fig. 7-7

Often, however, the term "curvature" is used to refer to the positive quantity $|\kappa|$, as when I said earlier that the curvature of a circle (to which no direction had been assigned) was $1/r$.[5] Curvature in that sense has the advantage of being independent of the direction of a curve, but as a mathematical function it is harder to work with than the signed κ.

7.1.5 Curvature as a derivative. An arc-length coordinate along the curve is obtained as in §7.1.2 by choosing an initial point P_0. Let the coordinate of P be s, and of Q, $s + \Delta s$; then $S = \Delta s$ and

$$\kappa(P) = \lim_{\Delta s \to 0} \frac{\Psi}{\Delta s}.$$

The curvature κ is a function of the variable s, since s determines P: $\kappa(s) = \kappa(P)$ is the curvature-number assigned to the point P with coordinate s.

Suppose now that the tangent line at P is not vertical—this will surely be the case if the curve is the graph of a differentiable function. Then in view of the smoothness of the curve the tangents at points near P are also non-vertical, so that the *angle of inclination* α of the tangent line to the horizontal is defined at all points of an arc containing P (§3.3.7). It is not hard to see that regardless of the direction of the curve, we have

$$\Psi = \alpha(Q) - \alpha(P),$$

[5]Similarly, "total curvature" is used to refer to $|\Psi|$.

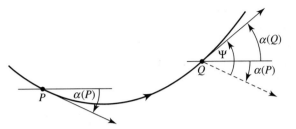

Fig. 7-8

as long as Q lies on the arc (Fig. 7-8). If α is considered as a function of s this becomes

$$\Psi = \alpha(s + \Delta s) - \alpha(s) = \Delta\alpha\,,$$

so that

$$\kappa(s) = \lim_{\Delta s \to 0} \frac{\Delta\alpha}{\Delta s} = \frac{d\alpha}{ds}\,,$$

assuming that the derivative exists. In words: the curvature is the rate of change of angle of inclination with respect to arc length.

In case there is a vertical tangent at P, $\kappa(s)$ can be represented in like manner as the derivative of the angle of inclination β of the tangent line to the vertical axis (defined analogously to α).

> If the arc containing P is concave, α is a monotonic function of s. Then although $d\alpha/ds$ cannot vanish along an s-interval, it may vanish at individual points (§3.9.1); see Example 3, §7.2.5. That is why in the last section it was said only that κ is *generally* positive or negative along a concave curve.

7.1.6   The circle of curvature and the radius of curvature. The limit by which curvature is defined, if that limit exists at a point P, may be zero, or finite but non-zero, or infinite ($\pm\infty$). In the last of these cases the curve is sometimes said to have **infinite curvature** at P ("$|\kappa| = \infty$"), although properly speaking κ does not exist there. Let us give our attention to the middle case, in which κ exists and $\kappa \neq 0$ at P.

At any such point a tangent circle can be found to match the curve in respect of curvature, for circles come in all non-zero curvatures. We define the **circle of curvature** at P to be the unique oriented circle through P that has the same *direction* and *curvature* as the curve at P. Its center lies on the concave side of the curve—to the left of P, as you proceed along the curve in the positive direction, if $\kappa > 0$, to the right if $\kappa < 0$ (Fig. 7-9). Its radius $\rho = 1/|\kappa|$ is called the **radius of curvature** at P, its center the **center of curvature** there.[6]

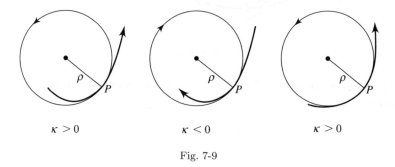

Fig. 7-9

As the tangent line is the *straight*, or *uniformly directed*, line that best matches a curve at a point, so the circle of curvature is the *uniformly curved* line that best matches it. Thus of all tangent curves that agree in curvature with the given arc at P, the circle of curvature is the simplest in respect of curvature; but it is not simplest in respect of equation, as we will see by and by (§8.1.2).

Where $\kappa = 0$ there is no circle of curvature; its role is taken by the tangent line, and we say that the radius of curvature is infinite. Where the curvature is infinite the radius of curvature is said to be zero; in that case the circle of curvature has degenerated to a point. Examples to illustrate these situations will be given in §7.2.5, after we have developed a formula for calculating κ.

7.1.7   Invariance of arc length and curvature.   We have defined arc length S or s as a limit of lengths of polygons (with a sign added) and κ as a limit of Ψ/S, Ψ being an angle between directed tangents. Neither definition relied in any way on a choice of coordinate axes in the plane. Arc length and curvature are **intrinsic** properties of a curve, like the area contained by a closed curve, and unlike such quantities as ordinate, slope, and area under

---

[6]The circle of curvature is also called the **osculating** (kissing) circle, as having closer contact with the curve than does the tangent (touching) line.

a curve. Hence the formulas for s and κ to be obtained in the next article, which *involve* quantities that do depend upon the coordinate axes, nevertheless *express* quantities that remain **invariant** if new axes are chosen. This point will be amplified in §7.4.6.

The definition of Ψ did rely upon the choice of counterclockwise rotation as positive; in addition, of course, S depends on the direction chosen for the curve. If we consider only $|S|$ and $|\kappa|$, these choices too are without effect.

7.2   LENGTH AND CURVATURE OF A GRAPH

7.2.1   The differential of arc length.   Arc length as an integral.
The length of the chord c of an arc from $P(x,y)$ to $Q(x + \Delta x, y + \Delta y)$ is determined by the Pythagorean theorem:

$$c^2 = (\Delta x)^2 + (\Delta y)^2$$

(Fig. 7-10(a)). This quantity c^2 is in general smaller than the square of arc length $(\Delta s)^2$. But if we pass to infinitesimals arc and chord can be identified, for by Proposition 2, §7.1.1, their ratio will be unity (Fig. 7-10(b)). Then the square of arc length is

$$ds^2 = dx^2 + dy^2 \,.$$

(It is customary to omit the parentheses in such squares as $(ds)^2$ and $(dx)^2$, since confusion with $d(s^2)$, etc., is unlikely.)

The identity of infinitesimals derived from quantities "ultimately in the ratio of equality" (§3.1.11.Q4) is consistent with the remarks in §2.4.8 to the

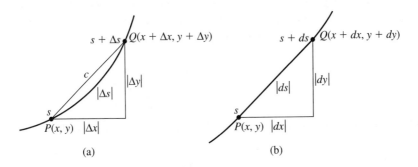

(a)   (b)

Fig. 7-10

effect that negligibility of infinitesimals is a matter of their ratios. Let a be a finite arc, c its chord, and δ the difference between them; then $a = c + \delta$, so $\delta/c = (a/c) - 1$. As $c \to 0$, therefore, $\delta/c \to 0$. For the corresponding infinitesimals c_1, δ_1 this means that δ_1/c_1 is an infinitesimal ratio: δ_1 is negligible in comparison to c_1.

If the direction from P to Q is the positive one, the differential of arc length is the positive square root of the sum of squares,

$$ds = \sqrt{dx^2 + dy^2} \, .$$

Returning to the finite arc from $P(s)$ to $Q(s + \Delta s)$, we then have

$$\Delta s = \int_s^{s+\Delta s} ds = \int_s^{s+\Delta s} \sqrt{dx^2 + dy^2} \, .$$

This can be regarded as the length of a polygon with infinitely many infinitesimal sides; thus it expresses a version of Proposition 1, §7.1.1. In a different notation the formula is

$$S = s_2 - s_1 = \int_{s_1}^{s_2} ds = \int_{s_1}^{s_2} \sqrt{dx^2 + dy^2} \, ,$$

the length of arc from s_1 to $s_2 > s_1$.

The sum on the right does not have the look of an integral, because ds is not expressed as the differential of a function of x (or y). Let us assume that the curve we are considering is the graph of a continuously differentiable function $y = f(x)$, and further that *the positive direction is that of increasing* x (towards the right). From the first assumption it follows that dy/dx is a continuous function of x, and from the second that ds and dx always have the same sign. The latter consequence implies that the relation

$$|ds| = \sqrt{dx^2 + dy^2} = \sqrt{1 + \left(\frac{dy}{dx}\right)^2} \, |dx| \, ,$$

which holds regardless of the signs of ds and dx, can be replaced by

$$ds = \sqrt{1 + \left(\frac{dy}{dx}\right)^2} \, dx \, .$$

Here the square root is a continuous function of x, so if x_1 and x_2 are the x-values corresponding to s_1 and s_2 we have[7]

$$S = s_2 - s_1 = \int_{s_1}^{s_2} ds = \int_{x_1}^{x_2} \sqrt{1 + \left(\frac{dy}{dx}\right)^2}\, dx\,.$$

This is the formula for length of arc of the graph of a function of x. In §7.2.3 it will be derived without infinitesimals.

By writing an indefinite integral instead of a definite one we can express the arc-length coordinate s as a function of x. Let x_0 be the x-value corresponding to $s = 0$, then

$$s(x) = \int_0^{s(x)} ds = \int_{x_0}^{x} \sqrt{1 + \left(\frac{dy}{dx}\right)^2}\, dx\,.$$

The Fundamental Theorem (Part I) shows that $s(x)$ is continuously differentiable and its derivative is positive everywhere. It then has an inverse $x(s)$ with the same properties.

In case the curve is the graph of a function $x(y)$ instead of $y(x)$, analogous formulas are obtained, e.g.

$$S = \int_{y_1}^{y_2} \sqrt{\left(\frac{dx}{dy}\right)^2 + 1}\, dy\,.$$

A curve such as a circle, which is neither the graph of a function of x nor of a function of y, can be dealt with by dividing it up into portions that are graphs (see Example 2). A formula that does not require such division will be developed in §7.4.3.

The application of the present arc-length formulas can be illustrated by a few examples.

7.2.2 Examples. EXAMPLE 1. Let $y = x^{3/2}$, $x \geq 0$ (Fig. 7-11; the virtue of this function is that its arc-length integral is easy to evaluate). We have

---

[7]The limits of integration are transformed according to the rule for a change of variable (§4.6.3).

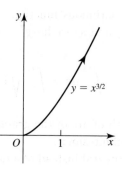

Fig. 7-11

$y' = \frac{3}{2}x^{1/2}$, so

$$ds = \sqrt{1 + y'^2}\, dx = \sqrt{1 + \left(\frac{3}{2}\right)^2 x}\, dx = \frac{3}{2}\left(x + \frac{4}{9}\right)^{1/2} dx\,.$$

Then the arc length from $x = a$ to $x = b$ is

$$\int_a^b \frac{3}{2}\left(x + \frac{4}{9}\right)^{1/2} dx = \left(x + \frac{4}{9}\right)^{3/2}\Big|_a^b = \left(b + \frac{4}{9}\right)^{3/2} - \left(a + \frac{4}{9}\right)^{3/2}.$$

The arc length measured from the origin to any $x \geq 0$ is

$$s = \int_0^x \sqrt{1 + y'^2}\, dx = \left(x + \frac{4}{9}\right)^{3/2} - \left(\frac{4}{9}\right)^{3/2} = \left(x + \frac{4}{9}\right)^{3/2} - \frac{8}{27}\,.$$

EXAMPLE 2. To calculate the length of the upper semicircle of the unit circle, we let $y = \sqrt{1 - x^2}$, $-1 \leq x \leq 1$ (Fig. 7-12). As x increases from -1 to 1, the curve is described *clockwise*; this is then the positive direction. From

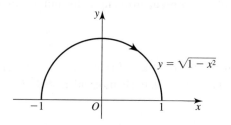

Fig. 7-12

$y' = -x/\sqrt{1-x^2}$ follows

$$ds = \sqrt{1 + y'^2}\, dx = \sqrt{1 + \left(\frac{x^2}{1-x^2}\right)}\, dx = \frac{1}{\sqrt{1-x^2}}\, dx,$$

and consequently the arc length is

$$\int_{-1}^{1} \frac{1}{\sqrt{1-x^2}}\, dx = \arcsin x \Big|_{-1}^{1} = \frac{\pi}{2} - \left(-\frac{\pi}{2}\right) = \pi$$

(§4.7.2, Example 4). The outcome cannot be called a surprise. Yet the integration is more than an idle exercise; for if π and the trigonometric functions are defined independently of geometry as in §4.7.4, their connection to the circle must be established by just such calculations as this.

    The use of improper integrals in this example and the next is justified because a length of arc from P_0 to P is the limit of the lengths of arc from P_0 to Q as $Q \to P$. (By Propositions 1 and 2, §7.1.1, both of which apply to these curves, arc $P_0 P = $ arc $P_0 Q + $ arc QP and arc $QP \to 0$.) Thus $s(x)$ here is continuous at $x = \pm 1$ even though ds/dx is not defined.

EXAMPLE 3. The upper half of the ellipse $\dfrac{x^2}{a^2} + \dfrac{y^2}{b^2} = 1$, $a > b$, has equation

$$y = \frac{b}{a}\sqrt{a^2 - x^2}\,,$$

so

$$y' = -\frac{b}{a}\frac{x}{\sqrt{a^2 - x^2}}$$

and

$$ds = \sqrt{1 + y'^2}\, dx = \sqrt{1 + \frac{b^2}{a^2}\left(\frac{x^2}{a^2 - x^2}\right)}\, dx = \sqrt{\frac{a^2 - k^2 x^2}{a^2 - x^2}}\, dx,$$

where I have put $k = \sqrt{1 - \dfrac{b^2}{a^2}}$, the **eccentricity** of the ellipse. The arc length from $x = 0$ to any x is then

$$s(x) = \int_{0}^{x} \sqrt{\frac{a^2 - k^2 x^2}{a^2 - x^2}}\, dx$$

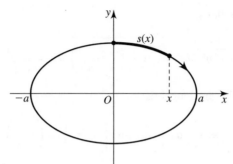

Fig. 7-13

(Fig. 7-13), or with the substitution $z = x/a$,

$$s(x) = a \int_0^{x/a} \sqrt{\frac{1 - k^2 z^2}{1 - z^2}} \, dz \, .$$

If the ellipse degenerates to a circle, $k = 0$ and

$$s(x) = a \int_0^{x/a} \frac{1}{\sqrt{1 - z^2}} \, dz = a \arcsin \frac{x}{a} \, ;$$

in particular, for $x = a$ we obtain $a \arcsin 1 = \frac{1}{2}\pi a$, the length of a quarter-circle of radius a. But if $k \neq 0$ the integral for $s(x)$ cannot be evaluated in terms of elementary functions. Called an **elliptic integral**, it defines a *new* function which generalizes the arc sine (cf. §4.7.4). The inverse of this function is one of the **elliptic functions**, which generalize the trigonometric or "circular" functions.

7.2.3   Limit derivation of the arc-length integral. Equivalence of Propositions 1 and 2.[8]   Let $y = f(x)$ as in §7.2.1, and consider the length of arc S of the graph of f from $x = a$ to $x = b > a$. Let the successive vertices of an inscribed polygon have x-coordinates $x_0 = a, x_1, x_2, \ldots, x_{n-1}, x_n = b$ (Fig. 7-14). The length of the ith side is

$$c_i = \sqrt{\Delta x_i^2 + \Delta y_i^2} = \sqrt{1 + \left(\frac{\Delta y_i}{\Delta x_i}\right)^2} \, \Delta x_i \, .$$

---

[8]This section can be omitted.

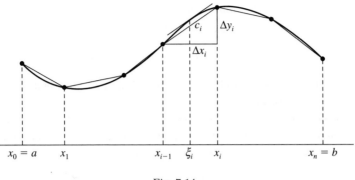

Fig. 7-14

According to the mean value theorem there is a point ξ_i between x_{i-1} and x_i such that

$$\frac{\Delta y_i}{\Delta x_i} = f'(\xi_i)$$

—at ξ_i the slope of the curve equals the slope of the ith side. Then

$$c_i = \sqrt{1 + f'(\xi_i)^2}\,\Delta x_i\,,$$

and the length of the polygon is

$$c_1 + c_2 + \cdots + c_n$$
$$= \sqrt{1 + f'(\xi_1)^2}\,\Delta x_1 + \sqrt{1 + f'(\xi_2)^2}\,\Delta x_2 + \cdots + \sqrt{1 + f'(\xi_n)^2}\,\Delta x_n\,.$$

The assertion of Proposition 1, §7.1.1, as applied to this graph, is that the limit of the sum of the c_i as all the c_i approach zero is the arc length S. But the definition of the integral tells us that the limit of this sum as all the Δx_i approach zero is $\displaystyle\int_a^b \sqrt{1 + f'(x)^2}\,dx$. If, therefore, we show that it is the same thing for all the Δx_i to approach zero as for all the c_i to do so, we will have proved that *the graph is rectifiable* (§7.1.1), and further that for the graph of a function of x, *Proposition 1 is equivalent to the arc-length formula*

$$S = \int_a^b \sqrt{1 + f'(x)^2}\,dx\,.$$

The required argument is as follows. Since Δx_i cannot be larger than c_i, it is clear that all the Δx_i approach zero when all the c_i do. To prove the converse, note that by the maximum–minimum property the continuous function $\sqrt{1 + f'(x)^2}$ has a greatest value M on the interval $[a, b]$, so that

$$c_i = \sqrt{1 + f'(\xi_i)^2}\,\Delta x_i \leq M\Delta x_i$$

for all i. As the Δx_i approach zero so do the $M\Delta x_i$, and hence also the c_i. This completes the proof.

In the preceding derivation we assumed $a < b$, but the integral formula obviously holds for $a \geq b$ as well. Its validity for arbitrary a and b means that *Proposition 1 is also equivalent to the formula*

$$s(x) = \int_{x_0}^{x} \sqrt{1 + f'(x)^2}\,dx\,,$$

where $s(x_0) = 0$.

Let us now consider Proposition 2. Assuming that "length of arc" has *some* meaning, but not necessarily the meaning assigned to it by Proposition 1, we postulate an increasing function $s(x)$ such that $s(x_2) - s(x_1)$ is the signed arc length of the graph of $y = f(x)$ from $x = x_1$ to $x = x_2$. Then the length of the arc between $P(x, y)$ and $Q(x + \Delta x, y + \Delta y)$ is $|\Delta s|$, while the length of chord PQ is $c = \sqrt{(\Delta x)^2 + (\Delta y)^2}$. Proposition 2 states that as $c \to 0$, $|\Delta s|/c \to 1$. But we proved a moment ago that $c \to 0$ if and only if $\Delta x \to 0$. Thus the assertion of Proposition 2 is that as $\Delta x \to 0$, $|\Delta s|/c \to 1$, or equivalently

$$\frac{\dfrac{\Delta s}{\Delta x}}{\sqrt{1 + \left(\dfrac{\Delta y}{\Delta x}\right)^2}} \to 1\,.$$

The denominator approaches the non-zero limit $\sqrt{1 + f'(x)^2}$, so the stated limit is equivalent to

$$\frac{\Delta s}{\Delta x} \to \sqrt{1 + f'(x)^2}\,.$$

In other words, *what Proposition 2 says is that $s(x)$ is differentiable, and its derivative is the continuous function $\sqrt{1 + f'(x)^2}$.*

An application of the Fundamental Theorem now demonstrates the equivalence of Propositions 1 and 2 for the graph of a smooth function. Part I of the theorem shows that Proposition 1 implies Proposition 2, while Part II shows the converse. Clearly the propositions are complementary as well as equivalent.

The graph of a function of y is analogously treated. Similar arguments apply to a curve which is not the graph of a function (§7.4.3).

7.2.4   The curvature of a graph in terms of derivatives. When it exists, the curvature of the graph of a continuously differentiable function $y = f(x)$ is given by $\kappa = d\alpha/ds$, where α is the angle of inclination (§7.1.5). By the chain rule

$$\frac{d\alpha}{ds} = \frac{d\alpha}{dx} \Big/ \frac{ds}{dx} \,,$$

provided that $d\alpha/dx$ exists, because ds/dx has the non-zero value $\sqrt{1 + y'^2}$ (§7.2.1). Now the slope of the graph is $\tan \alpha = y'$, so $\alpha = \arctan y'$; wherever y'' exists we then have

$$\frac{d\alpha}{dx} = \frac{y''}{1 + y'^2}$$

(cf. §4.4.9.Q9), and it follows that

$$\frac{d\alpha}{ds} = \frac{y''}{1 + y'^2} \Big/ \sqrt{1 + y'^2} \,,$$

or

(1) $$\kappa = \frac{y''}{(1 + y'^2)^{3/2}} \,.$$

As usual, I have taken the positive direction along the graph to be towards the right (in assuming $ds/dx > 0$).

This formula asserts that α differs from y'' by a continuous positive factor, namely $(1 + y'^2)^{-3/2}$; consequently κ and y'' are positive, negative, or zero together, and there is infinite curvature where $|y''| \to \infty$. Where there is a horizontal tangent ($y' = 0$) κ and y'' are the same; a geometrical explanation of this fact will be given in §7.3.2. As a rule, y'' is continuous and non-zero at all but isolated points, and so therefore is the curvature.

> Continuity of κ as a function of x, which is equivalent to continuity of y'', is also equivalent to continuity of κ as a function of s, because $s(x)$ is continuous and has a continuous inverse.

Several varieties of behavior of κ are illustrated in the following examples. The figures show a few circles of curvature together with the curves.

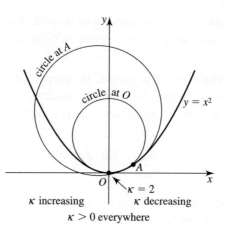

Fig. 7-15

7.2.5 Examples. EXAMPLE 1. Let $y = x^2$, then $\kappa = \dfrac{2}{(1 + 4x^2)^{3/2}}$
(Fig. 7-15). At $x = 0$ κ has its maximum value, which is 2.

EXAMPLE 2. Let $y = x^3$, then $\kappa = \dfrac{6x}{(1 + 9x^4)^{3/2}}$ (Fig. 7-16). The sign of
κ changes at the inflection point $x = 0$, where $\kappa = 0$. By calculating $d\kappa/dx$

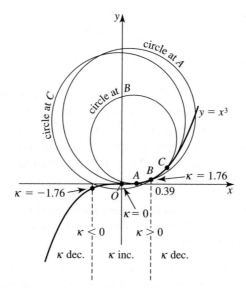

Fig. 7-16

we find that κ takes its least value at $x = -45^{-1/4} \approx -0.39$ and its greatest at $x = 45^{-1/4} \approx 0.39$; at the former point $\kappa \approx -1.76$, at the latter $\kappa \approx 1.76$. Thus $|\kappa|$ has its minimum value of 0 at $x = 0$ and its maximum value of 1.76 at $x \approx \pm0.39$. (This could have been shown directly by differentiating κ^2 instead of κ.)

EXAMPLE 3. Let $y = x^4$, then $\kappa = \dfrac{12x^2}{(1 + 16x^6)^{3/2}}$ (Fig. 7-17). At $x = 0$ the curve is so flat that its curvature is zero. The maximum curvature of 2.15 occurs where $d\kappa/dx = 0$, at $x = \pm56^{-1/6} \approx \pm0.51$.

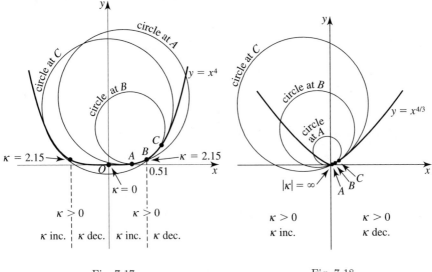

Fig. 7-17   Fig. 7-18

EXAMPLE 4. Let $y = x^{4/3}$, then $\kappa = \dfrac{12}{x^{2/3}(9 + 16x^{2/3})^{3/2}}$ (Fig. 7-18). The curve is so sharp at $x = 0$ that its curvature is infinite there: $\kappa \to \infty$ as $x \to 0$.

EXAMPLE 5. Let a curve be formed of arcs of two tangent circles of radius r, as shown in Fig. 7-19. Where they join, the curvature κ is undefined, because it switches abruptly from $-1/r$ to $1/r$. Although $|\kappa|$ is the same on both sides of this point, it too fails to exist at the point.

EXAMPLE 6. The upper half of the ellipse $\dfrac{x^2}{a^2} + \dfrac{y^2}{b^2} = 1$, $a > b$, is

$$y = \frac{b}{a}\sqrt{a^2 - x^2}$$

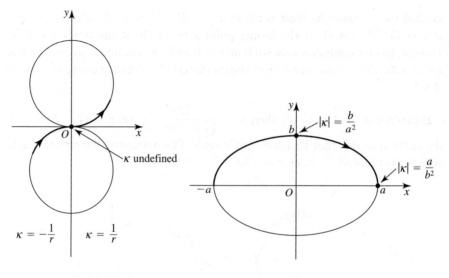

Fig. 7-19   Fig. 7-20

(Fig. 7-20), from which we calculate

$$y' = -\frac{b}{a}\,\frac{x}{\sqrt{a^2 - x^2}}\,,$$

$$y'' = -ab\,\frac{1}{(a^2 - x^2)^{3/2}}\,,$$

and

$$\kappa = -\frac{ab}{\left[a^2 - \left(1 - \dfrac{b^2}{a^2}\right)x^2\right]^{3/2}}\,.$$

In the notation of Example 3, §7.2.2, we then have

$$|\kappa| = \frac{ab}{(a^2 - k^2x^2)^{3/2}} = \frac{b/a^2}{(1 - k^2z^2)^{3/2}}\,;$$

by symmetry this is valid for the whole ellipse. By setting $k = 0$, or $b = a$, we recover the curvature of a circle of radius a, $|\kappa| = 1/a$. When $k \neq 0$, the minimum curvature occurs at $x = 0$, where $z^2 = 0$ and $|\kappa| = b/a^2$, and the maximum curvature at $x = \pm a$, where $z^2 = 1$ and $|\kappa| = a/b^2$.

The formula for κ written here, as simplified from $\dfrac{y''}{(1+y'^2)^{3/2}}$, makes it a continuous function of x, although y' and y'' are discontinuous at $x = \pm a$. This is correct, because $s(x)$ is continuous (see the remark following Example 2, §7.2.2) and κ is a continuous function of s at $x = \pm a$, as we can show by representing κ in terms of β (end of §7.1.5; see also §7.4.3).

7.2.6　QUESTIONS

Q1. (a) Use the arc-length integral to calculate the length of the line $y = cx$ from $x = 0$ to any $x > 0$. Verify by geometry that the answer is correct. (b) For any $y = f(x)$, what is $\sqrt{1+y'^2}$ in terms of the angle of inclination α? (c) In terms of α, what therefore is the ratio of κ to y''?

Q2. Find the arc-length function $s(x)$ for the graphs of the following functions, taking $x = 0$ as the starting point. (a) $y = \frac{1}{2}x^2$. (Use §4.6.5.Q5(b) or §4.6.8.Q4(b).) (b) $y = e^x$. (Let u equal the integrand of the integral with respect to x. After transforming, integrate by partial fractions.) (c) $y = \cosh x$

Q3. (a) What is the curvature of the earth's equator? (b) How many degrees per mile does the mast of a ship turn through as the ship sails along the equator?

Q4. (a) Show by a geometrical argument that the change in direction of travel as you go from A to B on a semicircle equals the central angle AOB. (b) Conclude directly (without referring to $2\pi r$ as in §7.1.4) that the curvature of a circle is $1/r$.

Q5. Let L be the length of $y = \cos x$ between $x = -\pi/4$ and $x = \pi/4$. Show that $\pi/2 < L < \sqrt{3/2}\,\pi/2$. (Use §3.7.3.Q4(b) or Proposition 3, §7.1.3.)

Q6. (a) Find the rate of change of direction with respect to length of arc along the graph of $y = \sin x$ at the point $P(\pi/6, 1/2)$. (b) Approximately how far must you move along the curve from P for your direction to have changed by $0.1°$?

Q7. (a) Find the curvature of the graph of $y = e^x$ as a function of x. (b) Show that it has a maximum value at a point $x_0 < 0$. Find x_0 and $\kappa(x_0)$.

Q8. (a) Following Example 6, §7.2.5, find the curvature $|\kappa|$ of the hyperbola $\dfrac{x^2}{a^2} - \dfrac{y^2}{b^2} = 1$ in terms of $z = x/a$ and $l = \sqrt{1+(b^2/a^2)}$, the eccentricity of the hyperbola. (b) Show that its maximum value agrees with the maximum curvature of the corresponding ellipse (Example 6). Where does it take the value of the minimum curvature of the ellipse?

Q9. (a) Prove that the curvature $|\kappa|$ of an ellipse has exactly four local extrema. (b) Is there a closed, convex, non-circular curve of which the curvature is continuous and has fewer than four local extrema?

Q10. In Fig. 7-21, PR is the tangent line to concave arc PQ at P and QR is another straight line. Is $PR > PQ$? (Let QR be vertical and cf. Q5.)

Q11. As a rule, the tangent line to a curve at a point does not cross the curve there, but sometimes it does. How does the circle of curvature at a point P behave in respect of crossing the curve at P?

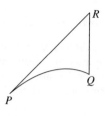

Fig. 7-21

Q12.   Use the expressions for κ and ds in terms of x to verify by integration that the total curvature Ψ of the graph of $y = f(x)$ from s_1 to s_2 is the integral of the curvature,

$$\Psi = \int_{s_1}^{s_2} \kappa \, ds \, .$$

Q13.   Show that any arc of the continuous curve of §4.2.9, Example 2, which includes the origin, does not have finite length. (Find a sequence of inscribed polygons whose lengths are comparable to the terms of the sequence $1, 1 + \frac{1}{2}, 1 + \frac{1}{2} + \frac{1}{3}, \ldots$, and use §6.2.10.Q12.)

7.3   LOCAL GEOMETRY

7.3.1   Local tangent and normal coordinates.

Curvature being independent of the axes of coordinates, we are at liberty to employ any convenient axes in its investigation. At a given point on a directed curve a natural choice for these is suggested by the curve itself, and with this choice we will be able to pursue by analytic methods the geometrical study of the relation of curvature at a point to the behavior of the curve near it.

Let a curve have continuous curvature at and near a point P, and suppose that the curvature is non-zero at P. I will assume it to be positive; the opposite case differs only in the direction assigned to the curve. Since the curvature is continuous, it is positive near point P as well as at it. As x-axis we take the tangent line at P, choosing the positive direction to agree with that of the curve; the y-axis is then the normal in the direction of the center of curvature (because the curve is concave towards the left with respect to its direction), and the situation is as in Fig. 7-22. Here the coordinate axes are drawn at an angle to emphasize that they are chosen to fit the curve at the arbitrary point P.

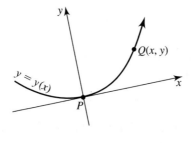

Fig. 7-22

In terms of these coordinates the curve has (near P) an equation $y = y(x)$ such that $y(0) = y'(0) = 0$ and y'' is continuous and positive (because κ is: §7.2.4). Let us denote by Q a variable point (x, y) of the curve, different from P; "y" and "y'" are to be understood to mean their values at Q, i.e. $y(x)$ and $y'(x)$. Further, let arc length be measured from P, so that $s = s(Q) = s(x)$. All limits will be taken as $Q \to P$, i.e. as $x \to 0$; thus

$$0 = y'(0) = \lim \frac{y}{x}, \quad y''(0) = \lim \frac{y'}{x}$$

—the symbol Δ can be omitted, because $\Delta y = y - y(0) = y$, etc.

7.3.2   Curvature in terms of local coordinates. As a first observation, which incidentally does not depend on the existence of κ, we have

$$\lim \frac{s}{x} = 1$$

(Fig. 7-23): the ratio of arc PQ to the tangent line PR cut off by perpendicular QR approaches unity. For the limit is

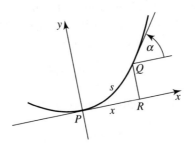

Fig. 7-23

$$\frac{ds}{dx}\bigg|_0 = \sqrt{1 + y'(0)^2} = 1\,.$$

This fact is a complement to Proposition 2, §7.1.1. (Cf. the end of §7.1.3.)
From $y'(0) = 0$ also follows

(1) $\kappa(P) = y''(0)\,,$

by the formula for curvature (§7.2.4(1)). The geometrical reason for this equality is that curvature is the rate of change of angle of inclination with respect to length of arc, while the second derivative is the rate of change of the *tangent* of angle of inclination (slope) with respect to length of the *tangent* line (abscissa), and in the limit the ratios of angle and arc to their respective tangents become unity. In symbols (Fig. 7-23): $\kappa(P) = \lim(\alpha/s)$; since $(\tan\alpha)/\alpha \to 1$ (§4.4.4) and $s/x \to 1$ this limit is the same as

$$\lim\left[\frac{\tan\alpha}{\alpha}\cdot\frac{\alpha}{s}\cdot\frac{s}{x}\right] = \lim\frac{\tan\alpha}{x} = \lim\frac{y'}{x} = y''(0)\,.$$

If y'' is constant there is a simple expression for $y''(0)$ in terms of x and y. We have $y'' = y''(0)$ everywhere, whence by integration $y = \frac{1}{2}y''(0)x^2$, or

$$y''(0) = \frac{2y}{x^2}\,.$$

In this case the curve is a parabola with vertex at P. If instead it is a circular arc, the same result holds in the limit:

(2) $y''(0) = \lim\frac{2y}{x^2}\,.$

For the reciprocal of $y''(0) = \kappa(P)$ is the radius r, and in Fig. 7-24 we clearly have $RS \to 2r$, while by elementary geometry $RS = x^2/y$; therefore $x^2/2y \to r$, and the result follows.

The limit expression that holds for the circle turns out to be valid generally. This is a consequence of the second-order mean value theorem (§3.8.7) with $a = 0$, which asserts the existence of a number ξ between 0 and x such that

$$y(x) = \frac{1}{2}y''(\xi)x^2$$

(there is only one term on the right because y and y' vanish at 0). Then

$$\frac{2y}{x^2} = y''(\xi) \to y''(0)\,,$$

by continuity of y''. (Another proof is given in the next section.)

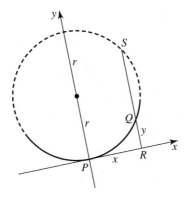

Fig. 7-24

Putting together our results (1) and (2) we have an expression for the curvature in terms of ordinate and abscissa:

$$(3) \qquad\qquad \kappa(P) = \lim \frac{2y}{x^2}.$$

7.3.3 Geometrical proof. The expression for $y''(0)$ as a limit can be obtained by an elementary argument reminiscent of the one that established the limit of $(\sin h)/h$ (§4.4.2).[9] The demonstration does not use the continuity of y'' (as the preceding proof did), but we have to assume that y'' is monotone on each side of 0—a mild hypothesis, as the limit involves only the immediate vicinity of the origin. To be definite, let y'' be increasing for $x \geq 0$, and consider the graph of y' (rather than of y) in that region. Since the derivative of y' is positive and increasing, while $y'(0) = 0$, the graph, represented in Fig. 7-25 by arc PS, rises from the origin and is concave up. Let SR be an ordinate, and draw secant PS and tangent PT.

Two representations of $y''(0)$ are afforded by the figure:

(i) as the slope of PT,

$$y''(0) = \frac{RT}{PR} = \frac{RT}{x};$$

(ii) as the limit of the slope of PS,

$$y''(0) = \lim \frac{RS}{PR} = \lim \frac{y'}{x}.$$

---

[9]The argument is from Newton's *Principia*, Book 1, Lemmas 9 and 10.

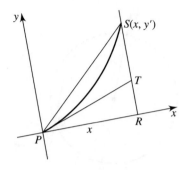

Fig. 7-25

We have

$$\triangle PRT < \text{area } PRS \text{ under arc } PS < \triangle PRS,$$

or

$$\frac{1}{2} PR \cdot RT < \int_0^x y' \, dx < \frac{1}{2} PR \cdot RS,$$

or, evaluating the integral and using (i),

$$\frac{1}{2} x^2 y''(0) < y < \frac{1}{2} xy',$$

and finally

$$y''(0) < \frac{2y}{x^2} < \frac{y'}{x}.$$

By (ii), then, necessarily $2y/x^2 \to y''(0)$.

A similar argument applies for $x \leq 0$, or if y'' is decreasing rather than increasing (Q1).

7.3.4　The second derivative as a limit.　In the derivation of (2) from the second-order mean value theorem continuity of y'' was needed only at 0, so we have the following result (as yet *proved* only under the different hypothesis of the last section).

> **Proposition 0.**　*If $f(x)$ is twice differentiable in a neighborhood of 0 and f'' is continuous at 0, and $f(0) = f'(0) = 0$, then as $x \to 0$*
>
> $$\frac{f(x)}{x^2} \to \frac{1}{2} f''(0).$$

It implies the apparently more general

Proposition a.   *If $f(x)$ is twice differentiable in a neighborhood of a and f'' is continuous at a, then as $x \to a$*

$$\frac{f(x) - [f(a) + f'(a)(x-a)]}{(x-a)^2} \to \frac{1}{2}f''(a).$$

Like Proposition 0, Proposition a is an immediate consequence of the second-order mean value theorem. To deduce it from Proposition 0, let

$$g(x) = f(a+x) - [f(a) + f'(a)x];$$

then g satisfies the hypotheses of Proposition 0, so as $x \to 0$, $g(x)/x^2 \to \frac{1}{2}g''(0) = \frac{1}{2}f''(a)$. Hence as $x \to a$, i.e. $x-a \to 0$, $g(x-a)/(x-a)^2 \to \frac{1}{2}f''(a)$; and this is the assertion of Proposition a, because

$$g(x-a) = f(x) - [f(a) + f'(a)(x-a)].$$

The significance of Proposition a will be explained in §8.1.2.

7.3.5   Geometrical interpretations and applications.   Let curves, coordinates, etc., be as in §7.3.1. The first example shows that curvature measures deviation from the tangent line.

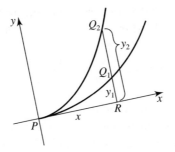

Fig. 7-26

EXAMPLE 1.   *The ratio of the curvatures at P of two curves (tangent at P) is the limit of the ratio of their ordinates.* For (Fig. 7-26) $\kappa_1(P) = \lim 2y_1/x^2$ and $\kappa_2(P) = \lim 2y_2/x^2$, hence

$$\frac{\kappa_1(P)}{\kappa_2(P)} = \lim \frac{2y_1/x^2}{2y_2/x^2} = \lim \frac{y_1}{y_2}.$$

A corollary is that if the curvatures are the same, the ratio of ordinates approaches unity. In particular, this is the case for any curve and its circle of curvature at P: in the limit, curve and circle depart alike from the tangent.

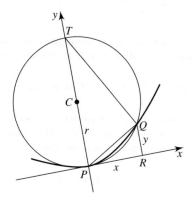

Fig. 7-27

EXAMPLE 2. *The circle of curvature of a curve at P is the limit of circles tangent to the curve at P and passing through a point Q of it.* That is, the radii of such circles approach the radius of curvature as a limit. For in circle PQT of radius $CP = r$ (Fig. 7-27) we have, by similar triangles, $PT = PQ^2/QR$. Since $PT = 2r$, $QR = y$, and x/PQ (ratio of tangent to chord, or equivalently $\cos \angle RPQ) \to 1$, this implies $\lim 2r = \lim x^2/y$, whence follows

$$\lim r = \lim \frac{x^2}{2y} = \frac{1}{\kappa(P)} = \rho(P).$$

This property can be taken as the *definition* of radius of curvature.[10] If that is done, the preceding argument shows that the radius so defined is the same as the radius $\rho = 1/\kappa$ defined by the method we have used.

EXAMPLE 3. *The center of curvature of a curve at P is the limiting position of the intersection of the normal to the curve at Q with the normal at P* (the y-axis). For (Fig. 7-28) the slope of normal QV is $-1/y'$, so $UV = x/y'$ and therefore $PV = y + (x/y')$. In the limit $y \to 0$ and $x/y' \to 1/y''(0) = 1/\kappa = \rho$; hence $PV \to \rho$.

---

[10] Cf. Newton's *Principia*, Book 1, Lemma 11 and Prop. 6, Cor. 3.

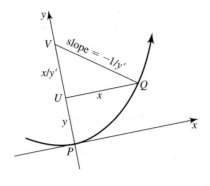

Fig. 7-28

Another characterization of the circle of curvature resembles the description of the tangent line at P as the limit of secants, or lines through P and one other point.

EXAMPLE 4. *The circle of curvature of a curve at P is the limit of circles passing through P and two points Q_1, Q_2 of the curve. That is, the center of curvature is the limiting position of the centers of such circles.*

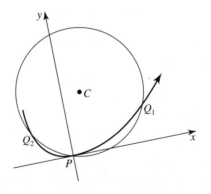

Fig. 7-29

See Fig. 7-29 (C is the center of circle Q_1PQ_2). A proof of the case in which Q_1 and Q_2 are on opposite sides of P is indicated in Q3.

7.3.6　QUESTIONS

Q1 (§7.3.3). Give the argument for the case when y'' decreases for $x \geq 0$.

Q2. Deduce Example 2 of §7.3.5 from Example 3. (By §4.2.7.Q11, there is a point Q' on the curve between P and Q (Fig. 7-27) such that CQ' is normal to the curve.)

Q3 (§7.3.5, Example 4). Supply the proof. (Find Q'_1 and Q'_2 as in Q2. Let CQ'_1 and CQ'_2 intersect the y-axis at C_1 and C_2 respectively, and use Example 3 to show that C approaches the center of curvature.)

7.4   PARAMETRIC REPRESENTATION

7.4.1   Parametric representation of curves. Time as parameter.

We have already met the idea of representing the motion of a particle in the plane by two functions $x(t)$ and $y(t)$ which together specify the coordinates of its position at any time t (§§4.2.4, 4.5.9). If a curve is thought of as traced out in time by such a motion (of a pen-point, if you like), it thereby acquires a description as the collection of points (x, y) determined by the equations

$$x = x(t), \quad y = y(t),$$

where t takes values in some finite or infinite interval of numbers. Thus, for example, the unit circle $x^2 + y^2 = 1$ can be represented by the pair of equations

$$x = \cos t, \quad y = \sin t,$$

since the point $(\cos t, \sin t)$ moves around it as t increases (§3.3.5). In this case the geometrical interpretation of t as a central angle, or the periodicity of the sine and cosine, shows that we must restrict t to an interval of length 2π if we are to trace out the curve once only.

For this kind of representation of a curve the essential element is the introduction of a third, independent, variable t that governs x and y. Regarding t as time may be helpful, but is not necessary. Whether, in a given situation, the independent variable t is interpreted as time, or as a geometrical quantity, or not at all, it is called a **parameter**.[11] We **parametrize** a curve by choosing a parameter t together with **parametric equations** $x = x(t)$, $y = y(t)$, and an interval of values of t; the result is a **parametric representation**, or **parametrization**, of the curve. (Needless to say, another letter than t may be used for the parameter.) Let us assume that each point (x, y) of the curve is determined by only one value t; thus a moving particle cannot

---

[11]After the parameter of a conic section—because varying t produces a family of points (a curve), as varying the Apollonian parameter produces a family of curves.

reverse its course, or return to the path it has traveled.[12] Then with the se-
lection of a parametrization a direction is assigned to the curve, the direction
of increasing t.

As this talk of choice implies, a parametrization of a given curve is by no
means unique. Indeed, a particle may traverse the curve at any constant or
variable speed, and during any time interval; corresponding to each individual
motion will be a set of parametric equations for the curve. A parametrization
available for any curve that is the graph of a function $y = f(x)$ is provided by
the equations

$$x = t, \quad y = f(t),$$

where the t-interval is the same as the x-interval. The upper semicircle of the
unit circle (Fig. 7-30), for example, in addition to the parametrization (a)

$$x = \cos t, \quad y = \sin t, \quad 0 \le t \le \pi,$$

has also the quite different one (b)

$$x = t, \quad y = \sqrt{1 - t^2}, \quad -1 \le t \le 1.$$

In the latter representation the positive direction is clockwise and the particle
P does not move at constant speed (Q2).

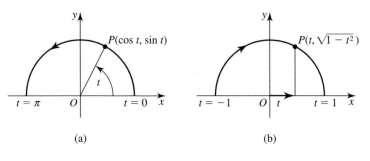

(a)   (b)

Fig. 7-30

The advantages that a well-chosen parametrization of a curve may have over
its representation by one or more functions $f(x)$ include the *simplicity* of the
parametric equations and the *unity* with which they express the whole curve.

---

[12] An exception can be made where a closed curve returns to its starting point; e.g., for
the circle above we can admit the interval $0 \le t \le 2\pi$.

The unit circle can be described by (i) $y = \sqrt{1 - x^2}$ (upper semicircle) and $y = -\sqrt{1 - x^2}$ (lower semicircle), or by (ii) $x = \cos t$ and $y = \sin t$; while the latter functions are smooth and define the whole curve as t varies from 0 to 2π, the former require the curve to be unnaturally divided in two, and they fail even to be differentiable at $x = \pm 1$, although the points $(-1, 0)$ and $(1, 0)$ of the curve are in no way exceptional.

If a curve is originally given by parametric equations $x = x(t)$, $y = y(t)$, we may be able to obtain a non-parametric equation for it by eliminating t from these equations. In the case of the unit circle, given $x = \cos t$, $y = \sin t$, we find $x^2 + y^2 = \cos^2 t + \sin^2 t = 1$, or simply $x^2 + y^2 = 1$.

7.4.2   Differentiation with respect to a parameter.   In order to study a curve which can be represented either by an equation $y = f(x)$ or by parametric equations $x = x(t)$, $y = y(t)$, we adopt a special notation for differentiation with respect to the parameter, namely a dot placed above a letter. The prime is then reserved for differentiation with respect to x. Thus we write, e.g., $dy/dx = y'$ but $dy/dt = \dot{y}$, $d^2y/dt^2 = \ddot{y}$. In what follows the functions $x(t)$ and $y(t)$ are supposed continuously differentiable, or twice continuously differentiable if necessary.

Given the equations just mentioned, at any t the values $x(t)$ and $y(t)$ satisfy $y = f(x)$; that is, $y(t) = f(x(t))$. Then by the chain rule

$$\frac{dy}{dt} = \frac{dy}{dx}\frac{dx}{dt},$$

or

$$\frac{dy}{dx} = \frac{dy}{dt} \bigg/ \frac{dx}{dt},$$

or, in our new notation,

$$y' = \frac{\dot{y}}{\dot{x}}.$$

Similarly, if x is expressed as a function of y we have $dx/dy = \dot{x}/\dot{y}$.

Relations such as these are useful where a curve given parametrically can be represented by a differentiable function $y(x)$ or $x(y)$, but we do *not* assume that that is the case at all points. We therefore have to derive the parametric arc-length formula independently of the formula found in §7.2.1.

7.4.3   Parametric expressions for arc length and curvature.   Let an arc-length coordinate s be chosen for a given curve. Since we have assumed that the positive direction is that of increasing t, ds is of the same sign as dt (or possibly zero). It therefore follows from

$$ds^2 = dx^2 + dy^2$$

(§7.2.1) that

$$ds = \sqrt{\left(\frac{dx}{dt}\right)^2 + \left(\frac{dy}{dt}\right)^2} \; dt \,,$$

or

$$ds = \sqrt{\dot{x}^2 + \dot{y}^2} \; dt \,.$$

Then corresponding to the arc-length integral with respect to x derived in §7.2.1, which is

$$s_2 - s_1 = \int_{x_1}^{x_2} \sqrt{1 + y'^2} \; dx \,,$$

we now have the parametric form

$$s_2 - s_1 = \int_{t_1}^{t_2} \sqrt{\dot{x}^2 + \dot{y}^2} \; dt \,,$$

where t_1 and t_2 are the t-values corresponding to s_1 and s_2. The arc length is expressed as a function of t by

$$s(t) = \int_{t_0}^{t} \sqrt{\dot{x}^2 + \dot{y}^2} \; dt \,,$$

where t_0 corresponds to $s = 0$. If the parametric equations define the motion of a particle, this is its *location along the curve* at time t.

The derivative of $s(t)$ is $\dot{s}(t) = \sqrt{\dot{x}^2 + \dot{y}^2}$. This is the rate of change of the location s of the particle with respect to time; in other words, it is the *speed at which the particle describes the curve*. If we assume that it never vanishes (the particle never stops), then at all times

$$\dot{s} = \sqrt{\dot{x}^2 + \dot{y}^2} > 0 \,.$$

In consequence $\dot{x}$ and $\dot{y}$, the velocities with respect to the separate axes, are never *both* zero: the particle is always moving with respect to one axis or the other. Where $\dot{x} = 0$ the motion is vertical; the curve has a vertical tangent, y' is not defined, and the formula $y' = \dot{y}/\dot{x}$ does not apply. Where $\dot{y} = 0$ the tangent is horizontal and dx/dy, which elsewhere equals $\dot{x}/\dot{y}$, fails to exist.

To calculate the curvature in terms of t we proceed as in §7.2.4. At a point where $\dot{x} \neq 0$ the angle of inclination is $\alpha = \arctan y' = \arctan(\dot{y}/\dot{x})$; then

$$\dot{\alpha} = \frac{d\alpha}{dt} = \frac{1}{1 + (\dot{y}/\dot{x})^2} \frac{\dot{x}\ddot{y} - \dot{y}\ddot{x}}{\dot{x}^2} = \frac{\dot{x}\ddot{y} - \dot{y}\ddot{x}}{\dot{x}^2 + \dot{y}^2}.$$

The denominator is $\dot{s}^2$, while by the chain rule $\kappa = d\alpha/ds = \dot{\alpha}/\dot{s}$; hence

$$\kappa = \frac{\dot{x}\ddot{y} - \dot{y}\ddot{x}}{(\dot{x}^2 + \dot{y}^2)^{3/2}} = \frac{\dot{x}\ddot{y} - \dot{y}\ddot{x}}{\dot{s}^3}.$$

In the same way, where $\dot{y} \neq 0$ let β be the angle of inclination to the y-axis; then $\beta = -\arctan(\dot{x}/\dot{y})$ and differentiation yields the same formula for $\kappa = d\beta/ds$ (Q4).

> The foregoing derivation of the expression for $s(t)$ can be made rigorous in the manner of §7.2.3. A difference is that the mean value theorem gives *two* numbers τ_i' and τ_i'' such that $c_i = \sqrt{\dot{x}(\tau_i')^2 + \dot{y}(\tau_i'')^2}$, so that the sum of the c_i does not quite have the form of the kind of approximating sum we are used to. Nevertheless its limit is the integral; the proof sketched in §2.4.5 covers this type of sum as well as the simpler one. In this way rectifiability is established for all curves that can be parametrized by smooth functions (see also the remark at the end of §7.4.4). If $\dot{s} \neq 0$, it is easy to see that the integral formula implies Proposition 2 of §7.1.1 and is implied by it.

7.4.4 Parametrization by arc length. The parametric formulas for arc length and curvature take their simplest form when the parameter is chosen to be arc length itself, $t = s$. In that case $\dot{s} = 1$: when time is distance, the particle moves at *unit* speed along the curve. The arc-length integral is just

$$s_2 - s_1 = \int_{s_1}^{s_2} ds,$$

while the curvature is

$$\kappa = \dot{x}\ddot{y} - \dot{y}\ddot{x}$$

(see also Q6). An important example is given by the straight line.

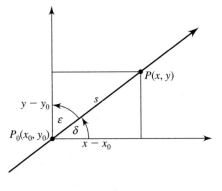

Fig. 7-31

EXAMPLE 0. A straight line is parametrized by arc length in the following way. Take a point P_0 on the line, and measure s from P_0 (Fig. 7-31). Then $x - x_0$ and $y - y_0$ are both proportional to s. Let p and q be the constants of proportionality,[13] so that $x - x_0 = ps$ and $y - y_0 = qs$; then the parametric equations are

$$x = x_0 + ps \,,$$

$$y = y_0 + qs \,.$$

The constants p and q are the cosines of the angles δ and ε which the positive direction of the line makes with the positive x-axis and y-axis respectively; for this reason they are called the **direction cosines** of the line.[14] They satisfy the relation $p^2 + q^2 = 1$, because $s^2 = (ps)^2 + (qs)^2$ (or because $\cos \varepsilon = \pm \sin \delta$). If the parameter is eliminated (by multiplying the parametric equations for x and y by q and p respectively and subtracting) the result is

$$q(x - x_0) - p(y - y_0) = 0 \,,$$

a non-parametric equation of the line. For $p \neq 0$ the slope of the line is q/p. The curvature $\kappa(s) = \dot{x}\ddot{y} - \dot{y}\ddot{x}$ is of course zero, since $\ddot{y} = \ddot{x} = 0$.

In this example the parametric equations show that $\dot{x} = p$ and $\dot{y} = q$. In general, assuming Proposition 2 of §7.1.1 to hold, $\dot{x}(s)$ and $\dot{y}(s)$ are the

---

[13] If the line is horizontal or vertical, one of these is zero.

[14] Any choice of angle δ between the positive x-direction and the positive direction of the line will yield the same cosine, for $\cos \delta = \cos(-\delta) = \cos(\delta + 2\pi)$, etc. Likewise for ε.

direction cosines of the tangent arrow to a curve at s (Q16). Then smoothness of the curve (its tangent turning continuously) corresponds to smoothness of the functions $x(s)$ and $y(s)$ (continuity of their derivatives). It follows that a curve defined by smooth parametric functions $x(t)$, $y(t)$ is smooth if ds/dt is nowhere zero. In the next example ds/dt vanishes at one point.

7.4.5   Examples.   Let us now apply parametric representation to the examples of §7.2.2.

EXAMPLE 1. The equation $y = x^{3/2}$, or $y = (\sqrt{x})^3$, $x \geq 0$, suggests introducing the parameter $t = \sqrt{x}$. Doing so, we obtain parametric equations $x = t^2$, $y = t^3$; these functions of t are better behaved than the original function of x, being everywhere defined and differentiable as many times as we like. As t increases from 0 towards ∞, x and y do the same; thus the interval $0 \leq t < \infty$ of the parameter corresponds to the given interval $0 \leq x < \infty$.

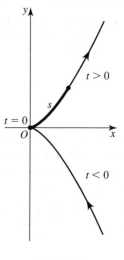

If we allow t to assume negative values, we double the graph of Fig. 7-11 (Fig. 7-32). By differentiating we obtain $\dot{x} = 2t$, $\dot{y} = 3t^2$; the speed of the moving particle is then

$$\dot{s} = \sqrt{\dot{x}^2 + \dot{y}^2} = \sqrt{4t^2 + 9t^4}$$

$$= |t|\sqrt{4 + 9t^2} = 3|t|\left(t^2 + \frac{4}{9}\right)^{1/2},$$

so the arc length from 0 to any $t \geq 0$ is

$$s = \int_0^t \dot{s}\, dt = \int_0^t 3t\left(t^2 + \frac{4}{9}\right)^{1/2} dt$$

$$= \left(t^2 + \frac{4}{9}\right)^{3/2} - \frac{8}{27},$$

Fig. 7-32

in agreement with the calculation in §7.2.2.

At $t = 0$ there is a cusp (§2.9.6). The direction of the curve changes abruptly by 180°, and the moving particle stops momentarily, for $\dot{s} = 0$. The curvature is not defined at this point, but everywhere else we have

$$\kappa = \frac{\dot{x}\ddot{y} - \dot{y}\ddot{x}}{\dot{s}^3} = \frac{(2t)(6t) - (3t^2)(2)}{|t|^3(4 + 9t^2)^{3/2}} = \frac{6}{|t|(4 + 9t^2)^{3/2}}$$

—always positive, because the curve is concave towards the left.

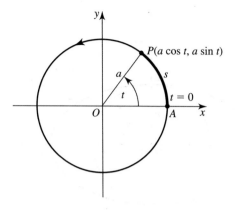

Fig. 7-33

EXAMPLE 2. The circle of radius a about the origin can be parametrized by its central angle t (Fig. 7-33), as we have already seen in the case $a = 1$ (§7.4.1). The parametric equations are $x = a\cos t$, $y = a\sin t$, with $0 \le t \le 2\pi$. Then $\dot{x} = -a\sin t$, $\dot{y} = a\cos t$, and the speed of the moving particle is

$$\dot{s} = \sqrt{\dot{x}^2 + \dot{y}^2} = a\,.$$

For the arc length from A to P we therefore have

$$s = \int_0^t \dot{s}\,dt = at\,.$$

Differentiating $\dot{x}$ and $\dot{y}$ gives $\ddot{x} = -a\cos t$, $\ddot{y} = -a\sin t$, so the curvature is

$$\kappa = \frac{\dot{x}\ddot{y} - \dot{y}\ddot{x}}{\dot{s}^3} = \frac{a^2\sin^2 t + a^2\cos^2 t}{a^3} = \frac{1}{a}\,.$$

If instead we parametrize the circle by arc length $s = at$, the equations become $x = a\cos(s/a)$, $y = a\sin(s/a)$, where $0 \le s \le 2\pi a$. The speed $\dot{s}$ is now 1, but calculation of κ gives the same result as before (Q3).

EXAMPLE 3. The first parametrization of the circle in Example 2 suggests a parametrization for the ellipse $\dfrac{x^2}{a^2} + \dfrac{y^2}{b^2} = 1$, namely $x = a\cos t$, $y = b\sin t$, $0 \le t \le 2\pi$. Here t has a geometric interpretation as the **eccentric angle** at the center (Fig. 7-34): the point P on the ellipse corresponding to t is determined as the intersection of the vertical UP and the horizontal VP drawn from the points U and V, on the circumscribed and inscribed circles respectively,

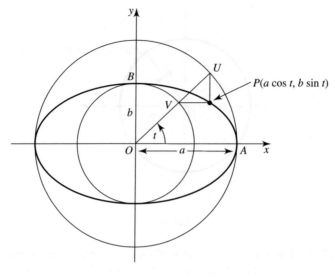

Fig. 7-34

which correspond to t as central angle (Q8). When b is set equal to a this parametrization reduces to the former one.

We find

$$\dot{s} = \sqrt{a^2 \sin^2 t + b^2 \cos^2 t}\,.$$

The length of the arc from B to P, which was expressed in Example 3 of §7.2.2 as an integral with respect to x, in terms of t is

$$-\int_{\pi/2}^{t} \sqrt{a^2 \sin^2 t + b^2 \cos^2 t}\; dt$$

(the minus sign is needed to make the quantity positive, because $t < \pi/2$). This is then another form of the elliptic integral of the earlier example.

The parametric formula for the curvature is

$$\kappa = \frac{ab}{(a^2 \sin^2 t + b^2 \cos^2 t)^{3/2}} = \frac{ab}{\dot{s}^3}$$

(cf. Example 6 of §7.2.5). Thus in this representation the particle travels at a speed inversely proportional to the cube root of the curvature.

7.4.6 Intrinsic equation of a curve.[15] I have already said (§7.1.7)
that in virtue of their definitions the arc length and curvature of a curve are
independent of what coordinate axes may be used to provide the curve with an
equation $y = f(x)$. With a change of axes the equation changes too, and often
in such a way that one would not at once know that the new equation and the
old represent the same curve; for instance, the standard equation of an ellipse,
$\dfrac{x^2}{a^2} + \dfrac{y^2}{b^2} = 1$, upon rotation and translation of axes becomes a rather general
equation of the second degree, which can be recognized as defining an ellipse of
semiaxes a and b only after some treatment. The form of parametric equations
$x = x(t)$, $y = y(t)$ likewise depends on the choice of axes, and also on the
choice of t. But arc length and curvature are (apart from sign) independent
of these choices, as can be verified directly by use of our formulas for s and
κ (Q10, Q11). We may then ask whether a given curve can be defined by
an equation involving only s and κ, and therefore independent of coordinate
axes—an equation truly characteristic of the curve itself.

It is plausible that an equation of the form

$$\kappa = f(s)$$

will determine a curve, since it specifies at each location s the rate at which
direction changes with respect to distance. This is in fact the case, and such
an equation is called a **natural** or **intrinsic** equation of the curve; s and κ
are natural or intrinsic coordinates. (For $\kappa \neq 0$ we can also write the equation
in the form $\rho = g(s)$, where $g(s) = 1/|f(s)|$; then both variables represent
distances.) In more detail, the assertion is this: given any continuous function
$\kappa = f(s)$ defined on some interval $[s_1, s_2]$, where s and κ are variables which as
yet have no geometrical meaning, there is a curve of which s is the arc length
and κ the curvature; moreover, the curve is unique, in the sense that any other
curve having the same s and κ differs from it only by a rotation or translation
or both.[16]

To prove this, let us first suppose that we have a curve parametrized by
arc length, whose parametric equations are $x = x(s)$, $y = y(s)$, and whose
curvature is a continuous function $\kappa(s)$. The interval of definition $[s_1, s_2]$ of
these functions may be assumed to contain 0 (if it does not, we choose a point

---

[15]This section and the next one can be omitted.

[16]The curve determined by f may be self-intersecting.

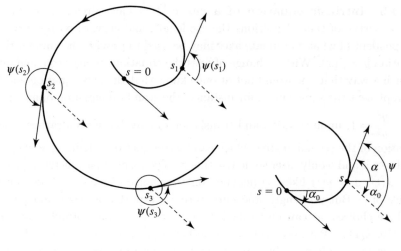

Fig. 7-35 Fig. 7-36

s_0 in the interval and replace s by $s - s_0$). The total, or rather cumulative, curvature from 0 to s (§§7.1.4–5; §7.2.6.Q12) is then

$$(1) \qquad \psi(s) = \int_0^s \kappa(s)\, ds$$

(Fig. 7-35). On an s-interval where $\dot{x} \neq 0$ this quantity differs only by a constant from the angle of inclination α, for both α and ψ have derivative κ.[17] Hence on such an interval we can write

$$\alpha = \psi + \alpha_0\,,$$

where α_0 is the angle of inclination at $s = 0$ (Fig. 7-36). Now since $\dot{y}/\dot{x} = \tan\alpha = \sin\alpha/\cos\alpha$ and $\dot{x}^2 + \dot{y}^2 = 1$ we must have either $\dot{x} = \cos\alpha$ and $\dot{y} = \sin\alpha$ or $\dot{x} = -\cos\alpha$ and $\dot{y} = -\sin\alpha$, and by continuity of $\dot{x}$ either one or the other possibility holds throughout the interval.[18] But the latter case can be turned into the former by adding π to α_0. Thus with an appropriate choice

---

[17]In the exceptional case that there is no such interval we can work with the angle of inclination to the y-axis.

[18]Here $\dot{x}$ and $\dot{y}$ are direction cosines of the tangent arrow (§7.4.4); the two cases correspond to its possible directions.

of constant γ we have

$$\dot{x} = \cos{(\psi + \gamma)},$$
$$\dot{y} = \sin{(\psi + \gamma)}.$$

Moreover, the formulas remain valid without the hypothesis $\dot{x} \neq 0$, for continuity of $\dot{y}$ prevents the signs of $\dot{x}$ and $\dot{y}$ from changing together at a point where $\dot{x} = 0$.

From these equations we recover $x(s)$ and $y(s)$ by integration:

(2)
$$x(s) = \int_0^s \cos{(\psi + \gamma)}\, ds + c_1,$$
$$y(s) = \int_0^s \sin{(\psi + \gamma)}\, ds + c_2.$$

Here $c_1 = x(0)$ and $c_2 = y(0)$. In these expressions ψ is determined by $\kappa(s)$, but the constants γ, c_1, and c_2 are not. It is clear that to change c_1 or c_2 is to translate the curve; and since $\psi + \gamma$ is the same as the angle of inclination (up to an additive multiple of π), to change γ is to rotate the curve. Arc length and curvature thus determine the curve up to translation and rotation.

It remains to show that if given an arbitrary continuous function $\kappa(s)$ (on an interval containing 0) we *define* a curve by the parametric equations (2), where ψ is defined by (1) and γ, c_1, and c_2 may be any constants, then s is its arc length and κ its curvature. By differentiation we obtain

$$\dot{x} = \cos{(\psi + \gamma)},$$
$$\ddot{x} = -[\sin{(\psi + \gamma)}]\,\dot{\psi} = -\kappa \sin{(\psi + \gamma)},$$
$$\dot{y} = \sin{(\psi + \gamma)},$$
$$\ddot{y} = [\cos{(\psi + \gamma)}]\,\dot{\psi} = \kappa \cos{(\psi + \gamma)}.$$

Then the arc length of the curve is

$$\int_0^s \sqrt{\dot{x}^2 + \dot{y}^2}\, ds = \int_0^s 1\, ds = s$$

and its curvature is

$$\dot{x}\ddot{y} - \dot{y}\ddot{x} = \kappa,$$

as expected. This concludes the proof.

Several examples are given in the next section.

Curves in three-dimensional space can likewise be characterized by natural coordinates. Besides arc length s and curvature κ a third coordinate is required, the **torsion** τ, which measures the *twisting* of a space curve (a kind of turning impossible in the plane). There is then a pair of natural equations, $\kappa = \kappa(s)$ and $\tau = \tau(s)$.

7.4.7   Examples.   EXAMPLE 1.   A straight line has $\kappa = 0$, and a circle of radius a, $\kappa = \pm 1/a$. These are then the *only* curves of constant curvature (i.e. constant for all s), since the intrinsic equation determines the curve up to translation and rotation.

Starting from the equation $\kappa = $ constant we can determine parametric equations for the line and circle by calculating the integrals (1) and (2); this work is to be carried out in Q12.

EXAMPLE 2.   Figure 7-37 shows a thread TP being unwound (without stretching) from a circular spool of radius a. The curve traced out by its end is called the **involute** of the circle. For $P \neq A$ it is plausible that T is the center of curvature at P. Assuming this to be so, tangent TP is normal to the curve at P, and therefore the total-curvature angle ψ formed by the tangent at P and the x-axis is equal to the central angle AOT. Then $\rho = TP = \operatorname{arc} AT = a\psi$. We always have $d\psi/ds = \kappa$, so $ds/d\psi = \rho = a\psi$; consequently if s is measured

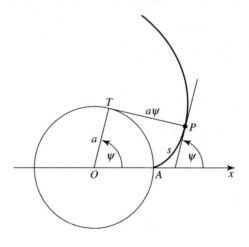

Fig. 7-37

from A, $s = \frac{1}{2}a\psi^2$, or $\psi = \sqrt{2s/a}$. Thus for $s > 0$

$$\kappa = \frac{1}{a\psi} = \frac{1}{\sqrt{2as}} :$$

the curvature of the involute is inversely proportional to the square root of the arc length. For a circle of unit diameter the radius of curvature is

$$\rho = \sqrt{s}.$$

Parametric equations are found in Q13, where in addition the hypothesis about the center of curvature is verified.

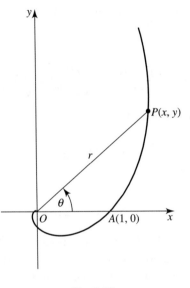

Fig. 7-38

EXAMPLE 3. Let us determine the intrinsic equation of the logarithmic spiral $r = e^{\theta}$ (Fig. 7-38; §3.3.12). Taking θ as parameter, we use the relations $x = r\cos\theta$, $y = r\sin\theta$ to obtain parametric equations

$$x = e^{\theta}\cos\theta,$$

$$y = e^{\theta}\sin\theta.$$

Then

$$\dot{x} = e^{\theta}(\cos\theta - \sin\theta),$$

$$\dot{y} = e^{\theta}(\sin\theta + \cos\theta),$$

$$\ddot{x} = -2e^{\theta}\sin\theta,$$

$$\ddot{y} = 2e^{\theta}\cos\theta.$$

Hence

$$\dot{s} = \sqrt{\dot{x}^2 + \dot{y}^2} = \sqrt{2}\,e^{\theta},$$

so the arc length measured from $A(1,0)$ is

$$s = \int_0^{\theta} \sqrt{2}\,e^{\theta}\,d\theta = \sqrt{2}\,(e^{\theta} - 1).$$

The curvature is

$$\kappa = \frac{\dot{x}\ddot{y} - \dot{y}\ddot{x}}{\dot{s}^3} = \frac{2e^{2\theta}}{2\sqrt{2}\,e^{3\theta}} = \frac{1}{\sqrt{2}\,e^{\theta}},$$

and the intrinsic equation is then

$$\kappa = \frac{1}{s + \sqrt{2}},$$

or

$$\rho = s + \sqrt{2}.$$

This takes an even simpler form if we adjoin to the curve the point it approaches as $\theta \to -\infty$, namely the pole $(r = 0)$, and then measure arc length from that point. It is possible to do this, because although the spiral winds about the pole infinitely many times for $\theta \le 0$, its length from A to the pole has the finite value $\int_{-\infty}^0 \dot{s}\,d\theta = \sqrt{2}$. Then

$$s = \int_{-\infty}^{\theta} \dot{s}\,d\theta = \sqrt{2}\,e^{\theta},$$

and the intrinsic equation becomes

$$\rho = s.$$

At the pole the curvature is infinite.

The general case of a logarithmic spiral is addressed in Q15.

7.4.8 QUESTIONS

Q1.　Find parametric equations with respect to arc length of the line $y = 2x - 3$, when it is given each of its two possible directions. (In Example 0, §7.4.4, determine p and q from $p^2 + q^2 = 1$ and $q/p = 2$.)

Q2.　Calculate the speed $\dot{s}$ of the particle for each of the parametrizations of the semicircle illustrated in Fig. 7-30. What happens at the endpoints in (b)?

Q3 (§7.4.5, Example 2). Verify this.

Q4 (§7.4.3).　Establish the formula for β, and thereby verify that $\kappa = d\beta/ds$ is the same as before. (First define "slope" relative to the y-axis.)

Q5.　(a) Show that if t is not an odd multiple of $\pi/2$ the point $P(x = \sec t, y = \tan t)$ lies on the hyperbola $x^2 - y^2 = 1$. (b) Parametrize the right-hand branch of the hyperbola by finding a t-interval such that as t traverses that interval, P traverses the branch. (c) Parametrize the left-hand branch. (d) Parametrize the hyperbola by using hyperbolic functions (§4.1.5.Q7) instead of trigonometric ones.

Q6.　(a) Show that if parametrization is by arc length, $\dot{x}\ddot{x} + \dot{y}\ddot{y} = 0$. (Of what is the left side the derivative?) (b) Deduce that $\kappa = \ddot{y}/\dot{x}$ if $\dot{x} \neq 0$, $\kappa = -\ddot{x}/\dot{y}$ if $\dot{y} \neq 0$.

Q7.　The formula $\dot{s} = \sqrt{\dot{x}^2 + \dot{y}^2}$ is symmetrical in x and y; that is, $\dot{s}$ is not affected if x and y are interchanged. The same is not true of $\kappa = (\dot{x}\ddot{y} - \dot{y}\ddot{x})/\dot{s}^3$. What is the reason for this difference?

Q8 (§7.4.5, Example 3).　Verify that the construction described produces the indicated parametrization.

Q9.　In Example 3, §7.4.5, show that $\dot{s}$ takes its least value at A and its greatest value at B, and find those values.

Q10.　Let $x'y'$-axes (the primes do not signify differentiation) be obtained from xy-axes by rotation about the origin through an angle η. The xy- and $x'y'$-coordinates of any point P are related by the equations

$$x' = (\cos\eta)x + (\sin\eta)y\,,$$

$$y' = -(\sin\eta)x + (\cos\eta)y\,,$$

or equivalently

$$x = (\cos\eta)x' - (\sin\eta)y'\,,$$

$$y = (\sin\eta)x' + (\cos\eta)y'\,.$$

(a) Prove this by showing that under these equations (i) $x = 0$, $y = 0$ corresponds to $x' = 0$, $y' = 0$, (ii) $x = \cos\eta$, $y = \sin\eta$ corresponds to $x' = 1$, $y' = 0$, (iii) $x = -\sin\eta$, $y = \cos\eta$ corresponds to $x' = 0$, $y' = 1$, and (iv) the distance between any two points is unaffected by the transformation of coordinates. (b) Show that the formulas for arc length and curvature are unaltered in form by the transformation. (c) The same as (b) for the transformation of coordinates corresponding to a translation of axes.

By (b) and (c) together, arc length and curvature are independent of the coordinate axes.

Q11.   (a) A parametrization is **reversed** if the parameter t is replaced by the parameter $u = -t$. What is the effect upon the values of the quantities $\dot{x}$, $\dot{y}$, $\ddot{x}$, $\ddot{y}$, $\dot{s}^2$, $\dot{s}$, s, and κ at a given point? (b) Let parameters t and u be related by twice differentiable invertible functions $u = u(t)$, $t = t(u)$. Show by direct calculation from the formulas that s and κ have the same values whether calculated in terms of t or in terms of u, except that their signs are reversed if $du/dt < 0$.

Q12 (§7.4.7, Example 1).   Determine parametric equations from the natural equation, and compare with Examples 0 of §7.4.4 and 2 of §7.4.5.

Q13 (§7.4.7, Example 2).   Determine parametric equations from the argument given, and confirm that the curve they define is the involute.

Q14.   Parametrize the parabola $y = \frac{1}{2}x^2$ by $t = x$, calculate ψ from §7.4.6(1) (taking $s = 0$ at the origin), and verify that the result is the same as the angle of inclination found by differentiating $y = \frac{1}{2}x^2$.

Q15.   Find the intrinsic equation $\rho = \rho(s)$ of the general logarithmic spiral $r = r_0 e^{\alpha\theta}$, $\alpha > 0$, where s is measured (a) from $\theta = 0$, (b) from the pole. (Follow §7.4.7, Example 3.)

Q16 (§7.4.4).   Give the proof.

7.5   LINE INTEGRALS AND CURVILINEAR MOTION

7.5.1   Line integrals. Integration with respect to a parameter. Any quantity defined at the points of a curve with arc-length coordinate s can be considered a function of s. In many geometrical and physical situations we are interested in the integral of such a function with respect to s, called a **line integral** ("line" signifying what we have called a "curve"), or integral **along the curve**. A definite line integral has the general form

$$\int_{s_1}^{s_2} f(s)\, ds\,.$$

As we know, if the curve is the x-axis we can take s to be x itself (§7.1.2); thus the line integral generalizes the ordinary integral with respect to x,

$$\int_{x_1}^{x_2} f(x)\, dx\,.$$

In fact there would seem to be nothing new in the idea of line integral, inasmuch as the letter used to represent the variable of integration does not matter. However, $f(s)$ is not as a rule given explicitly as a function of s; rather, it is given as a function of another parameter t, or as a function of x and y (Art. 4.5), which in turn have expressions in terms of t. Regarding s as a function of t (§7.4.3) and putting $s_1 = s(t_1)$, $s_2 = s(t_2)$, we can then write the line integral as

$$\int_{t_1}^{t_2} f(s(t))\dot{s}(t)\,dt\,,$$

or

$$\int_{t_1}^{t_2} f(x(t), y(t))\sqrt{\dot{x}(t)^2 + \dot{y}(t)^2}\,dt\,.$$

The complicated-looking expressions $f(s(t))$ and $f(x(t), y(t))$ are in practice just "f expressed as a function of t." In particular, if f is given as a function of x and the curve admits an equation $y = y(x)$ the integral becomes

$$\int_{x_1}^{x_2} f(x)\sqrt{1 + y'(x)^2}\,dx\,.$$

If an oriented curve is designated by a single letter, say C, the line integral in the positive direction along its full length is denoted by

$$\int_C f(s)\,ds$$

(cf. the notation for an integral over a region, §§4.8.2–3). When the curve is closed, this symbol means the integral once around C.

A few examples will indicate the usefulness of line integrals and illustrate their evaluation by integration with respect to a parameter.

7.5.2   Examples. A. Length, curvature, line density. EXAMPLE 1. If $f(s) = 1$, the line integral is the arc length:

$$\int_{s_1}^{s_2} ds = s_2 - s_1\,.$$

We have already calculated such integrals, using x and t as variables of integration (§§7.2.2, 7.4.5). More generally, the line integral of any constant c is the product of c and the arc length:

$$\int_{s_1}^{s_2} c\,ds = c(s_2 - s_1)\,.$$

This trivial case is often encountered.

As a numerical example, the line integral of the constant 3 in the positive direction around a circle of radius 2 is $3 \cdot 2\pi(2) = 12\pi$.

EXAMPLE 2. The cumulative curvature ψ of a curve is the line integral of the curvature κ, as we saw in §7.4.6 (equation (1)). There we went on to obtain parametric formulas for x and y as line integrals of functions of ψ (equations (2)).

EXAMPLE 3. The concept of line density, whether of mass or charge or some other quantity, applies as well to a curved as to a straight line. Let a curved wire have line density $\lambda(s)$ kg/m; then its mass between s_1 and $s_2 > s_1$ is $\int_{s_1}^{s_2} \lambda(s)\,ds$; similarly for charge.

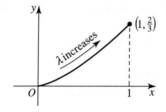

Fig. 7-39

Suppose, for example, that the line density of the arc of $y = \frac{2}{3}x^{3/2}$ from the origin to $(1, \frac{2}{3})$ is proportional to the abscissa, $\lambda(x) = 5x$ (Fig. 7-39; cf. §7.2.2, Example 1). Here λ is given as a function of x rather than of s. We have $y' = x^{1/2}$, so the mass (or other total quantity) is

$$\int_0^1 \lambda(x)\sqrt{1 + y'^2}\,dx = \int_0^1 5x\sqrt{1 + x}\,dx = \frac{4}{3}\left(\sqrt{2} + 1\right) \approx 3.22$$

(the integral can be evaluated by making the substitution $u = \sqrt{1 + x}$).

Alternatively, we can represent the curve parametrically by $x = t^2$, $y = \frac{2}{3}t^3$ (cf. §7.4.5, Example 1). Then $\lambda(t) = 5t^2$ and the mass is

$$\int_0^1 \lambda(t)\sqrt{\dot{x}^2 + \dot{y}^2}\,dt = \int_0^1 5t^2\sqrt{4t^2 + 4t^4}\,dt = \int_0^1 10t^3\sqrt{1 + t^2}\,dt,$$

which is equivalent to the other integral.

In contrast, if the straight segment of the x-axis from 0 to 1 has the same line density $\lambda(x) = 5x$, its mass is

$$\int_0^1 \lambda(x)\, dx = \int_0^1 5x\, dx = \frac{5}{2} = 2.5\,.$$

7.5.3   B. Work.   When a particle moves along a curve under the influence of a force $f(s)$ directed along the curve, the work done on it by the force from s_1 to s_2 is

$$W_{12} = \int_{s_1}^{s_2} f(s)\, ds$$

(cf. §§5.3.2–3). This is then also the change in the particle's kinetic energy due to the force. The next three examples are of line integrals representing work.

EXAMPLE 4. If the path of the particle is a radial straight line from a center of inverse-square force, we have the case treated in Art. 5.4. There we called the arc-length coordinate x; one would be more likely to call it r, following the notation of polar coordinates. As a variation on the earlier treatment, suppose that a movable proton (positively charged particle) Q approaches a fixed one P (Fig. 7-40); like charges repel, so the moving proton slows as it passes from $r = r_1$ to $r = r_2 < r_1$; how much work is done on it? The charge of a proton is 1.6×10^{-19} coulomb, and Coulomb's inverse-square law says that the force of repulsion (in newtons) is proportional to the quotient of the product of the charges by the square of the distance, with the constant of

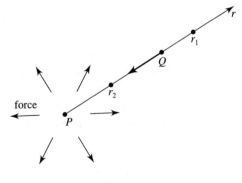

Fig. 7-40

proportionality as shown:

$$f(r) = (9.0 \times 10^9)\frac{(1.6 \times 10^{-19})^2}{r^2} = \frac{2.3 \times 10^{-28}}{r^2},$$

so the work done is

$$W_{12} = \int_{r_1}^{r_2} \frac{2.3 \times 10^{-28}}{r^2}\, dr = 2.3 \times 10^{-28}\left(\frac{1}{r_1} - \frac{1}{r_2}\right) \text{ newton m},$$

a negative quantity because the motion is against the force: $dr < 0$ during the motion from r_1 in to r_2.

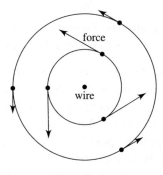

Fig. 7-41

EXAMPLE 5. A different kind of force is the magnetic force felt by a north magnetic pole placed in the vicinity of a straight wire carrying electric current (flow of charge). In any plane perpendicular to the wire it acts tangentially to all circles concentric with the wire (either counterclockwise everywhere or clockwise everywhere, depending on the direction of the current), and its magnitude is inversely proportional to the radius r of such circles (Fig. 7-41). Without concerning ourselves with how the direction and strength of the force are determined, we can write $f(s) = c/r$ and calculate in terms of c the work done by this force on a trip around a circle of radius a in the positive direction. Since the force has the constant value c/a along the circle, by Example 1 the work is

$$\frac{c}{a} \cdot 2\pi a = 2\pi c,$$

a quantity *independent of the radius.*

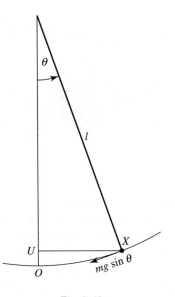

Fig. 7-42

EXAMPLE 6. In Example 4 the path was straight, and in Example 5 the force was constant. Consider now the motion of the pendulum bob of §5.2.4 from X to O (Fig. 7-42). We parametrize circle OX by θ, so that $s = l\theta$ and $ds = l\,d\theta$. The gravitational force on the particle at X directed along the curve is $f(s) = -mg\sin\theta$, so the work done on it from X to O is

$$W = \int_{\theta}^{0} (-mg\sin\theta)l\,d\theta = mgl\cos\theta\Big|_{\theta}^{0} = mgl(1 - \cos\theta) = mg \cdot OU.$$

This is exactly the same as the work that would be done by gravity on the particle if it descended in a straight line from U to O.

It is also the work that would be done if the particle moved along the longer path XUO, since along XU no gravitational force acts in the direction of motion. Thus the work done from X to O is the same whether the path taken is arc XO or the two straight segments XUO. This is a special case of a very general principle of the highest importance: for a so-called **conservative** force such as uniform or inverse-square gravitation, the work done along a path from A to B is *independent of the path*. In consequence a potential energy function can be defined, not just along a straight line but throughout space, so as to satisfy (2) and (3) of §5.3.3. No such function can exist for the magnetic force of Example 5, which is not conservative (Q9).

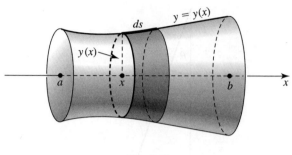

Fig. 7-43

7.5.4   C. Integration over a surface of revolution.   A final geometrical application is to a **surface of revolution**, which is the curved surface of a solid of revolution (§4.8.6): the surface obtained by revolving a graph $y = y(x)$ about the x-axis. As the volume of a solid of revolution is the sum of the volumes of infinitely thin *slices*, so the area of its curved surface is the sum of the areas of the infinitely narrow *bands*, or circular strips, which form the curved surfaces of these slices. A typical one of these surface elements is the band pictured in Fig. 7-43, which can be viewed as the surface of a frustum of a right circular cone with radius $y(x)$ at one end and infinitesimal slant height ds; its area is therefore $dS = 2\pi y(x)\,ds$.[19]

> This is a consequence of the formula $dA = 2\pi r\,dr$ of §4.8.3, Example 4, for the area of a plane ring element. Let a right circular conical surface be slit along a generator (straight line from the vertex) and unrolled to form a circular sector (Fig. 7-44). If

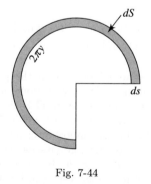

Fig. 7-44

---

[19]S here stands for "surface," and is different from the arc length S introduced in §7.1.2 and used subsequently.

whole ring = inner circumference × width,

surely

partial ring = inner partial circumference × width,

i.e. $dS = 2\pi y \times ds$. Similarly for the outer circumference. Where $y' = 0$ the band lies on a cylinder instead of a cone, and unrolls to a straight strip.

The area of the whole surface from $x = a$ to $x = b$ is then the line integral

$$S = \int_{s(a)}^{s(b)} 2\pi y(x)\,ds = \int_a^b 2\pi y\sqrt{1 + y'^2}\,dx\,.$$

Following §4.8.3 we can say further that if the surface is made of metal, and its surface density σ can be considered constant on each area element dS, then its total mass will be given by

$$M = \int_a^b 2\pi y\sigma\sqrt{1 + y'^2}\,dx\,;$$

here σ, like y, is necessarily a function of x. Other total quantities associated with the surface are expressed in the same way.

The following example corresponds to Example 1 of §4.8.6. (For the analogue of Example 2 of that section, see Q5.)

EXAMPLE 7. The area of the sphere of radius ρ obtained by revolving $y = \sqrt{\rho^2 - x^2}$ about the x-axis is

$$\int_{-\rho}^{\rho} 2\pi\sqrt{\rho^2 - x^2}\sqrt{1 + \frac{x^2}{\rho^2 - x^2}}\,dx = 2\pi\int_{-\rho}^{\rho} \rho\,dx = 4\pi\rho^2\,.$$

Consequently, if a surface density σ (or other specific quantity) is constant over the whole spherical surface, the mass (or other total quantity) is $4\pi\rho^2\sigma$. Integration of a non-constant σ is requested in Q8.

Our formulas here express integrals over a certain special kind of curved surface as line integrals, on the basis of a decomposition of the surface into circular elements. As with the plane regions of §§4.8.2–4, other decompositions of surfaces can be used to perform other integrations. A rather general approach is to consider a surface S that lies over a region R of the xy-plane (the graph of a function of two variables (§4.5.1)) as analogous to a curve lying above the straight x-axis (the graph of a function of one variable). Then in the same way as the element of arc length ds is expressed by multiplying dx by a suitable factor (namely $\sqrt{1 + y'^2}$), the element of surface area dS is expressed by multiplying dR by a suitable factor (Q11); here dR is an element infinitesimal in diameter, as described in §4.8.4.

7.5.5   Velocity and acceleration along a curve.   Let a particle move in the positive direction along a curve in the plane, so that its location $(x, y) = (x(t), y(t))$ gives a parametric representation of the curve in terms of time t. If we regard the location of the particle as the intersection of perpendiculars to the axes, we can associate with its curvilinear motion two rectilinear motions, namely the motions along the axes of the feet of those perpendiculars (Fig. 7-45). Their equations are respectively $x = x(t)$ and $y = y(t)$, the parametric equations of the curve.

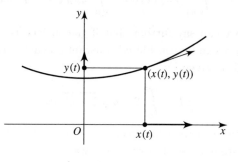

Fig. 7-45

The velocity of the particle moving through the plane is at every moment directed along the curve, and its magnitude is the derivative $\dot{s}$ of arc length with respect to time. Along the axes the corresponding velocities are $\dot{x}$ and $\dot{y}$, called the **component** velocities of the particle. Considered with reference to its motion parallel to one or the other axis, the particle moves with a component velocity: as it moves northeast, say, at velocity $\dot{s}$, it is also moving east at $\dot{x}$ and north at $\dot{y}$ (cf. §4.2.4).

Let us now draw tangent and normal axes at a point P, as in §7.3.1 (Fig. 7-46), and for the sake of convenience set $t = 0$ at P and measure arc length

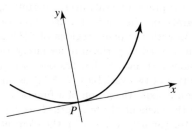

Fig. 7-46

s from P. Denoting differentiation with respect to x by a prime, we have $s' = \sqrt{1 + y'^2}$, hence $s'' = y'y''/\sqrt{1 + y'^2}$, and therefore

$$(1) \qquad\qquad s'(0) = 1, \quad s''(0) = 0.$$

Since at P the particle is moving in the x-direction only, it should have component velocities

$$\dot{x}(0) = \dot{s}(0), \quad \dot{y}(0) = 0.$$

To check this, we calculate as follows by the chain rule: $\dot{s} = s'\dot{x}$, so

$$\dot{s}(0) = s'(0)\dot{x}(0) = \dot{x}(0)$$

(because $s'(0) = 1$); likewise $\dot{y} = y'\dot{x}$, so

$$\dot{y}(0) = y'(0)\dot{x}(0) = 0$$

(because $y'(0) = 0$). Thus the speed at which the particle moves in the tangent direction is the same as the speed at which it moves along the curve; with respect to the normal direction it is stationary.

The acceleration of the particle along the curve—the rate at which its speed $\dot{s}$ is changing—is given by $\ddot{s}$, and we might by analogy with the preceding result expect to have $\ddot{x}(0) = \ddot{s}(0)$ and $\ddot{y}(0) = 0$. Differentiation of $\dot{s} = s'\dot{x}$ gives

$$\ddot{s} = s''\dot{x}^2 + s'\ddot{x},$$

from which by (1) does follow

$$(2) \qquad\qquad \ddot{x}(0) = \ddot{s}(0).$$

But $\ddot{y}(0)$ is not in general zero, for differentiation of $\dot{y} = y'\dot{x}$ yields

$$\ddot{y} = y''\dot{x}^2 + y'\ddot{x},$$

whence

$$(3) \qquad\qquad \ddot{y}(0) = y''(0)\dot{x}(0)^2.$$

Since $y''(0) = \kappa(0)$ (§7.3.2(1)) and $\dot{x}(0) = \dot{s}(0)$, this can be written

$$(4) \qquad\qquad \ddot{y}(0) = \kappa(0)\dot{s}(0)^2.$$

So while acceleration of the particle in the tangent direction is the same as its acceleration along the curve, there is also an *acceleration normal to the curve*, equal to the product of the curvature and the square of the speed.

We can derive (3) also from the representation of the second derivative as a limit given in §7.3.2(2). The function $y(t)$ satisfies $y(0) = \dot{y}(0) = 0$, hence

$$\ddot{y}(0) = \lim_{t \to 0} \frac{2y}{t^2} = \lim_{t \to 0} \frac{2y}{x^2} \left(\frac{x}{t}\right)^2 ;$$

but as t and $x \to 0$, $2y/x^2 \to y''(0)$ and $x/t \to \dot{x}(0)$, from which (3) follows.

Having established (2) and (4) at an arbitrary point P, let us change our notation so as to express these results without explicit reference to coordinates. Let v denote the speed of the particle, and let a_T and a_N denote its accelerations in the tangent and normal directions respectively. Then what we have proved is

(5)
$$a_T = \dot{v} ,$$
$$a_N = \kappa v^2 .$$

The former quantity is independent of the shape of the curve; the latter varies with shape and speed, but is independent of the rate at which the speed is changing. Wherever curvature is non-zero, even a particle moving at constant speed has normal acceleration.

7.5.6   Uniform circular motion.   In the case of uniform motion in a circle of radius r and center C, we have $a_T = 0$ and

(6)
$$|a_N| = \frac{v^2}{r} ;$$

the normal acceleration is directed toward C (Fig. 7-47). To produce this acceleration a force is required; by Newton's law $f = ma$ its magnitude is

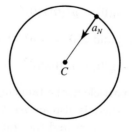

Fig. 7-47

$$f = \frac{mv^2}{r}.$$

The force may be supplied by a tense string, or by gravitational attraction, or in some other way.

EXAMPLE.[20] What holds the moon in its nearly circular orbit is terrestrial gravitation. Good evidence for this fact is provided by showing that if the two sides of (6) are computed separately, the left side being determined by gravity and the right side by the moon's motion, their values come out equal.

Let r be the mean radius of the moon's orbit; it is determined by observation to be about $60R = 3.8 \times 10^8$ m, where R is the radius of the earth. According to the inverse-square law (§5.4.1), if the acceleration of gravity (or f/m) is $g = 9.8$ m/sec^2 on earth, it is smaller than g by the ratio $(R/r)^2 = (1/60)^2$ in the vicinity of the moon.[21] That is,

$$|a_N| = 9.8 \times \left(\frac{1}{60}\right)^2 = 2.7 \times 10^{-3} \text{ m/sec}^2.$$

The moon circles the earth, traveling the distance $2\pi r$ in about 27 days, so at a velocity of about 1.0×10^3 m/sec. Hence

$$\frac{v^2}{r} = \frac{(1.0 \times 10^3)^2}{3.8 \times 10^8} = 2.6 \times 10^{-3} \text{ m/sec}^2.$$

There is agreement within our limits of accuracy.

7.5.7   The curve determined by accelerations. It follows from (5) that the tangential and normal accelerations of any motion of a particle along a curve determine the arc length s and curvature κ of the curve as functions of t by the formulas

$$v = \int a_T \, dt,$$

$$s = \int v \, dt,$$

---

[20] Newton's *Principia*, Book 3, Proposition 4.

[21] In symbols: $\dfrac{f}{m} = \dfrac{GM}{r^2} = \dfrac{GM}{R^2}\left(\dfrac{R}{r}\right)^2 = g\left(\dfrac{1}{60}\right)^2.$

$$\kappa = \frac{a_N}{v^2},$$

provided only that v is never zero. Upon eliminating t from the second and third equations we obtain the natural equation $\kappa = \kappa(s)$ of the curve (§7.4.6). Thus we can say that the curve itself is determined by a_T and a_N, regardless of how the moving particle's speed may vary.

7.5.8   QUESTIONS

Q1.   The line density of the arc of $y = \frac{1}{2}x^2$ from $x = 1$ to $x = 2$ is given by $\lambda(x) = x$. Find the mass of the arc.

Q2.   A car travels on a frictionless circular rail, propelled along it in the counter-clockwise direction by a fluctuating force of magnitude $\cos^2 \theta$, where θ is the central angle. If it starts from rest at $\theta = 0$, (a) how much work is done on it in the course of one revolution? (b) How fast is it moving at the end of the revolution? (c) The force acts tangentially, yet there is normal as well as tangential acceleration—why? (d) By Newton's second law, normal acceleration implies a normal force on the car. What exerts it?

Q3.   Find the line integral of xy from (x_0, y_0) to $(x_0 + 2k, y_0 + 2l)$ along the straight line $x = x_0 + kt$, $y = y_0 + lt$, where x_0, y_0, k, and l are constants, and t is a parameter. (Cf. §7.4.4, Example 0; here, however, t need not be arc length.)

Q4.   (a) By treating it as a surface of revolution, find the area of the curved surface of a right circular cone of base radius r and height h. (b) Express the result in terms of the slant height l (length of a generator) and verify it by the method of unrolling illustrated in Fig. 7-44.

Q5.   Find the surface area up to $x = b$ of the paraboloid of §4.8.6, Example 2.

Q6.   Compare the normal and tangential accelerations at a point P on a curve of (i) a particle traveling on the curve and (ii) a particle traveling at *constant* speed on the circle of curvature at P, whose speed and direction at P agree with those of the first particle.

Q7.   A particle moves around an ellipse, its location at time t being given by $x = a \cos t$, $y = b \sin t$ (§7.4.5, Example 3). Write formulas for a_T and a_N, evaluate them at A and B, and comment on the results. (Refer to the example cited and §7.4.8.Q9.)

Q8.   A hollow metal sphere has two poles, N and S. Its surface density at any P is given by $\sigma = 1 + d$, where d is the distance of P from axis NS. Find its mass.

Q9.   Show that the magnetic force of §7.5.3, Example 5, is not conservative (see the remark after Example 6).

Q10.   Let the origin of xy-coordinates be at the surface of the (flat) earth, and let the y-axis be vertical (positive direction up). At $t = 0$ a particle of mass m is projected horizontally in the positive x-direction from $(0, 1)$ at unit velocity (Fig. 7-48); no horizontal force acts on it thereafter. (a) (Galileo) Assuming that horizontal and

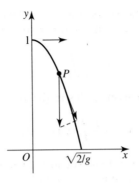

Fig. 7-48

vertical motions are determined independently by horizontal and vertical forces, show that the particle follows the parabola $y = 1 - \frac{1}{2}gx^2$ to the ground. (Find parametric equations, then eliminate t. Cf. §5.1.3.Q8.) (b) Apply the method of perpendicular projection onto the tangent used in §5.2.4 to determine the gravitational force in the direction of motion at any point P. (c) As in §7.5.3, Example 6, show by integration that the work done on the particle is the same as if it had fallen straight down.

Q11 (§7.5.4).   What is the factor?

Q12.   (a) Let a particle accelerate from rest at the origin in the positive direction along the x-axis, the motion beginning at time $t = 0$. Show that the location function $x(t)$ satisfies $\ddot{x}(0) = \lim\limits_{t \to 0} \dfrac{2x}{t^2}$. (Cf. the remark following (4), §7.5.5.) (b) Assuming $\ddot{x}(0) \neq 0$, what can be said about the ratio $x(t_1)/x(t_2)$ of displacements of the particle from the origin at different times?[22]

Q13.   The tangential and normal accelerations of a particle moving in the plane are both constant. What curve is it traveling on? (Distinguish zero and non-zero cases, and find the intrinsic equation.)

Q14.   (a) A closed convex curve C, oriented counterclockwise, is parametrized by t, $\alpha \leq t \leq \beta$. Show that its area is given by $A = -\int_\alpha^\beta y\dot{x}\, dt$, and also by $A = \int_\alpha^\beta x\dot{y}\, dt$. (First assume that the curve is located in the first quadrant, and refer to §1.5.5.Q2(b).) (b) Deduce that

$$A = -\frac{1}{2}\int_\alpha^\beta (y\dot{x} - x\dot{y})\, dt\,.$$

---

[22]Newton's *Principia*, Book 1, Lemma 10.

(c) Apply formula (b) to calculate the area of an ellipse. (d) Use formula (b) to show that if in polar coordinates C has equation $r = r(\theta)$, $0 \leq \theta \leq 2\pi$, its area is given by

$$A = \frac{1}{2} \int_0^{2\pi} r^2 \, d\theta \, .$$

Compare §4.8.5.Q9. (e) By canceling dt's in (b) we obtain

$$A = -\frac{1}{2} \int_C (y \, dx - x \, dy) \, ,$$

where the C indicates an integration around the curve in the positive direction. Is $y \, dx - x \, dy$ the differential of a function $f(x, y)$ (§4.5.7)? (Consider mixed partials.) (f) The last expression for A is independent of the parametrization. Show that it is also independent of the choice of xy-axes (use §7.4.8.Q10).

CHAPTER 8

Representation of Functions by Infinite Taylor Series

INTRODUCTION

The polynomials are a large and varied class of functions formed by the simplest arithmetic operations and presenting no obstacle to the performance of differentiation and integration. Although a natural wish that every function might be a polynomial goes unrealized, we will see in this chapter that most ordinary functions can be arbitrarily well *approximated* by polynomials of successively higher degrees, and this in a systematic way. There results a uniform representation of functions as polynomial-like expressions of infinitely many terms, somewhat analogous to the decimal system by which numbers are expressed as infinite "sums" of powers of ten with coefficients. The benefits for both understanding and calculation are very great.

In the first article this representation by the Taylor series, as it is called, is derived from polynomial approximations, and evidence of its value is provided by an uncritical look at the series associated with the exponential function and the sine and cosine. The second article begins with a theorem by which the Taylor series can be justified, and goes on to establish the validity of several important series by means of the theorem and certain special arguments.

8.1 POLYNOMIAL APPROXIMATION AND THE TAYLOR SERIES

8.1.1 The first three approximating polynomials. A. Constant and linear approximation. I will begin with a review of the approximations considered in §§3.7.11 and 3.8.6–8, with somewhat different emphasis.[1]

---

[1] It is advisable to read or re-read those sections.

Suppose that the value of a continuous function $f(x)$ is known at $x = a$, but unknown elsewhere. Because of continuity, it is reasonable to estimate its value at points x near a by

$$f(x) \approx f(a).$$

The estimate may be poor for a given x, but it can be made as accurate as we please by taking x close enough to a, because $f(x) \to f(a)$ as $x \to a$. Our estimation amounts to approximating $f(x)$ in the vicinity of a by the constant function $f(a)$. Geometrically speaking, we are approximating the height of a curve $y = f(x)$ near the point $(a, f(a))$ by the height of the horizontal line $y = f(a)$ through the point.

If f is differentiable and both $f(a)$ and $f'(a)$ are known we can do better, in two respects. To see this, first recall that as $x \to a$,

(1) $$\frac{f(x) - f(a)}{x - a} \to f'(a),$$

which is the same as

(2) $$\frac{f(x) - f(a)}{x - a} = f'(a) + \delta_1,$$

where $\delta_1 \to 0$ as $x \to a$. Let us write this equation in two other ways:

(3) $$f(x) = f(a) + (f'(a) + \delta_1)(x - a),$$

(4) $$f(x) = f(a) + f'(a)(x - a) + \delta_1(x - a).$$

The first of these latter formulas gives us an idea of how accurate the approximation $f(x) \approx f(a)$ is—the accuracy is measured by the added *error* term $(f'(a) + \delta_1)(x - a)$. Provided that $f'(a) \neq 0$, when x is near enough to a the quantity δ_1 is very small in comparison to $f'(a)$, and the added term is almost equal to $f'(a)(x - a)$; that is, the error is roughly proportional to the distance of x from a, the constant of proportionality being the derivative at a (up to sign). If $f'(a) = 0$ the error is even smaller.

The second formula suggests a new estimate:

$$f(x) \approx f(a) + f'(a)(x - a).$$

At the same time it gives us a reason to prefer this estimate to the former one. Indeed, the error term is $\delta_1(x - a)$, which for x near a is in general smaller than the error term calculated a moment ago. In fact, if $f'(a) \neq 0$ the ratio of the new error to the old approaches zero as $x \to a$:

$$\frac{\delta_1(x-a)}{(f'(a)+\delta_1)(x-a)} = \frac{\delta_1}{f'(a)+\delta_1} \to 0 \,.$$

In the exceptional case when $f'(a) = 0$ the two are equal.

It is better, then, to estimate $f(x)$ by the *linear* function

$$y = f(a) + f'(a)(x-a)$$

than by the constant function $y = f(a)$. The graph of the linear function is the tangent line at $x = a$, whose superiority to the horizontal line as an approximation to the graph of f near a is plain to see (Fig. 8-1). The error

$$(f'(a)+\delta_1)(x-a) = f'(a)(x-a) + \delta_1(x-a)$$

in the first estimate

$$f(x) \approx f(a)$$

had a part proportional to $x - a$ and a smaller part (much smaller as $x \to a$); what we did in devising the second estimate

$$f(x) \approx f(a) + f'(a)(x-a)$$

was to incorporate the larger part into the estimating function. That function agrees with $f(x)$ both in its value at a and in the value of its first derivative

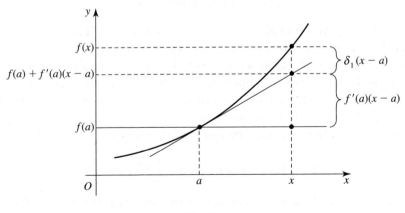

Fig. 8-1

there; in other words, at $x = a$ the estimating graph has the same height, and the same slope—or, equivalently, the same inclination—as the graph of f.

8.1.2   B. Quadratic approximation.   For the next step in approximation we assume f twice differentiable and f'' continuous at a, and suppose $f(a)$, $f'(a)$, and $f''(a)$ all known. With a different hypothesis we proved the following corollary of the second-order mean value theorem in §§7.3.3–4 (Prop. a): as $x \to a$,

$$(5) \qquad \frac{f(x) - [f(a) + f'(a)(x - a)]}{(x - a)^2} \to \frac{1}{2} f''(a) \,,$$

or

$$(6) \qquad \frac{f(x) - [f(a) + f'(a)(x - a)]}{(x - a)^2} = \frac{1}{2} f''(a) + \delta_2 \,,$$

where $\delta_2 \to 0$ as $x \to a$. Consequently there are analogues of (3) and (4):

$$(7) \qquad f(x) = f(a) + f'(a)(x - a) + \left(\frac{1}{2} f''(a) + \delta_2 \right)(x - a)^2 \,,$$

$$(8) \qquad f(x) = f(a) + f'(a)(x - a) + \frac{1}{2} f''(a)(x - a)^2 + \delta_2(x - a)^2 \,.$$

Reasoning exactly as before, we obtain from (7) a new estimate of the error in the second approximation

$$f(x) \approx f(a) + f'(a)(x - a) \,,$$

namely that it goes roughly as $(x - a)^2$ for x near a. This is more precise than the earlier representation of that error as $\delta_1(x - a)$ (in which we knew only that the factor δ_1 approached 0, not that it was roughly proportional to $x - a$), and since $(x - a)^2$ is much smaller than $|x - a|$ for x near a, we are confirmed in our preference for the second approximation over its predecessor $f(x) \approx f(a)$, whose error went as $x - a$.

Continuing in the same way, we find that (8) suggests a third estimate

$$f(x) \approx f(a) + f'(a)(x - a) + \frac{1}{2} f''(a)(x - a)^2$$

and shows it to be better than the last, in the sense that if $f''(a) \neq 0$ the ratio of the error in (8) to that in (7) approaches zero as $x \to a$. This time the

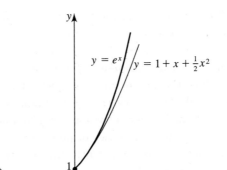

Parabola of curvature of e^x at $x = 0$

Fig. 8-2

estimating function

$$y = f(a) + f'(a)(x - a) + \frac{1}{2} f''(a)(x - a)^2$$

is *quadratic*, a second-degree polynomial (when $f''(a) \neq 0$), and has the same value, first derivative, and second derivative, as $f(x)$ at $x = a$. Its graph (Fig. 8-2) is the unique parabola with vertical axis that matches the graph of f at a in height, inclination, and curvature; for these three quantities determine and are determined by $f(a)$, $f'(a)$, and $f''(a)$ (for curvature see §7.2.4(1)). The equation of this **parabola of curvature**[2] is simpler than that of the circle of curvature, as it is quadratic in x only.

8.1.3   The Taylor polynomial of order n.   So far we have found three polynomials associated with $f(x)$:

$$p_0(x) = f(a),$$

$$p_1(x) = f(a) + f'(a)(x - a),$$

$$p_2(x) = f(a) + f'(a)(x - a) + \frac{1}{2} f''(a)(x - a)^2.$$

---

[2]Or *osculating* parabola.

If $f'(a)$ and $f''(a)$ are non-zero, the three polynomials are distinct; otherwise two or all three are the same. In any case, for each number $n = 0$, 1, or 2, *the value of $p_n(x)$ at $x = a$, and the values of its first n derivatives there, are the same as the corresponding values of $f(x)$ and its first n derivatives.* This property determines the polynomials uniquely, for if n is any non-negative integer a polynomial $p(x)$ of degree $\leq n$ is uniquely determined by its value and the values of its first n derivatives at any given point a (§3.8.3.Q7). I will now show how to continue the series of polynomials having this property for $n > 2$; it will be shown in Art. 8.2 that in this way the series of *approximating* polynomials is also continued.

Suppose first that $f'''(a) \neq 0$. Looking at how p_1 was obtained from p_0 and p_2 from p_1, we may hypothesize that

$$p_3(x) = p_2(x) + cf'''(a)(x - a)^3,$$

where c is some constant. The added term and its first two derivatives vanish at $x = a$, so we have

$$p_3(a) = p_2(a) = f(a),$$
$$p_3'(a) = p_2'(a) = f'(a),$$
$$p_3''(a) = p_2''(a) = f''(a).$$

Further, since $p_2'''(x) = 0$,

$$p_3'''(a) = \frac{d^3}{dx^3} \left[cf'''(a)(x - a)^3 \right] \Big|_a = 6cf'''(a).$$

If this is to equal the non-zero $f'''(a)$ we must have $c = \frac{1}{6}$, so that

$$p_3(x) = f(a) + f'(a)(x - a) + \frac{1}{2} f''(a)(x - a)^2 + \frac{1}{6} f'''(a)(x - a)^3.$$

Differentiation verifies $p_3'''(a) = f'''(a)$; and the formula for $p_3(x)$ is obviously valid also if $f'''(a) = 0$, in which case p_3 is the same as p_2.

Succeeding polynomials are obtained by the same method. In general,

$$p_n(x) = p_{n-1}(x) + cf^{(n)}(a)(x - a)^n,$$

and (when $f^{(n)}(a) \neq 0$) c is determined by

$$f^{(n)}(a) = p_n^{(n)}(a) = \frac{d^n}{dx^n} \left[cf^{(n)}(a)(x-a)^n \right] \Big|_a = n!\, cf^{(n)}(a),$$

where by definition $n!$, or n **factorial**, means the product $1 \cdot 2 \cdot 3 \cdots n$, which is the nth derivative of $(x-a)^n$. The result is[3]

$$p_n(x) = f(a) + \frac{f'(a)}{1!}(x-a) + \frac{f''(a)}{2!}(x-a)^2$$

$$+ \frac{f'''(a)}{3!}(x-a)^3 + \cdots + \frac{f^{(n)}(a)}{n!}(x-a)^n.$$

This is the **Taylor polynomial of order n** of $f(x)$ at $x = a$. It is of simplest form when $a = 0$:

$$p_n(x) = f(0) + \frac{f'(0)}{1!}x + \frac{f''(0)}{2!}x^2 + \frac{f'''(0)}{3!}x^3 + \cdots + \frac{f^{(n)}(0)}{n!}x^n.$$

Note that regardless of the value of a, if $f(x)$ is itself a polynomial of degree $\leq n$ it is equal to $p_n(x)$, in consequence of the remark on uniqueness in the first paragraph of this section.

8.1.4   The Taylor series.   In §§8.1.1–2 all statements about the accuracy with which the polynomials $p_n(x)$ approximated $f(x)$ had to be qualified by "as $x \to a$." On such a basis we cannot say that $p_n(x)$ is near $f(x)$, or approaches it as n increases, for any *given fixed* value of x other than a itself. As I will prove in the next article, however, the fact is that in many cases of interest we do have

$$p_n(x) \to f(x) \quad \text{as} \quad n \to \infty,$$

either for all x or for x in a neighborhood of a. (I am, of course, assuming that $f(x)$ has derivatives of all orders at a.) Moreover, limits on the size of the error in the approximation can often be determined quite easily; then the polynomials can serve as practical approximations to the function. We have already seen something of this in the sections of Chapter 3 cited at the beginning of §8.1.1.

---

[3]By convention $0! = 1$, so $0!$ could be added as denominator to the first term of the expression on the right.

To say that $p_n(x) \to f(x)$ is to say that the error or **remainder** term $R_n(x)$ defined by

$$f(x) = p_n(x) + R_n(x)$$

$$= f(a) + \frac{f'(a)}{1!}(x-a) + \frac{f''(a)}{2!}(x-a)^2 + \cdots$$

$$+ \frac{f^{(n)}(a)}{n!}(x-a)^n + R_n(x)$$

approaches zero as $n \to \infty$. This state of affairs is symbolized by writing

$$(9) \quad f(x) = f(a) + \frac{f'(a)}{1!}(x-a) + \frac{f''(a)}{2!}(x-a)^2 + \frac{f'''(a)}{3!}(x-a)^3 + \cdots .$$

In a similar way we understand the equation

$$\frac{1}{3} = 0.3333 \cdots ,$$

which is shorthand for

$$(10) \qquad \frac{1}{3} = 3 \cdot \frac{1}{10} + 3 \cdot \frac{1}{10^2} + 3 \cdot \frac{1}{10^3} + 3 \cdot \frac{1}{10^4} + \cdots ,$$

to mean that the terms of the sequence

$$3 \cdot \frac{1}{10}, \quad 3 \cdot \frac{1}{10} + 3 \cdot \frac{1}{10^2}, \quad 3 \cdot \frac{1}{10} + 3 \cdot \frac{1}{10^2} + 3 \cdot \frac{1}{10^3}, \cdots$$

eventually differ arbitrarily little from $\frac{1}{3}$.

The general name of **infinite series** is given to an expression such as the right side of (9) or (10); we call $f(x)$ or $\frac{1}{3}$ the **sum** of the infinite series. An infinite series of ascending powers of $x - a$ with coefficients is a **power series** in $x - a$, and the power series of (9) is called a **Taylor series**. We say that in (9) $f(x)$ has been **expanded** in a Taylor series about $x = a$. The statement that (9) is valid, i.e. that for some value or class of values of x we have $p_n(x) \to f(x)$, or equivalently $R_n(x) \to 0$, can be expressed by saying that the series **converges** to, or **represents**, $f(x)$ at the x-values in question.[4]

For $a = 0$ the Taylor series has the form

$$f(x) = f(0) + \frac{f'(0)}{1!}x + \frac{f''(0)}{2!}x^2 + \frac{f'''(0)}{3!}x^3 + \cdots .$$

---

[4]It is not enough simply to say that the series converges, for there are Taylor series that converge to functions other than the ones from which they are formed.

8.1.5   Examples: e^x, $\sin x$, $\cos x$.   The following examples of expansion about $x = 0$ will display some of the uses of the Taylor series, with no pretense of demonstrating their legitimacy.

EXAMPLE 1. Let us begin with $f(x) = e^x$. We have $f'(x) = e^x$, $f''(x) = e^x$, etc., so every $f^{(n)}(0) = 1$ and the series is

$$e^x = 1 + \frac{x}{1!} + \frac{x^2}{2!} + \frac{x^3}{3!} + \cdots .$$

In the next article it will be possible to show that this representation is valid for all x, and to determine the accuracy of approximation by a finite number of terms (i.e. by a Taylor polynomial). This will permit us to compute values of the function to any degree of accuracy; in particular, to calculate e itself:

$$e = 1 + \frac{1}{1!} + \frac{1}{2!} + \frac{1}{3!} + \cdots .$$

If we naively assume that a power series can be differentiated term by term like a polynomial, the result is gratifying:

$$\frac{d}{dx} e^x = 0 + \frac{1}{1!} + \frac{2x}{2!} + \frac{3x^2}{3!} + \cdots$$

$$= 1 + \frac{x}{1!} + \frac{x^2}{2!} + \cdots = e^x .$$

Term-by-term integration therefore works equally well (only one constant of integration is added for the whole series):

$$\int e^x \, dx = x + \frac{x^2}{2 \cdot 1!} + \frac{x^3}{3 \cdot 2!} + \cdots + \text{constant}$$

$$= 1 + \frac{x}{1!} + \frac{x^2}{2!} + \frac{x^3}{3!} + \cdots + C = e^x + C .$$

And the law of exponents is verified by straightforward multiplication:

$$e^a e^b = \left(1 + \frac{a}{1!} + \frac{a^2}{2!} + \frac{a^3}{3!} + \cdots \right) \left(1 + \frac{b}{1!} + \frac{b^2}{2!} + \frac{b^3}{3!} + \cdots \right)$$

$$= 1 + \left(\frac{a}{1!} + \frac{b}{1!} \right) + \left(\frac{a^2}{2!} + \frac{ab}{1!1!} + \frac{b^2}{2!} \right)$$

$$+ \left(\frac{a^3}{3!} + \frac{a^2 b}{2!1!} + \frac{ab^2}{1!2!} + \frac{b^3}{3!} \right) + \cdots$$

$$= 1 + \frac{a+b}{1!} + \frac{(a+b)^2}{2!} + \frac{(a+b)^3}{3!} + \cdots$$

$$= e^{a+b}.$$

Justification can be given for these and other manipulations of power series, although I will not attempt it here.[5] The success of such operations shows that the Taylor series offers more than the mere appearance of polynomial simplicity.

To venture a little further, let us put ourselves in the position of §6.2.1, where we had to solve the differential equation

$$y' = y$$

with initial condition

$$y(0) = 1$$

(the notation is different here). Assuming the existence of a solution $y(x)$ that can be represented by its Taylor series about $x = 0$, we have

$$y = 1 + \frac{y'(0)}{1!} x + \frac{y''(0)}{2!} x^2 + \frac{y'''(0)}{3!} x^3 + \cdots ;$$

the initial term of the series has been specified by the initial condition on y. Differentiating term by term we obtain a power series for y':

$$y' = y'(0) + \frac{y''(0)}{1!} x + \frac{y'''(0)}{2!} x^2 + \frac{y^{(4)}(0)}{3!} x^3 + \cdots .$$

Now it is a fact that a power series in $x - a$ which represents a function at points besides $x = a$ is necessarily the Taylor expansion of that function about a. In the present situation $a = 0$, and we are taking the differentiated series to represent y' everywhere. Then if $y' = y$ we must have $y'(0) = 1$, $y''(0) = y'(0)$, etc., so that all the coefficients are 1 and the unique solution is

$$y = 1 + \frac{x}{1!} + \frac{x^2}{2!} + \frac{x^3}{3!} + \cdots .$$

This calculation illustrates an effective general method for solving differential equations. In the case at hand the series serves to identify the solution as a function already known; but we could, if we wished, *define* the exponential function by this series (rather than via the logarithm), and deduce the

---

[5]The argument for the law of exponents is justified in Q17.

law of exponents and other properties of the function directly from the series definition. Both of these uses of a series solution have their applications.

EXAMPLE 2. The sine and cosine can also be expanded about $x = 0$. The values at 0 of $\sin x$ and its first three derivatives are 0, 1, 0, -1, after which this sequence is repeated. Thus

$$\sin x = x - \frac{x^3}{3!} + \frac{x^5}{5!} - \frac{x^7}{7!} + \cdots .$$

Similarly,

$$\cos x = 1 - \frac{x^2}{2!} + \frac{x^4}{4!} - \frac{x^6}{6!} + \cdots .$$

Each series can be derived from the other by means of term-by-term differentiation.

It is obvious from the two series that $\sin x$ is an odd function and $\cos x$ an even one; in general, odd or even functions are those whose expansions about 0 contain respectively only odd or only even powers of x (§3.8.3.Q9(b)). The addition formulas can be established by the kind of calculation we used to obtain the exponential law in Example 1 (Q6). On the other hand, nothing about the series suggests that they represent *periodic* functions. This fact can be proved directly, although not without effort; it would have been a considerable discovery if the functions had originally been known only through their series representations.

Like the series for the exponential function, these series are readily obtained as solutions to a differential equation, in this case

$$y'' = -y$$

with the appropriate values of y and y' at 0 as initial conditions (Q9(a)).

8.1.6   Euler's formula. If it were not for the negative signs, the series for $\cos x$ would consist of the even-degree terms, and the series for $\sin x$ of the odd-degree terms, of

$$1 + \frac{x}{1!} + \frac{x^2}{2!} + \frac{x^3}{3!} + \cdots = e^x .$$

The considerations that follow reveal a deep meaning in this curious coincidence.

As is usually pointed out in elementary algebra, it is possible to adjoin to the system of real numbers a new, so-called **imaginary**, number i, having the property that $i^2 = -1$, in such a way that the four operations of arithmetic apply as before to the enlarged system, the **complex** numbers. (One reason for the great utility of this extension is that every polynomial has a complex root, whereas there are polynomials with no real roots.) The first four powers of i are

$$i\,,$$
$$i^2 = -1\,,$$
$$i^3 = i^2 \cdot i = -i\,,$$
$$i^4 = (i^2)^2 = 1\,,$$

and the sequence of powers continues by repeating this pattern. Assuming that we can substitute ix for x in the Taylor series of e^x, by doing so we get

$$e^{ix} = 1 + \frac{ix}{1!} - \frac{x^2}{2!} - \frac{ix^3}{3!} + \frac{x^4}{4!} + \frac{ix^5}{5!} - \cdots .$$

If now we rearrange the terms and factor out i (always without justification) we obtain

$$e^{ix} = \left(1 - \frac{x^2}{2!} + \frac{x^4}{4!} - \cdots\right) + i\left(x - \frac{x^3}{3!} + \frac{x^5}{5!} - \cdots\right),$$

or

(11) $$e^{ix} = \cos x + i \sin x\,.$$

This extraordinary formula expresses an intimate connection between functions originally defined in entirely different ways.

These series will be examined further in §8.2.7.

8.2   TAYLOR'S THEOREM AND THE CONVERGENCE OF TAYLOR SERIES

8.2.1   Taylor's theorem.[6]   The following theorem gives explicit form to the remainder term R_n by which a function differs from its Taylor polynomial of order n.

---

[6]In order to understand the applications in §§8.2.7–8 it suffices to read this section and §8.2.4.

Taylor's Theorem.   *If $f^{(n)}(x)$ is continuous on $[a, b]$ and differentiable for $a < x < b$ there is a number ξ, $a < \xi < b$, such that*

$$f(b) = f(a) + \frac{f'(a)}{1!}(b-a) + \frac{f''(a)}{2!}(b-a)^2 + \cdots$$

$$+ \frac{f^{(n)}(a)}{n!}(b-a)^n + \frac{f^{(n+1)}(\xi)}{(n+1)!}(b-a)^{n+1}.$$

The same formula holds if $b < a$, provided that $f^{(n)}$ is continuous on $[b, a]$ and differentiable for $b < x < a$; in that case $b < \xi < a$.

The case $n = 0$ is the mean value theorem,

$$f(b) = f(a) + f'(\xi)(b-a),$$

and the case $n = 1$ is the second-order mean value theorem,

$$f(b) = f(a) + f'(a)(b-a) + \frac{1}{2}f''(\xi)(b-a)^2.$$

As in those theorems, we may replace b by x:

(1) $$f(x) = f(a) + \frac{f'(a)}{1!}(x-a) + \frac{f''(a)}{2!}(x-a)^2 + \cdots$$

$$+ \frac{f^{(n)}(a)}{n!}(x-a)^n + R_n(x),$$

where

(2) $$R_n(x) = \frac{f^{(n+1)}(\xi)}{(n+1)!}(x-a)^{n+1}.$$

And as in the applications of those theorems in Chapter 3 (§§3.7.11, 3.8.7), a bound on the size of $f^{(n+1)}$ over the interval between a and x yields a bound on the size of $R_n(x)$, even though ξ remains unknown.

8.2.2   Proof of Taylor's theorem. A. Rolle's theorem generalized.

The second part of the theorem is an easy consequence of the first part (Q11); for the proof we can therefore assume $a < b$. The proof begins with a generalization of Rolle's theorem.

Lemma. *Suppose that $F^{(n)}(x)$ is continuous on $[a,b]$ and differentiable for $a < x < b$. If $F(a) = F(b) = 0$ and $F'(a) = F''(a) = \cdots = F^{(n)}(a) = 0$, there is a number ξ, $a < \xi < b$, such that $F^{(n+1)}(\xi) = 0$.*

The geometrical meaning of this lemma for $n = 1$ can be understood as follows. Recall that Rolle's theorem, which is the case $n = 0$ of the lemma, asserts that if the graph of a function $F(x)$ meets the x-axis at a and b it must have a horizontal tangent at a point ξ_0 in between (Fig. 8-3). The case $n = 1$ adds the condition of a horizontal tangent at a. In order to return to the horizontal at ξ_0 after leaving a, the curve evidently must have a point of inflection, where $F'' = 0$, at some ξ_1 between a and ξ_0 (Fig. 8-4)—unless, of course, it coincides with the x-axis along the interval from a to ξ_0, in which case $F'' = 0$ at all points of that interval. In effect, we are applying Rolle's theorem to the slope function F', which is zero at a and ξ_0.

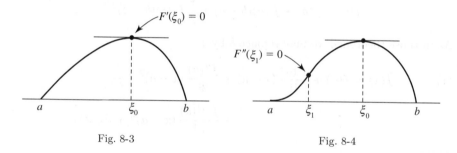

Fig. 8-3   Fig. 8-4

Hence the proof of the lemma is carried out by successive applications of Rolle's theorem.

Step 0. Since $F(a) = F(b) = 0$, there is a number ξ_0, $a < \xi_0 < b$, such that $F'(\xi_0) = 0$.

Step 1. Since $F'(a) = F'(\xi_0) = 0$, there is a number ξ_1, $a < \xi_1 < \xi_0$, such that $F''(\xi_1) = 0$.

Continue in this way to step n:

Step n. Since $F^{(n)}(a) = F^{(n)}(\xi_{n-1}) = 0$, there is a number ξ, $a < \xi < \xi_{n-1}$, such that $F^{(n+1)}(\xi) = 0$. Clearly $\xi < b$, so the lemma is proved.

8.2.3   B. Taylor's theorem proved.   Taylor's theorem can now be established by an argument closely related to the treatment of approximating polynomials in Art. 8.1. As we know, the polynomial

$$p_n(x) = f(a) + \frac{f'(a)}{1!}(x-a) + \frac{f''(a)}{2!}(x-a)^2 + \cdots + \frac{f^{(n)}(a)}{n!}(x-a)^n$$

matches $f(x)$ at $x = a$ in its value and the values of its first n derivatives. With an eye to the form of the remainder term proposed by the theorem, we observe that this match is unaffected by the addition to $p_n(x)$ of a term $\lambda(x-a)^{n+1}$, λ being any constant; for the added term and *its* first n derivatives are all zero at $x = a$. Let us choose λ so that at $x = b$ the added term *is* the remainder; that is, define λ by

$$f(b) = p_n(b) + \lambda(b-a)^{n+1},$$

or

$$\lambda = \frac{f(b) - p_n(b)}{(b-a)^{n+1}}.$$

It then follows immediately that the function

$$F(x) = f(x) - \left[p_n(x) + \lambda(x-a)^{n+1}\right],$$

which represents the difference between $f(x)$ and the augmented polynomial, satisfies the hypotheses of the lemma; consequently there exists ξ, $a < \xi < b$, such that $F^{(n+1)}(\xi) = 0$. Now $p_n(x)$ being of degree $\leq n$, its $(n+1)$st derivative is zero, while the $(n+1)$st derivative of $(x-a)^{n+1}$ is $(n+1)!$. Therefore

$$F^{(n+1)}(x) = f^{(n+1)}(x) - \lambda(n+1)!,$$

and substituting the value ξ for x,

$$0 = f^{(n+1)}(\xi) - \lambda(n+1)!,$$

or

$$\lambda = \frac{f^{(n+1)}(\xi)}{(n+1)!}.$$

Thus

$$f(b) = p_n(b) + \frac{f^{(n+1)}(\xi)}{(n+1)!}(b-a)^{n+1},$$

q.e.d.

8.2.4 Corollaries of Taylor's theorem. From (2) we have, for $x \neq a$,

$$\frac{R_n(x)}{(x-a)^{n+1}} = \frac{f^{(n+1)}(\xi)}{(n+1)!}.$$

There is then the following corollary to Taylor's theorem, by which the limits §8.1.1(1) and §8.1.2(5) are generalized.

Corollary 1. *If in addition to the hypotheses of Taylor's theorem* $f^{(n+1)}(x)$ *exists at a and is continuous there, then*

$$\lim_{x \to a} \frac{R_n(x)}{(x-a)^{n+1}} = \frac{f^{(n+1)}(a)}{(n+1)!}.$$

For as $x \to a$, also $\xi \to a$, and continuity ensures that $f^{(n+1)}(\xi) \to f^{(n+1)}(a)$.

Just as with §8.1.1(2) and §8.1.2(6) we can understand the corollary to say that $R_n(x)$ *approaches proportionality with* $(x-a)^{n+1}$ *if* $f^{(n+1)}(a) \neq 0$, *or tends to zero even more rapidly if* $f^{(n+1)}(a) = 0$. Even without the explicit form of the remainder given by Taylor's theorem, this result would qualify the Taylor polynomials as approximating polynomials in the sense of §§8.1.1–2.

It is possible to prove the conclusion of the corollary under the weaker hypothesis that $f^{(n)}$ exists on $[a, b]$ and is differentiable at a: see §151 of the work of G. H. Hardy cited in §4.7.4.

If the remainder approaches zero in near-proportionality to $(x-a)^{n+1}$ (or even faster), *a fortiori* it does so more rapidly than $(x-a)^n$. The formal deduction from the corollary goes as follows:

$$\lim_{x \to a} \frac{R_n(x)}{(x-a)^n} = \lim_{x \to a} \left[\frac{R_n(x)}{(x-a)^{n+1}} (x-a) \right] = \frac{f^{(n+1)}(a)}{(n+1)!} \cdot 0 = 0.$$

This property of the Taylor approximation actually characterizes it: no other approximation by a polynomial of degree $\leq n$ has a remainder that falls off so rapidly. That is the content of the second part of the next corollary.

Corollary 2. *With the hypotheses of Corollary 1,*

(3)
$$\lim_{x \to a} \frac{R_n(x)}{(x-a)^n} = 0.$$

Moreover, if $q_n(x)$ is any polynomial of degree $\leq n$ such that $f(x) = q_n(x) + S_n(x)$ and $\lim\limits_{x \to a} \dfrac{S_n(x)}{(x-a)^n} = 0$, *then $q_n = p_n$ and $S_n = R_n$.*

The uniqueness of the Taylor representation will be used in §8.2.8 to identify
a Taylor series (Example 6).

8.2.5   Proof of uniqueness.   To prove the second part we first reduce
it to an assertion about polynomials. From

$$f(x) = p_n(x) + R_n(x)$$

and

$$f(x) = q_n(x) + S_n(x)$$

follows

$$(p_n(x) - q_n(x)) + (R_n(x) - S_n(x)) = 0 \,.$$

Let

$$r_n(x) = p_n(x) - q_n(x) \,,$$

a polynomial of degree $\leq n$, and let

$$T_n(x) = R_n(x) - S_n(x) \,.$$

We have

$$r_n(x) + T_n(x) = 0 \,,$$

and

$$\lim_{x \to a} \frac{T_n(x)}{(x-a)^n} = \lim_{x \to a} \frac{R_n(x)}{(x-a)^n} - \lim_{x \to a} \frac{S_n(x)}{(x-a)^n} = 0 \,;$$

hence

(4)
$$\lim_{x \to a} \frac{r_n(x)}{(x-a)^n} = 0 \,.$$

I claim that if a polynomial $r_n(x)$ of degree $\leq n$ satisfies (4), it must be the
zero polynomial. Proof of this fact will clearly establish the corollary.

By the argument used to derive (3) from Corollary 1 we can extend (4) to

(5)
$$\lim_{x \to a} \frac{r_n(x)}{(x-a)^k} = 0 \,, \quad 0 \leq k \leq n \,.$$

Now we know from §8.1.3 (at the end) that $r_n(x)$ is equal to its own nth-order Taylor polynomial,

$$r_n(x) = c_0 + c_1(x - a) + c_2(x - a)^2 + \cdots + c_n(x - a)^n$$

—the form of the coefficients does not concern us here, only the expression of $r_n(x)$ in terms of powers of $x - a$. As $x \to a$ the right side approaches c_0; hence

$$c_0 = \lim_{x \to a} r_n(x) = \lim_{x \to a} \frac{r_n(x)}{(x - a)^0} = 0,$$

by (5) with $k = 0$. Therefore

$$r_n(x) = c_1(x - a) + c_2(x - a)^2 + \cdots + c_n(x - a)^n$$

$$= (x - a)[c_1 + c_2(x - a) + \cdots + c_n(x - a)^{n-1}],$$

or

$$\frac{r_n(x)}{x - a} = c_1 + c_2(x - a) + \cdots + c_n(x - a)^{n-1}.$$

Then as before

$$c_1 = \lim_{x \to a} \frac{r_n(x)}{x - a} = 0,$$

by (5) with $k = 1$. Continuing in this way we show that all the c_k are zero, so that $r_n(x)$ is the zero polynomial.

8.2.6   Convergence of Taylor series.

The magnitude of the remainder term $R_n(x)$ in formula (1), §8.2.1, measures the accuracy with which $p_n(x)$ approximates $f(x)$. If for a given x-value $R_n(x) \to 0$ as $n \to \infty$, then at that point the Taylor series converges to $f(x)$,

$$f(x) = f(a) + \frac{f'(a)}{1!}(x - a) + \frac{f''(a)}{2!}(x - a)^2 + \cdots.$$

This is certain to be so if there exists a single constant M such that

(6) $$|f^{(n+1)}(\xi)| \leq M$$

for all numbers ξ between a and x and all values of n, because in that case

$$|R_n(x)| \leq M \frac{|x - a|^{n+1}}{(n + 1)!},$$

by §8.2.1(2), and the following lemma shows that for fixed x, $\dfrac{|x-a|^{n+1}}{(n+1)!} \to 0$ as $n \to \infty$.

Lemma. *If c is any number,* $\lim\limits_{n\to\infty} \dfrac{c^n}{n!} = 0.$

It is proved as follows.

Since $\left|\dfrac{c^n}{n!}\right| = \dfrac{|c|^n}{n!}$ we can assume for the proof that $c \geq 0$ (as in the present situation). Choose a fixed integer m so large that $\dfrac{c}{m+1} < \dfrac{1}{2}$; then also $\dfrac{c}{m+2} < \dfrac{1}{2}$, $\dfrac{c}{m+3} < \dfrac{1}{2}$, etc. For any integer $n > m$ we have

$$\frac{c^n}{n!} = \frac{c^m}{m!} \cdot \frac{c}{m+1} \cdot \frac{c}{m+2} \cdot \ldots \cdot \frac{c}{n} < \frac{c^m}{m!}\left(\frac{1}{2}\right)^{n-m} = \frac{(2c)^m}{m!} \cdot \frac{1}{2^n}.$$

The first factor is constant and the second approaches zero as $n \to \infty$.

While (6) is a sufficient condition that the remainder approach zero, it is by no means a necessary one. To determine convergence each series must be looked at individually.

I will now return to the examples of §8.1.5 and present several others as well. In each example the value of a will be 0.[7]

8.2.7 Examples of convergent Taylor series: A. e^x, $\sin x$, $\cos x$.

EXAMPLE 1. The exponential function has $f^{(n+1)}(\xi) = e^\xi$ for every ξ and n. Then if x is positive, for ξ between 0 and x

$$|f^{(n+1)}(\xi)| = e^\xi < e^x,$$

because the exponential function is increasing. Likewise if x is negative, $e^0 = 1$ bounds $|f^{(n+1)}(\xi)|$. This means that for any given x, e^x or 1 can serve as M; hence

(i) the Taylor series represents the function at all values of x,

$$e^x = 1 + \frac{x}{1!} + \frac{x^2}{2!} + \cdots,$$

---

[7] As to this limitation see the beginning of Example 3.

and

(ii) the remainder is smaller in magnitude[8] than

$$\frac{e^x}{(n+1)!}\, x^{n+1},$$

if $x > 0$, or

$$\frac{1}{(n+1)!}\, |x|^{n+1},$$

if $x < 0$.

Let us apply (i) and (ii) to calculate the number e. Taking $x = 1$, we have by (i)

$$e = 1 + \frac{1}{1!} + \frac{1}{2!} + \frac{1}{3!} + \cdots .$$

Obviously $e > 2$; moreover, we can easily show that $e \leq 3$. For $\frac{1}{1!} = 1$, $\frac{1}{2!} = \frac{1}{2}$, $\frac{1}{3!} < \frac{1}{2^2}$, $\frac{1}{4!} < \frac{1}{2^3}$, and so on. Then

$$\frac{1}{1!} = 1,$$

$$\frac{1}{1!} + \frac{1}{2!} = 1 + \frac{1}{2},$$

$$\frac{1}{1!} + \frac{1}{2!} + \frac{1}{3!} < 1 + \frac{1}{2} + \frac{1}{2^2},$$

$$\frac{1}{1!} + \frac{1}{2!} + \frac{1}{3!} + \frac{1}{4!} < 1 + \frac{1}{2} + \frac{1}{2^2} + \frac{1}{2^3},$$

and so on. It is well known, and will be proved in Example 3, that the numbers on the right approach 2, the sum of the geometric series

$$1 + \frac{1}{2} + \frac{1}{2^2} + \frac{1}{2^3} + \cdots ;$$

in consequence the limit of the numbers on the left, which is $e - 1$, cannot exceed 2.

---

[8]The inequality for $|f^{(n+1)}(\xi)|$ was strict, unlike the one in (6).

By (ii), then, the magnitude of the remainder is less than $\dfrac{3}{(n+1)!}$. To estimate e with an error of less than 0.0002, say, we have only to choose n so large that

$$\frac{3}{(n+1)!} \leq 0.0002$$

and sum the first n terms of the series. The value $n = 7$ suffices, so the estimate

$$e \approx 1 + \frac{1}{1!} + \frac{1}{2!} + \frac{1}{3!} + \frac{1}{4!} + \frac{1}{5!} + \frac{1}{6!} + \frac{1}{7!} = 2.71825\ldots$$

is accurate to at least three decimal places.[9] Then to three places

$$e \approx 2.718.$$

Thus the series provides a systematic procedure for calculating e to within any specified degree of accuracy, with certainty that the accuracy has been achieved. We could not expect a stronger result from any calculation, for I will now prove, by means of the exact expression for the remainder, that e is an irrational number.

Suppose the contrary, that $e = p/q$, where p and q are positive integers; we can assume $q > 1$. Taking $x = 1$ and $n = q$ we have

$$e = 1 + \frac{1}{1!} + \frac{1}{2!} + \cdots + \frac{1}{q!} + \frac{e^\xi}{(q+1)!},$$

where $0 < \xi < 1$. Hence

$$q!e = \left[q! + \frac{q!}{1!} + \frac{q!}{2!} + \cdots + \frac{q!}{q!} \right] + \frac{e^\xi}{q+1}.$$

Here $q!e = q!(p/q) = (q-1)!p$ is an integer, and so is the sum in brackets (because for each $k \leq q$, $k!$ is a factor of $q!$); so therefore is their difference $\dfrac{e^\xi}{q+1}$. But this is impossible, for $e^\xi < e < 3$ implies $\dfrac{e^\xi}{q+1} < \dfrac{3}{q+1} \leq 1$, and thus $0 < \dfrac{e^\xi}{q+1} < 1$.

---

[9] Compare the method of §6.2.3, which requires a very large value of n to achieve the same accuracy.

EXAMPLE 2. The sine and cosine functions obviously satisfy $|f^{(n+1)}(\xi)| \leq 1$ for every value of ξ and n. Hence for all x,

$$\sin x = x - \frac{x^3}{3!} + \frac{x^5}{5!} - \cdots$$

and

$$\cos x = 1 - \frac{x^2}{2!} + \frac{x^4}{4!} - \cdots .$$

In both cases the remainder after a term $\pm x^n/n!$, which is $R_n(x) = R_{n+1}(x)$, does not exceed $\dfrac{|x|^{n+1}}{(n+1)!}$ or $\dfrac{|x|^{n+2}}{(n+2)!}$ in magnitude; and since we can always choose x between 0 and $\pi/2$, the upper limit of error for practical calculation is $\dfrac{(\pi/2)^{n+2}}{(n+2)!}$ (the smaller of the two quantities).

From this result we find that for any x between 0 and $\pi/2$ we can estimate $\sin x$ or $\cos x$ to within 0.0005 by taking four terms of the sine series or five terms of the cosine series, because 7 is the least value of n such that $\dfrac{(\pi/2)^{n+2}}{(n+2)!} <$ 0.0005. For the particular x-value 0.01, two terms of the sine series give us the value of the function to an accuracy of $\dfrac{(0.01)^5}{5!} < 9 \times 10^{-13}$, or 11 decimal places:

$$\sin 0.01 \approx 0.01 - \frac{(0.01)^3}{3!} \approx 0.00999983333 .$$

8.2.8   B. $1/(1+x)$, $\log(1+x)$, $(1+x)^\alpha$, arctan x. The series representations in the preceding examples were valid for all x. The remaining ones will hold only in a finite interval about $x = 0$.

EXAMPLE 3. The function $1/x$ cannot be expanded in a Taylor series about $x = 0$, where it is undefined, but can be expanded about $x = 1$. To keep the notation simple I will work with $f(x) = 1/(1+x)$ instead of $1/x$, and expand $f(x)$ about $x = 0$; it comes to the same thing (as will become clear at the end of the example).

The function and its successive derivatives are

$$(1+x)^{-1}, \quad -(1+x)^{-2}, \quad 2(1+x)^{-3}, \quad -2 \cdot 3(1+x)^{-4}, \ldots ,$$

and at $x = 0$,

$$1, \ -1!, \ 2!, \ -3!, \ldots .$$

The Taylor series is therefore

$$1 - x + x^2 - x^3 + \cdots .$$

This is a geometric series with ratio $-x$,

$$1 + (-x) + (-x)^2 + (-x)^3 + \cdots .$$

It has an algebraic derivation, which will furnish a better form of the remainder than the expression supplied by Taylor's theorem.

We begin with the formula for the sum of a finite geometric series,

$$1 + u + u^2 + \cdots + u^n = \frac{1 - u^{n+1}}{1 - u}, \quad u \neq 1,$$

which is verified by multiplication by $1 - u$. Rearrangement gives

$$\frac{1}{1 - u} = 1 + u + u^2 + \cdots + u^n + \frac{u^{n+1}}{1 - u},$$

and putting $u = -x$ we get

$$(7) \qquad \frac{1}{1 + x} = [1 - x + x^2 - \cdots + (-1)^n x^n] + (-1)^{n+1} \frac{x^{n+1}}{1 + x},$$

where $x \neq -1$. The sum in brackets is the Taylor polynomial of order n of $1/(1 + x)$, so the term added to it is the remainder,

$$R_n(x) = (-1)^{n+1} \frac{x^{n+1}}{1 + x}.$$

Notice that this is an exact expression in x alone: it does not depend on an indeterminate number ξ.

As $n \to \infty$ the remainder approaches zero when x^{n+1} does; that is, if, and only if, $-1 < x < 1$ (§3.3.3.Q6). Then for such x we have

$$\frac{1}{1 + x} = 1 - x + x^2 - x^3 + \cdots .$$

Equivalently, for $0 < x < 2$ we have

$$\frac{1}{x} = 1 - (x - 1) + (x - 1)^2 - (x - 1)^3 + \cdots .$$

EXAMPLE 4. For the same reason that we considered $1/(1+x)$ rather than $1/x$ in the last example, we will investigate the logarithm in the form $\log(1+x)$. That function has derivative $1/(1+x)$, so with the aid of the same example we can immediately write down its value and the values of its successive derivatives at $x = 0$:

$$0,\ 1,\ -1!,\ 2!,\ -3!,\ \dots .$$

The Taylor series is then

$$x - \frac{x^2}{2} + \frac{x^3}{3} - \frac{x^4}{4} + \cdots .$$

The remainder can be derived from that of the geometric series. If in (7) we replace x by t and n by $n-1$ we obtain

$$(8) \qquad \frac{1}{1+t} = 1 - t + t^2 - \cdots + (-1)^{n-1}t^{n-1} + (-1)^n \frac{t^n}{1+t} ,$$

where $t \neq -1$. Let $x > -1$, and integrate both sides from 0 to x; the result is

$$\log(1+x) = \left[x - \frac{x^2}{2} + \frac{x^3}{3} - \cdots + (-1)^{n-1}\frac{x^n}{n} \right] + (-1)^n \int_0^x \frac{t^n}{1+t}\, dt .$$

Here the sum in brackets is the Taylor polynomial of order n, so the added term is the remainder,

$$R_n(x) = (-1)^n \int_0^x \frac{t^n}{1+t}\, dt .$$

I will now show that $R_n(x) \to 0$ if $-1 < x \leq 1$.

There are two cases.

(i) Let $0 \leq x \leq 1$. Then in the integral $t \geq 0$, so

$$0 \leq \frac{t^n}{1+t} \leq t^n ,$$

and therefore

$$0 \leq \int_0^x \frac{t^n}{1+t}\, dt \leq \int_0^x t^n\, dt .$$

Hence

$$|R_n(x)| = \int_0^x \frac{t^n}{1+t}\, dt \leq \int_0^x t^n\, dt = \frac{x^{n+1}}{n+1} \leq \frac{1}{n+1} \to 0$$

(the second inequality holds because $0 \leq x \leq 1$), and this implies $R_n(x) \to 0$.

(ii) If $-1 < x < 0$, let $\bar{x} = -x$, so that $0 < \bar{x} < 1$, and transform $R_n(x)$ by the substitution $u = -t$:

$$R_n(x) = (-1)^n \int_0^x \frac{t^n}{1+t}\, dt = -\int_0^{\bar{x}} \frac{u^n}{1-u}\, du\,.$$

In the integral $0 \le u \le \bar{x} < 1$, so $0 < 1 - \bar{x} \le 1 - u$ and we have

$$0 \le \frac{u^n}{1-u} \le \frac{u^n}{1-\bar{x}}\,.$$

Hence

$$|R_n(x)| = \int_0^{\bar{x}} \frac{u^n}{1-u}\, du \le \int_0^{\bar{x}} \frac{u^n}{1-\bar{x}}\, du$$

$$= \frac{1}{1-\bar{x}} \int_0^{\bar{x}} u^n\, du = \frac{1}{1-\bar{x}} \cdot \frac{\bar{x}^{n+1}}{n+1} < \frac{1}{1-\bar{x}} \cdot \frac{1}{n+1} \to 0\,,$$

which again implies $R_n(x) \to 0$.

While if $-1 < x \le 1$ we thus have

$$\log(1+x) = x - \frac{x^2}{2} + \frac{x^3}{3} - \frac{x^4}{4} + \cdots\,,$$

for other values of x the sum of the series is not defined; i.e., the sequence of polynomials x, $x - \dfrac{x^2}{2}$, $x - \dfrac{x^2}{2} + \dfrac{x^3}{3}$, $\ldots$, does not have a finite limit (Q16). Nevertheless we can calculate *all* values of the logarithm by means of the series. For it gives us $\log u$ for $0 < u \le 2$; if now u is an arbitrary number > 2, let m be the integer such that $m < u \le m+1$ and write

$$u = \frac{u}{m} \cdot \frac{m}{m-1} \cdot \frac{m-1}{m-2} \cdots \cdots \frac{3}{2} \cdot 2\,.$$

Each factor is between 1 and 2, and

$$\log u = \log \frac{u}{m} + \log \frac{m}{m-1} + \cdots + \log \frac{3}{2} + \log 2\,.$$

The series for $\log 2$ ($x = 1$) is remarkable:

$$\log 2 = 1 - \frac{1}{2} + \frac{1}{3} - \frac{1}{4} + \cdots\,.$$

EXAMPLE 5. The **binomial series** is the Taylor expansion of $(1+x)^\alpha$ about $x = 0$, where α may be any number. Calculation of the derivatives shows the expansion to be

$$(1+x)^\alpha = 1 + \frac{\alpha}{1!}x + \frac{\alpha(\alpha-1)}{2!}x^2 + \frac{\alpha(\alpha-1)(\alpha-2)}{3!}x^3 + \cdots.$$

The case $\alpha = -1$ is the series of Example 3, which we proved to represent the function when $|x| < 1$. It can be shown that when $|x| < 1$ the representation is valid for *every* value of α; I omit the proof, which is somewhat more difficult than the arguments that have sufficed for the preceding examples.[10]

If α is a non-negative integer n, the function is a polynomial of degree $\leq n$ and the series reduces to the finite nth-order Taylor polynomial,

$$(1+x)^n = 1 + \frac{n}{1!}x + \frac{n(n-1)}{2!}x^2 + \frac{n(n-1)(n-2)}{3!}x^3 + \cdots + x^n$$

(the last coefficient is $n!/n! = 1$), valid of course for all x. This is a special case of the *binomial theorem*, by which $(a+b)^n$ is expanded; the general case, as given in §3.2.3.Q6, is deducible from it in virtue of the relation

$$(a+b)^n = a^n \left(1 + \frac{b}{a}\right)^n.$$

As an example of the series for a non-integral exponent, here it is for $\alpha = -\frac{1}{2}$:

$$\frac{1}{\sqrt{1+x}} = 1 - \frac{1}{2}x + \frac{1\cdot 3}{2\cdot 4}x^2 - \frac{1\cdot 3\cdot 5}{2\cdot 4\cdot 6}x^3 + \frac{1\cdot 3\cdot 5\cdot 7}{2\cdot 4\cdot 6\cdot 8}x^4 - \cdots$$

$$= 1 - \frac{1}{2}x + \frac{3}{8}x^2 - \frac{5}{16}x^3 + \frac{35}{128}x^4 - \cdots.$$

EXAMPLE 6. Although its higher derivatives are not easy to calculate, the function $\arctan x$ can be treated similarly to the logarithm. We replace t by t^2 in (8) (Example 4) to get

$$\frac{1}{1+t^2} = 1 - t^2 + t^4 - \cdots + (-1)^{n-1}t^{2n-2} + (-1)^n \frac{t^{2n}}{1+t^2},$$

and then integrate both sides from 0 to x; the result is

$$\arctan x = \left[x - \frac{x^3}{3} + \frac{x^5}{5} - \cdots + (-1)^{n-1}\frac{x^{2n-1}}{2n-1}\right] + (-1)^n \int_0^x \frac{t^{2n}}{1+t^2}\, dt.$$

---

[10]Proofs can be found on pp. 337 and 406 of the work of R. Courant cited in §2.4.5.

Let

$$S_{2n-1}(x) = (-1)^n \int_0^x \frac{t^{2n}}{1+t^2} \, dt \, .$$

In order for the sum in brackets to be the Taylor polynomial of order $2n - 1$, it is sufficient, by Corollary 2 of Taylor's theorem (§8.2.4), that

$$\lim_{x \to 0} \frac{S_{2n-1}(x)}{x^{2n-1}} = 0 \, .$$

An argument like the one in Example 4 (case (i)) shows that this is actually the case, and moreover that for any fixed x-value in the interval $[-1, 1]$ the remainder approaches zero as $n \to \infty$.

The function $S_{2n-1}(x)$ is odd (§4.6.5.Q4), so we need consider only non-negative values of x. We have

$$0 \le \frac{t^{2n}}{1+t^2} \le t^{2n} \, ,$$

whence for $x \ge 0$

(9) $$|S_{2n-1}(x)| = \int_0^x \frac{t^{2n}}{1+t^2} \, dt \le \int_0^x t^{2n} \, dt = \frac{x^{2n+1}}{2n+1} \, ;$$

then

$$\left| \frac{S_{2n-1}(x)}{x^{2n-1}} \right| = \frac{|S_{2n-1}(x)|}{x^{2n-1}} \le \frac{x^2}{2n+1} \, .$$

For any given n, the final expression approaches zero with x, while for fixed x, $-1 \le x \le 1$, the final expression in (9) approaches zero as $n \to \infty$.

Thus we have established the validity of the Taylor series

$$\arctan x = x - \frac{x^3}{3} + \frac{x^5}{5} - \frac{x^7}{7} + \cdots$$

for $-1 \leq x \leq 1$. Since the arc tangent of 1 is $\pi/4$, putting $x = 1$ gives

$$\frac{\pi}{4} = 1 - \frac{1}{3} + \frac{1}{5} - \frac{1}{7} + \cdots .$$

8.2.9   Envoi.   With this formula I conclude the examples of Taylor series, and this book on calculus. It is curious how many of our purely mathematical themes are suggested by these few symbols. On the left is a quantity originally geometrical, the ratio of the area or circumference of a circle to the corresponding attribute of its circumscribed square; here it appears as a value of an inverse trigonometric function. On the right there is an infinite series, obtained by integration, the terms of which incorporate values of successively higher derivatives; through approximating polynomials it expresses as a limit a number quite remote from the integers of which it is composed. Curve and straight line, function and number, approximation and limit, integral and derivative—much has been put at the service of calculation. As it happens, the formula before us is not practical at all: a great many terms of the series would be needed to determine with any accuracy the number it represents. By this point, however, we can feel some confidence that calculus is able to provide a more efficient means of computing $\pi/4$. Accordingly we are free to contemplate the elegant formula with pleasure, recognizing in it the depth that yields the finest fruits of reason.

8.2.10   QUESTIONS

Q1.   Draw on one set of axes the graphs of e^x and its Taylor polynomials of orders 0, 1, 2, and 3 about $x = 0$. (Order 2 is illustrated in Fig. 8-2.) Which polynomial graphs cross the exponential curve at $x = 0$?

Q2.   (a) Draw on one set of axes the graphs of $y = \cos x$ and its first two distinct Taylor polynomials about $x = 0$. (b) Do the same for $y = \sin x$. Which polynomial graphs cross the given curves at $x = 0$?

Q3.   Calculate the Taylor series of $f(x) = x^3 - x^2 + x - 1$ about $x = 1$ and verify that it equals the function.

Q4.   What is the Taylor expansion of SIN x (defined in §4.4.3.Q1) about $x = 0$?

Q5.   (a) Show that the Taylor series about $x = a$ of $f(x) + g(x)$ is the term-by-term sum of the Taylor series of $f(x)$ and $g(x)$, and that the Taylor series of $cf(x)$, c constant, is the term-by-term product of the Taylor series of $f(x)$ by c. (b) Show that if the Taylor series of $f(x)$ converges to $f(x)$ and the Taylor series of $g(x)$ to $g(x)$, then the Taylor series of $f(x) + g(x)$ converges to $f(x) + g(x)$ and the Taylor series of $cf(x)$ to $cf(x)$.

Q6 (§8.1.5, Example 2).   Check the addition formula for $\cos{(u+v)}$ by calculation with the first few terms of Taylor series.

Q7.   (a) Show that at $x = \frac{1}{2}$ the remainder after $n + 1$ terms of the Taylor series of e^x about $x = 0$ is smaller than $\dfrac{1}{2^n (n + 1)!}$ in magnitude. (b) Hence compute $\sqrt{e}$ to within 0.0001.

Q8.   Calculate $\cos{(\pi/6)}$ by using two terms of the Taylor series about $x = 0$. To how many places is the calculation certain to be accurate? Check it against the known value.

Q9.   (a) (§8.1.5, Example 2) Derive these series from the differential equation. (b) Use the series alone to find the limits as $x \to 0$ of $(\sin x)/x$ and $(1 - \cos x)/x$ (as was done by other means in §4.4.2). What assumptions must you make to complete the argument? (c) If the sine and cosine are defined by their series, how do you suppose π is defined?

Q10.   (a) Show that $e^{\pi i} = -1$ and $e^{2\pi i} = 1$. (b) Find a complex number z such that $z^2 = i$. (Square $z = e^{ix}$, then apply formula (11), §8.1.6, to the result.)

Q11 (§8.2.2).   Supply the argument.

Q12.   Prove that $\sin 1$ is irrational. (Use the method of §8.2.7, Example 1.)

Q13.   In §8.2.8, Example 4, it is shown that the values of $\log{(1+x)}$ for $-1 < x \leq 1$ suffice to calculate all values of the logarithm. Show that the values for $0 \leq x \leq 1$ suffice.

Q14.   Write the first few terms of the binomial series in the following cases. (a) $1/(1 + x)^2$ (b) $\sqrt{1 + x}$

Q15.   Find the Taylor series of $\sinh x$ and $\cosh x$ (§4.1.5.Q7) at 0 (a) from the definitions of the functions and the series for e^x, and (b) directly by differentiation.

Q16 (§8.2.8, Example 4).   Prove this. (For $x = -1$ the series is the negative of the *harmonic series*

$$1 + \frac{1}{2} + \frac{1}{3} + \cdots ,$$

concerning which see §6.2.10.Q12. For $x > 1$ estimate the integral $|R_n(x)|$.)

Q17.   Let $f(x)$ and $g(x)$ be functions whose derivatives of all orders (including zero) are bounded in absolute value by a number M in some neighborhood of $x = 0$ (cf. §8.2.6(6)). (a) Show that the nth derivative of fg is bounded in absolute value by $2^n M^2$. (Use Leibniz's rule (§3.8.3.Q10) and the binomial expansion of $(1 + 1)^n$.) (b) Show that the Taylor series of fg at 0 is obtained by multiplying the series of f and g at 0, as in §8.1.5, Example 1.

Q18.   *L'Hospital's rule* (generalized). Suppose that (i) $f(x)$ and $g(x)$ satisfy the hypotheses of Corollary 1 of Taylor's theorem (§8.2.4), (ii) their derivatives of all orders (including zero) up to n vanish at a, and (iii) $g^{(n+1)}(a) \neq 0$. Prove that

$$\lim_{x \to a} \frac{f(x)}{g(x)} = \frac{f^{(n+1)}(a)}{g^{(n+1)}(a)}.$$

(b) Apply the rule to find $\lim_{x \to 0} \dfrac{x^2}{1 - \cos x}$.

Q19.   Let $f(x)$ satisfy the hypotheses of Corollary 1 of Taylor's theorem (§8.2.4). Show that if $f^{(n+1)}(a) \neq 0$, the graph of p_n crosses the graph of f at $x = a$ if n is even, but does not cross it if n is odd. (Consider the sign of $f - p_n$.)

Topical Summary

This summary collects the most important terms, propositions, rules and methods, formulas, and equations. Section references in parentheses are to good places to look first; for more direction on any topic use the cross-references found in those places or consult the index.

Contents

1. BASIC TERMS AND OPERATIONS

1A. General terms

Quantity
    Total quantity, M, etc. (§§1.1.2, 1.3.4, 1.6.2)
    Specific quantity, λ, etc. (§§1.1.2, 1.3.4, 1.6.2)
    Cumulative quantity, m, etc. (§1.6.3)
    Uniform relation of quantities (§§1.1.2, 3.6.5, 6.1.1)

Variable (§1.6.1)
    Independent variable, x, t (§§1.6.1, 1.6.2, 1.6.5)
    Dependent variable, y (§§1.6.2, 1.6.5)

Function, $y = f(x)$ (§§1.6.2, 1.6.5–7, 1.6.9)
    Monotone (increasing, decreasing) function (§1.6.8)
    Composite function, $(f \circ g)(x) = f(g(x))$ (§4.1.1)
    Inverse function, $y = f^{-1}(x)$ (§§3.3.11, 4.3.1–2)

Limit, $x_n \to a$, $x \to a$, $\lim\limits_{x \to a} f(x)$, etc. (§§2.2.1–2, 3.1.1–3)

Infinitesimal (§§2.2.5, 2.4.8)
    Differential, dx, dy (§2.2.7)

Continuous function, $\lim\limits_{x \to a} f(x) = f(a)$ (§3.1.5)

Integral (of a continuous function)
    Definite integral, $\int_a^b f(x)\,dx$ (§§2.2.7–9, 2.4.3)
    Indefinite integral, Primitive, $\int_a^x f(x)\,dx$, $\int f(x)\,dx$ (§§2.5.1, 3.5.6–7)

Derivative (of a differentiable function), $f'(x) = \lim\limits_{\Delta x \to 0} \dfrac{f(x + \Delta x) - f(x)}{\Delta x}$,

$y' = \dfrac{dy}{dx} = \lim\limits_{\Delta x \to 0} \dfrac{\Delta y}{\Delta x}$ (§§2.7.2–4, 2.9.1, 2.9.9–10)

    Higher-order derivatives, $y'' = f''(x) = \dfrac{d^2 y}{dx^2}$, etc. (§3.8.1)

1B. Particular functions

Elementary functions (§3.3.1)
    Power functions, x^α (§§3.3.2, 6.2.6)

Trigonometric functions, $\sin x$, $\cos x$, $\tan x$, etc. (§§3.3.4–5, 3.3.7, 4.7.4)
    Inverse trigonometric functions, $\arcsin x$, etc. (§§4.3.2, 4.4.7–8, 4.7.4)
  Exponential and logarithmic functions, a^x, $\log_a x$ (§§3.3.10–11, 6.2.1, 6.2.5)

Functions constructed from these include the polynomials (§§3.1.10, 3.2.1, 3.2.4), among which are the linear functions (§3.6.5); the rational functions (§§3.1.10, 4.6.9); and the hyperbolic functions and their inverses (§4.1.5.Q7, §4.4.9.Q10). The functions $x^n \sin(1/x)$ furnish examples (§4.2.9).

1C.   Limits and continuity

If $\lim\limits_{x \to a} f(x) = l$ and $\lim\limits_{x \to a} g(x) = m$ then the limits as $x \to a$ of $f + g$, $f - g$, fg, and $f/g \, (g \neq 0)$ are respectively $l + m$, $l - m$, lm, and $l/m \, (m \neq 0)$. (§3.1.9)

If $f(x) \leq g(x)$ for all x near a, then $l \leq m$. (§3.1.9)

If f and g are continuous so are $f + g$, $f - g$, fg, and $f/g \, (g \neq 0)$, and also the composite $f \circ g$ and, if f is monotone, the inverse f^{-1}. (§§3.1.10, 4.1.1, 4.3.2, 4.3.7)

2.   THEOREMS

2A.   Fundamental Theorem

I. $\dfrac{d}{dx} \displaystyle\int_a^x f(x)\, dx = f(x)$ and   II. $\displaystyle\int_a^x \dfrac{d}{dx} F(x)\, dx = F(x) - F(a)$ (§§3.5.3, 3.5.6)

Hence the primitives of $f(x)$ are the functions $F(x) = \int_a^x f(x)\, dx + C$, where a and C are arbitrary constants. If two functions have the same derivative, they differ by a constant. (§3.5.6)

2B.   Mean value theorem and Taylor's theorem

There is ξ between a and b such that:
  (Differential) mean value theorem: $f(b) - f(a) = f'(\xi)(b - a)$. (§3.7.5)
  Integral mean value theorem: $\int_a^b f(x)\, dx = f(\xi)(b - a)$. (§3.7.15)

The differential form has a second-order analogue (§3.8.7), and generalizes further to Taylor's theorem,

$$f(b) = f(a) + \frac{f'(a)}{1!}(b-a) + \frac{f''(a)}{2!}(b-a)^2 + \cdots$$

$$+ \frac{f^{(n)}(a)}{n!}(b-a)^n + \frac{f^{(n+1)}(\xi)}{(n+1)!}(b-a)^{n+1}. \quad (\S8.2.1)$$

A consequence is the Taylor series of a function,

$$f(x) = f(a) + \frac{f'(a)}{1!}(x-a) + \frac{f''(a)}{2!}(x-a)^2 + \frac{f'''(a)}{3!}(x-a)^3 + \cdots .$$

$(\S\S8.1.4\text{--}5)$

A number of functions have easily computable Taylor series. $(\S\S8.2.6\text{--}8)$

These theorems have application to approximation and estimation. $(\S\S3.7.11, 3.8.6\text{--}8, 8.1.1\text{--}4, 8.2.7)$

3. DIFFERENTIATION

3A. Rules

Sum: $(f+g)' = f' + g'$ (extends to more summands)

Difference: $(f-g)' = f' - g'$

Product: $(fg)' = f'g + fg'$ (extends to more factors),

and a special case: $(cf)' = cf'$

Quotient: $\left(\dfrac{f}{g}\right)' = \dfrac{f'g - fg'}{g^2}, \quad g \neq 0,$

and a special case: $\left(\dfrac{1}{g}\right)' = -\dfrac{g'}{g^2}$ $(\S\S3.2.4, 3.4.2\text{--}3)$

Differentiability implies continuity. $(\S3.1.7)$

Chain rule: $(f \circ g)'(x) = f'(g(x))g'(x)$, $\dfrac{dy}{dx} = \dfrac{dy}{du}\dfrac{du}{dx}$ $(\S4.1.3\text{--}4)$

Implicit differentiation: from an equation in x and y determine y' in terms of x and y by differentiating with the aid of the chain rule. $(\S\S4.2.3\text{--}4, 4.5.10)$

Inverse function: for $y = f^{-1}(x)$, $(f^{-1})'(x) = \dfrac{1}{f'(y(x))}$, $\dfrac{dy}{dx} = \dfrac{1}{\dfrac{dx}{dy}}$ (§§4.3.4–5)

3B.   Derivatives of particular functions

The derivative of a constant is 0. (§2.9.2)

$(x^\alpha)' = \alpha x^{\alpha-1}$ (§§3.2.1, 3.3.2, 6.2.6)

The derivative of a polynomial is a polynomial. (§3.2.4)

$(\sin x)' = \cos x$, $(\cos x)' = -\sin x$, $(\tan x)' = \sec^2 x$, etc. (§§3.3.7, 4.4.1)

$(\arcsin x)' = \dfrac{1}{\sqrt{1-x^2}}$, $(\arccos x)' = -\dfrac{1}{\sqrt{1-x^2}}$, $(\arctan x)' = \dfrac{1}{1+x^2}$, etc.

(§§4.4.7–8)

$(a^x)' = (\log a)a^x$, and the special case $(e^x)' = e^x$;

$(\log_a x)' = \left(\dfrac{1}{\log a}\right)\dfrac{1}{x}$, and the special case $(\log x)' = \dfrac{1}{x}$ (§§3.3.10–11, 4.3.5,

6.2.1, 6.2.5)

$(\sinh x)' = \cosh x$, $(\text{arsinh}\, x)' = \dfrac{1}{\sqrt{x^2+1}}$, etc.   (§4.1.5.Q7, §4.4.9.Q10)

4.   INTEGRATION

4A.   Rules and methods

$$\int_a^b f(x)\,dx + \int_b^c f(x)\,dx = \int_a^c f(x)\,dx \quad (§2.4.4)$$

$$\int_b^a f(x)\,dx = -\int_a^b f(x)\,dx \quad (§2.10.2)$$

If $F(x) = \displaystyle\int f(x)\,dx$, $\displaystyle\int_a^b f(x)\,dx = F(b) - F(a)$. (§3.6.1)

$$\int (f(x) + g(x))\, dx = \int f(x)\, dx + \int g(x)\, dx,$$

$$\int (f(x) - g(x))\, dx = \int f(x)\, dx - \int g(x)\, dx,$$

$$\int cf(x)\, dx = c \int f(x)\, dx,$$

and likewise for definite integrals.   (§3.6.3)

If $g(x) \le h(x)$ everywhere in $[a, b]$, then

$$\int_a^b g(x)\, dx \le \int_a^b h(x)\, dx .   (\S3.7.2)$$

Methods of integration:

Substitution, $\displaystyle \int f(u(x)) \frac{du}{dx}\, dx = \int f(u)\, du$ (§4.6.2) and

$$\int_a^b f(u(x)) \frac{du}{dx}\, dx = \int_{u(a)}^{u(b)} f(u)\, du   (\S4.6.3)$$

Parts, $\displaystyle \int h(x)k(x)\, dx = H(x)k(x) - \int H(x)k'(x)\, dx$, where

$$H(x) = \int h(x)\, dx   (\S4.6.6)$$

Partial fractions, for rational functions (§4.6.9)

Improper integrals, as limits, e.g. $\displaystyle \int_1^\infty \frac{1}{x^2}\, dx = \lim_{b \to \infty} \int_1^b \frac{1}{x^2}\, dx$ (§§4.7.1–2)

Integration over regions:

In the plane, $\displaystyle \int_R \sigma\, dA$, by strips and rings   (§§4.8.2–3)

In space, $\displaystyle \int_R \delta\, dV$, by slices   (§4.8.6)

On a surface of revolution, by bands, $\displaystyle \int_a^b 2\pi y \sigma \sqrt{1 + y'^2}\, dx$ (§7.5.4)

4B. Integrals of particular functions

$$\int x^\alpha \, dx = \frac{x^{\alpha+1}}{\alpha+1} + C, \quad a \neq 1;$$

$$\int \frac{1}{x} \, dx = \log|x| + C \quad (\S 3.6.2)$$

$$\int \sin x \, dx = -\cos x + C,$$

$$\int \cos x \, dx = \sin x + C \quad (\S 3.6.1)$$

$$\int \tan x \, dx = -\log|\cos x| + C \quad (\S 4.6.4, \text{ Example 3})$$

$$\int \frac{1}{\sqrt{1-x^2}} \, dx = \left\{ \begin{array}{l} \arcsin x + C, \\ -\arccos x + C, \end{array} \right\} \quad -1 < x < 1;$$

$$\int \frac{1}{1+x^2} \, dx = \arctan x + C, \quad -\infty < x < \infty \quad (\S 4.4.8)$$

$$\int e^x \, dx = e^x + C \quad (\S 3.6.4)$$

$$\int \log x \, dx = x \log x - x + C \quad (\S 4.6.7, \text{ Example 2})$$

Other integrals are calculated as well (mostly in Art. 4.6), among them $\int \sqrt{1-x^2} \, dx$ (§4.6.4, Example 8) and $\int \sqrt{x^2 \pm 1} \, dx$ (§4.6.5.Q5, §4.6.8.Q4), $\int \frac{1}{\sqrt{x^2 \pm 1}} \, dx$ and $\int \frac{1}{1-x^2} \, dx$ (§4.4.9.Q10), $\int \sqrt{\frac{x}{a \pm x}} \, dx$ (§4.6.4(3), §4.6.5.Q6), $\int x e^x \, dx$ (§4.6.7, Example 1), $\int e^x \sin x \, dx$ (§4.6.7, Example 3), $\int \cos^n x \, dx$ (§4.6.7, Example 4), $\int \arcsin x \, dx$ (§4.6.8.Q3), and integrals of rational functions (§4.6.9). Many more can be found by consulting a published "table of integrals" or a computer integration program.

5. PROPERTIES OF FUNCTIONS AND GRAPHS

The derivative can be interpreted as the slope of the graph. (§2.9.4)

The definite integral can be interpreted as the area under the graph. (§2.4.1)

The derivative of $y = f(x)$ measures the rate of change of y with respect to x. (§2.9.1)

If $f' > 0$ on an interval f is increasing; if $f' < 0$, decreasing. (§3.9.1)

If $f'' > 0$ on an interval the graph of f is concave up; if $f'' < 0$, concave down. (§3.9.6)

At an interior point of an interval on which f' exists, if f has a local maximum or minimum, $f' = 0$. (§§3.7.7, 3.9.2)

If $f' = 0$ at an interior point of an interval on which f' exists, and at that point f' changes sign from negative to positive, or $f'' > 0$, there is a local minimum there. If f' changes from positive to negative or $f'' < 0$ there is a local maximum. (§§3.9.2, 3.9.6)

6. FUNCTIONS OF SEVERAL VARIABLES

A function $z = f(x, y)$ has partial derivatives $f_x = \dfrac{\partial f}{\partial x}$, $f_y = \dfrac{\partial f}{\partial y}$, and higher partial derivatives. (§§4.5.1, 4.5.4–5)

Its differential is $df = \dfrac{\partial f}{\partial x}\, dx + \dfrac{\partial f}{\partial y}\, dy$. (§4.5.7)

For a compound function $F(t) = f(x(t), y(t))$ there is the chain rule $\dfrac{dF}{dt} = \dfrac{\partial f}{\partial x}\dfrac{dx}{dt} + \dfrac{\partial f}{\partial y}\dfrac{dy}{dt}$. (§4.5.9)

A function of two variables $\sigma(x, y)$ has an integral $\int_R \sigma\, dA$ over a plane region R; a function of three variables $\delta(x, y, z)$ has an integral $\int_R \delta\, dV$ over a region R of space. (§§4.8.3–4, 4.8.6)

7. DIFFERENTIAL EQUATIONS

Uniform gravitation, $\dfrac{d^2x}{dt^2} = -g$.

Solution: $x(t) = -\frac{1}{2}gt^2 + v_0t + x_0$ (§5.1.1)

Simple harmonic motion, $\dfrac{d^2x}{dt^2} = -\omega^2x$.

Solution: $x(t) = A\cos(\omega t + \delta)$ (§§5.2.1, 5.2.3)

The special case $\omega = 1$ (with initial conditions) characterizes the sine and cosine and leads to their Taylor series. (§8.1.5, Example 2)

Motion in terms of potential energy, $\dfrac{dx}{dt} = \pm\sqrt{\dfrac{2}{m}\left(E - \phi(x)\right)}$. (§5.3.6)

Inverse-square gravitation, $\dfrac{d^2x}{dt^2} = -\dfrac{GM}{x^2}$, $\quad x > 0$.

Solutions take several forms. (§§5.4.1, 5.4.4, §5.4.5.Q12 & Q13)

Extrinsic uniform (linear) growth, $\dfrac{du}{dt} = a$.

Solution: $u(t) = u_0 + at$ (§6.1.1)

Intrinsic uniform (exponential) growth, $\dfrac{dq}{dt} = \alpha q$.

Solution: $q(t) = q_0 e^{\alpha t}$ (§§6.1.2, 6.2.1, 6.2.5)

The special case $\alpha = 1$ (with an initial condition) characterizes the exponential function and leads to its Taylor series. (§8.1.5, Example 1)

8. CURVES

A curve is given a positive direction. (§7.1.1)

Arc length s is the limit of the lengths of inscribed polygons, with a sign. The ratio of arc to chord approaches unity as the length of the chord approaches 0. (§§4.4.4–5, 7.1.1–2, 7.2.3)

Curvature κ at a point is the limit of mean curvature (change of direction per arc length) as arc length approaches 0, with a sign. (§§7.1.4–5)

A curve may be given by parametric equations $x = x(t)$, $y = y(t)$; a dot signifies differentiation with respect to the parameter t. If $y = f(x)$, $y' = \dot{y}/\dot{x}$. (§§7.4.1–2)

Arc length: $ds = \sqrt{dx^2 + dy^2} = \sqrt{1 + \left(\dfrac{dy}{dx}\right)^2}\, dx = \sqrt{\dot{x}^2 + \dot{y}^2}\, dt$,

$$S = s_2 - s_1 = \int_{s_1}^{s_2} ds = \int_{s_1}^{s_2} \sqrt{dx^2 + dy^2}, \text{ etc. } (\S\S7.2.1,\ 7.4.3)$$

Curvature: $\kappa = \dfrac{y''}{(1 + y'^2)^{3/2}} = \dfrac{\dot{x}\ddot{y} - \dot{y}\ddot{x}}{(\dot{x}^2 + \dot{y}^2)^{3/2}}$ (§§7.2.4, 7.4.3)

In local coordinates x, y, at P, $\kappa(P) = \lim \dfrac{2y}{x^2}$. (§7.3.2)

Intrinsic equation of a curve: $\kappa = f(s)$ (§7.4.6)

Velocity and tangential and normal acceleration along a curve:

$$a_T = \dot{v}, \ a_N = \kappa v^2; \ v = \int a_T\, dt, \ s = \int v\, dt, \ \kappa = \frac{a_N}{v^2}. \ (\S\S7.5.5,\ 7.5.7)$$

APPENDIX 1

Conversion to CGS Units

The International System of units (SI), used in this book, employs the meter and the kilogram as the basic measures of length and mass. The older cgs system uses the centimeter and the gram. Both measure time in seconds. To convert from SI to cgs, multiply lengths measured in the SI unit by 100 and masses by 1000:

$$\text{length: } 1 \text{ m} = 10^2 \text{ cm}$$
$$\text{mass: } 1 \text{ kg} = 10^3 \text{ g}$$

This affects derived units correspondingly:

$$\text{area: } 1 \text{ m}^2 = 10^4 \text{ cm}^2$$
$$\text{volume: } 1 \text{ m}^3 = 10^6 \text{ cm}^3$$
$$\text{line density: } 1 \text{ kg/m} = 10 \text{ g/cm}$$
$$\text{surface density: } 1 \text{ kg/m}^2 = 10^{-1} \text{ g/cm}^2$$
$$\text{(volume) density: } 1 \text{ kg/m}^3 = 10^{-3} \text{ g/cm}^3$$
$$\text{velocity: } 1 \text{ m/sec} = 10^2 \text{ cm/sec}$$
$$\text{acceleration: } 1 \text{ m/sec}^2 = 10^2 \text{ cm/sec}^2$$
$$\text{volume flow rate: } 1 \text{ m}^3/\text{sec} = 10^6 \text{ cm}^3/\text{sec}$$
$$\text{mass flow rate: } 1 \text{ kg/sec} = 10^3 \text{ g/sec}$$
$$\text{momentum: } 1 \text{ kg m/sec} = 10^5 \text{ g cm/sec}$$

The cgs units of force and energy are the **dyne** and the **erg**.

$$\text{force: } 1 \text{ newton} = 1 \text{ kg m/sec}^2 = 10^5 \text{ g cm/sec}^2 = 10^5 \text{ dyne}$$

pressure: 1 newton/m^2 = 10 dyne/cm^2

impulse: 1 newton sec = 10^5 dyne sec

energy: 1 joule = 1 kg m^2/sec^2 = 10^7 g cm^2/sec^2 = 10^7 erg

work: 1 newton m = 10^7 dyne cm

Thus the acceleration of gravity g is 9.8×10^2 cm/sec^2, and the gravitational constant G (§5.4.1) works out to 6.7×10^{-8} dyne cm^2/g^2. For each of the units above, to convert a value we multiply by a power of 10 (move a decimal point), keeping the digits the same. But for the cgs "electrostatic unit" of charge, the **esu**, we have:

electric charge: 1 coulomb $\approx 3 \times 10^9$ esu

In Coulomb's law (§7.5.3) the cgs constant of proportionality is 1 (instead of the SI value of approximately 9×10^9).

APPENDIX 2

The Epigraphs

The first epigraph, in ancient Greek *ouk esti basilikē atrapos epi geōmetrian*, has the familiar English translation "There is no royal road to geometry." It is attributed to Euclid, a Greek mathematician of around 300 BCE who wrote the *Elements*, a systematic deductive treatise on plane and solid geometry, ratio and proportion, and number theory. (His book is cited in §4.4.5.) Euclid probably studied in Athens, where mathematics had long flourished, but he taught at Alexandria in Egypt. The remark is said to have been his response to a question of the Egyptian king Ptolemy I concerning geometry, whether there was any shorter road than that of the *Elements*.

In his question Ptolemy used the regular word for a road, *hodos*. Euclid's word *atrapos*, not very frequent in the literature that survives, is usually translated "path," sometimes "shortcut." It had been used more than a century before Euclid by Herodotus, in his history of the wars between the Greeks and the Persians, for the mountain path by which invading troops of the Persian king Xerxes, guided by a false Greek, got around to the rear of the defenders who had held the narrow pass at Thermopylae against every assault. This longer route was for Xerxes a shortcut that opened the way to Athens. In a reference to the celebrated affair of Thermopylae a generation after Herodotus, the Athenian historian Thucydides spoke of "the *atrapos*" used by the Persians, seemingly as a thing well known. Perhaps then Euclid's reply is more pointed than appears at first. To the mind of the Eastern monarch it may suggest one of his highways fit for a royal progress, but the geometer, remembering Athens, means that there is no king's pathway this time: you have to meet Greek mathematics head-on.

For Euclid's remark see *Selections Illustrating the History of Greek Mathematics*, ed. Ivor Thomas (Loeb Classical Library), Vol. 1, pp. 154–155. It is reported by the commentator Proclus in indirect discourse, here turned into direct. The second edition of the *Oxford English Dictionary* translates it with "royal short cut" (article "Road," I.6.c). On Euclid see *The Thirteen Books of Euclid's Elements*, ed. Thomas L. Heath, Vol. 1, pp. 1–6. The passages in the historians are: Herodotus 7.175, 212ff; Thucydides 4.36. Authors after Euclid's time (Diodorus Siculus, Strabo, Pausanias) also call the path at Thermopylae an *atrapos*.

Stobaeus (2.31.115) has a similar story of Menaechmus and Alexander, in which the geometer tells the king that "through the country there are roads for private citizens and roads for kings (royal roads), but in geometry there is one road for all." Like Ptolemy, Menaechmus uses *hodos*. Note that his road is *in* geometry while Euclid speaks of a path *to* it. If the version in Stobaeus derives from the other, it may exhibit the same misunderstanding of Euclid's wit as led to the English translation "royal road."

In Athens Euclid would have studied under students of Plato. The occurrences of *atrapos* in Plato's *Phaedo* (66b) and *Statesman* (258c) are interesting in connection with these speculations. See also Herodotus 5.49ff (on the Persian royal road) and Isaiah 40.3–4.

The second epigraph is attributed to Jean le Rond d'Alembert (1717–83), French mathematician, scientist, and man of letters, who among other things advocated the concept of limit as a foundation for calculus and observed that neither momentum nor kinetic energy was the "true" measure of a moving body's impetus, but that each (the integral over time and the integral over space, Art. 1.4) had its uses. The story is that a young man tackling calculus and finding contradictions in it (no doubt having to do with limits and infinitesimals) made bold to consult d'Alembert, who advised him, "Go on, faith will come to you." The first words can also be translated "Go forward"—that is, advance along your path. Calculus, which is everyman's shortcut to geometry (§1.5.4), had been discovered the century before.

D'Alembert's remark is an ironical echo of religious advice, surely intended as such by the anti-clerical *philosophe*. He may have had Pascal in mind. Pascal was a founder of the mathematical theory of probability, which d'Alembert worked on in some perplexity, and had used its language in a famous *pensée* to urge belief in God as a bet on a sure thing. Impressed by the argument but unsatisfied, his interlocutor protests that he cannot believe. He is told: "You want to go to faith and you don't know the road there." You should do as others before you, who risked "acting just as if they believed," and then "at each step you take on this road, you will see so much certainty of gain, and so much nothingness in what you are risking, that in the end you will recognize that you have bet on a thing that is certain, infinite, for which you have given nothing." That is not unlike the briefer counsel of d'Alembert.

D'Alembert's reply, and the story of it, are in *D'Alembert*, by Joseph Bertrand (Paris, 1889), p. 56. The quotation is also found with *et* ("and") instead of a comma between the clauses. On the concept of limit see articles by d'Alembert in the *Encyclopédie* of Diderot and d'Alembert (1751–72): "The theory of limits is the basis of the true Metaphysics of differential calculus" (article "*Limite*"); "Differential calculus does not deal with infinitely small quantities, as is still commonly said; it deals solely with limits of finite quantities. ... We employ the term *infinitely small* only for the sake of brevity of expression (*pour abréger les expressions*)" (article "*Différentiel*"). On momentum and kinetic energy see d'Alembert's *Traité de Dynamique* (1st ed., Paris, 1743, and reprints), *Préface*, pp. xvi–xxii. The wager *pensée* is no. 418 in Lafuma's ordering (Brunschvicg 233, Chevalier = Pléiade 451); a related *pensée* is no. 816 (240, 457).

Answers

CHAPTER 1

§1.1.3

Q1. (a) 3 kg/m (b) 32 kg; 8 kg **Q2.** λ takes its minimum value at the midpoint and rises to its maximum value at the ends. One thing to say about mass is that a segment of given length (much shorter than the wire) has least mass when it contains the midpoint, most when it includes an endpoint. **Q3.** Line density at P is determined by the masses of the segments that contain P; but not by the mass of any one such segment, or any finite number of segments. (This point will be taken up in §2.6.1.)

§1.2.2

Q1. (a) 9 m^3 of water enters the tank in 3 sec of uniform flow; the growth rate is 3 m^3/sec. (b) Volume grows uniformly at 4 m^3/sec for 8 sec; 32 m^3 of water is added; in 2 sec 8 m^3 is added. **Q2.** Growth rate at time t is determined by the volumes added during time intervals that contain t, but not by the volume added during any finite number of such intervals. **Q3.** (a) $\gamma_1 + \gamma_2$; $U_1 + U_2$ (b) If at a given instant the growth due to inlet 1 alone proceeds at γ_1 m^3/sec, and that due to inlet 2 alone at γ_2 m^3/sec, then at that same instant the growth rate with both open is $\gamma_1 + \gamma_2$. If in a given time interval inlet 1 adds U_1 m^3 and inlet 2 U_2 m^3, then in that same interval a total of $U_1 + U_2$ m^3 is added.

§1.3.2

Q1. (Cf. the analogous expressions in §1.1.2.) **Q2.** (a) 3.0 sec (b) Not necessarily all the time, but at some moment it must have been moving at exactly that speed.

§1.3.6

Q1. (a) In m/sec: 9.8; 19.6 (or 20 if rounded to two digits), 29.4 (or 29), 9.8n; $9.8(n+\frac{1}{2}) = 9.8n+4.9$. (b) A first observation is that the distance added in *any* time interval from t_1 to t_2 is greater than the velocity at t_1 would produce, if continued for the whole interval. **Q2.** Let train A always accelerate uniformly. At time t_1 let train B be traveling at a constant velocity less than the velocity of A; thus at t_1 B has zero acceleration. Let B increase its speed smoothly in such a way that by a later time t_2 it is traveling faster, and ever faster, than A. At t_1, A's speed is increasing relative to B's—a passenger on B sees A accelerating. At t_2, B's speed is increasing relative to A's, or (what is the same thing) A's is decreasing relative to B's—the passenger on B sees A decelerating. At a certain moment in between, A's speed is neither increasing nor decreasing relative to B's—the perspicacious passenger detects agreement in rate of increase of velocity.

§1.4.5

Q1. 50 kg m/sec   **Q2.** 8 m/sec   **Q3.** (a) Since force is momentum per time, $f_1 = (20 \text{ kg m/sec})/(10 \text{ sec}) = 2$ newton, $f_2 = (30 \text{ kg m/sec})/(10 \text{ sec}) = 3$ newton. (b) The ratio of the impulse of f_1 to the impulse of f_2 is 2/3. While f_1 may exceed f_2 during sōme part of the 10 seconds, it cannot always be greater than f_2.   **Q4.** By §1.3.5 the acceleration of gravity is 9.8 m/sec^2. Applying $f = ma$ with $m = 1$ kg gives $f = 9.8$ newton.   **Q5.** (i) means that Q/x depends only on y, i.e. is independent of x. Likewise (ii) means that Q/y is independent of y. Hence $Q/xy = (1/x)(Q/y)$ is independent of y (since each factor is), and also $Q/xy = (1/y)(Q/x)$ is independent of x (for the same reason). Then Q/xy is independent of x and y, i.e. Q is proportional to xy.   **Q6.** Multiplication of total and specific quantity by the same constant is equivalent to changing the scale of each by the same factor. On the wire, for example, changing from kilograms to grams, and from kilograms per meter to grams per meter, is the same thing as multiplying both M and λ by 1000. But the notion of total and specific quantity is independent of the units of measurement, which need only be consistent with one another.   **Q7.** $V = aT$ and $mV = fT$ together imply $f = ma$.

§1.4.10

Q1. The work done is 1 newton m, which equals the kinetic energy of the particle, $\frac{1}{2}mv^2 = \frac{1}{2}v^2$. Hence $v = \sqrt{2} \approx$ (is approximately equal to) 1.4 m/sec. **Q2.** (a) If the distance is L, the work done is $9.8mL$ newton m, which equals the kinetic energy $\frac{1}{2}mv^2 = \frac{1}{2}m(5^2) = 12.5m$. Hence $L \approx 1.3$ m. (b) Gravitational force is proportional to mass. **Q3.** The velocities in m/sec are approximately 0, 4.47, 24.6, 29.1. In the first case, $K = \frac{1}{2}mv_2^2 - \frac{1}{2}mv_1^2 = \frac{1}{2}m(v_2^2 - v_1^2) = \frac{1}{2}(1400)(4.47^2 - 0^2) \approx 1.4 \times 10^4$ joule. In the second, $K \approx 1.7 \times 10^5$ joule—more than ten times larger. **Q4.** It is work to push on an object that will not move. It is harder work to do a job quickly than to do it slowly.

§1.5.5

Q1. A is the difference of two triangles: $A = \frac{1}{2}b \cdot \mu b - \frac{1}{2}a \cdot \mu a = \frac{1}{2}\mu(b^2 - a^2)$. **Q2.** (a) Find the area of a semicircle. (b) In Fig. A-1, $A =$ area under $C_1 -$ area under

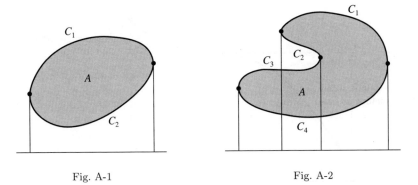

Fig. A-1   Fig. A-2

C_2. (c) In Fig. A-2, A = area under C_1 − a.u. C_2 + a.u. C_3 − a.u. C_4. **Q3.** Fig. A-3, etc. (five in all, besides the wire). **Q4.** As in Q1, $v = \mu t$ for some number μ. Area in that question corresponds to distance here, so the distance covered is $\frac{1}{2}\mu(b^2 - a^2)$.

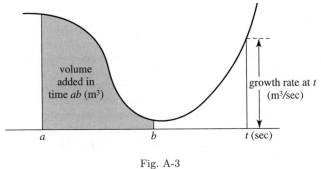

Fig. A-3

§1.5.10

Q1. Draw one vertically above the other to compare them. Notice first that the latter is steeper where the former is higher. **Q2.** Fig. A-4, etc. (five in all, besides the wire). **Q3.** None. **Q4.** Angle increases continuously from $0°$ to $90°$, slope begins at 0 and increases continuously without limit; slope is not defined at R. Both increase more and more rapidly as x approaches R.

§1.6.4

Q1. The rate, in m^3/sec, at which the volume of water in the tank is growing at time 3.5 sec; the distance in m covered by the particle between $t = 5$ sec and $t = 6$ sec; etc. **Q2.** (a) $m(x) = 3x$ (kg) (b) For $x \geq 1$, $m_1(x) = 3x - 3$. **Q3.** $m_0(x) = m(x) - m(x_0)$. **Q4.** The area under the curve from $x = 0$ to the variable point x. **Q5.** If motion begins at $t = 0$, so that $v(0) = 0$, then $p(0) = 0$ and hence

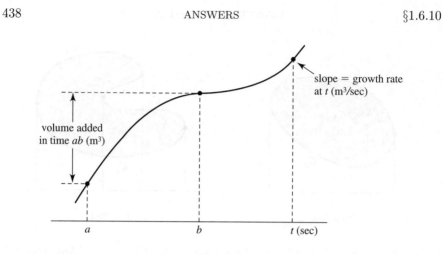

Fig. A-4

$p(t) = p(t) - p(0) = P([0, t])$ is the cumulative function corresponding to the total function P.

§1.6.10

Q1. (a) Yes, provided that "mother of" is unambiguous—e.g., biological mother. (b) No, because not every woman has exactly one daughter. **Q2.** The former is a number, regarded as a value taken by the variable y; the latter is a law which assigns the number 2 to every value of some unnamed independent variable (possibly x). **Q3.** Yes, *if* (i) its inputs are restricted to non-negative numbers, and (ii) "square root" is interpreted to mean "non-negative square root." ("Non-positive square root" would also work.) Regarding (i) and (ii) see §1.6.6. **Q4.** The curve can be divided into parts, each of them the graph of a function. Cf. §1.5.5.Q2, with Figs. A-1 and A-2, and see Art. 7.4. **Q5.** $f(x) = \begin{cases} 1 \text{ if } x \leq 0 \\ 2 \text{ if } x > 0 \end{cases}$ (Fig. A-5).

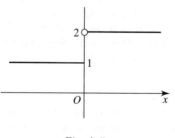

Fig. A-5

CHAPTER 2

§2.1.3

Q1. (a) $n = 1$: (i) 1 kg, (ii) $\frac{3}{2}$, (iii) 2. $n = 2$: (i) $\frac{5}{4}$, (ii) $\frac{3}{2}$, (iii) $\frac{7}{4}$. $n = 3$: (i) $\frac{4}{3}$, (ii) $\frac{3}{2}$, (iii) $\frac{5}{3}$. As n increases, estimates (i) increase, and estimates (iii) decrease, toward the constant intermediate estimates (ii). (b) $\frac{3}{2}$ kg, the same as estimates (ii). **Q2.** Since $\lambda(\xi_1) \geq \lambda(a)$ and $\lambda(\xi_2) > \lambda(a)$, $\lambda(\xi_1)(x_1 - a) + \lambda(\xi_2)(b - x_1) > \lambda(a)(x_1 - a) + \lambda(a)(b - x_1) = \lambda(a)(b - a)$. Similarly, the estimate is $< \lambda(b)(b - a)$.

§2.1.5

Q1. (a) $\xi_1 = 1$, $\xi_2 = 3$, so $\lambda(\xi_1) = 1 + 0.01(1^2) = 1.01$, $\lambda(\xi_2) = 1 + 0.01(3^2) = 1.09$, and the estimate is $1.01 \times 2 + 1.09 \times 2 = 4.20$. (b) In the earlier case the line density increased uniformly with x. In the present case it increases more rapidly than that, so the line density at $\xi_2 = 3$ exceeds the line density at 2 by more than the line density at $\xi_1 = 1$ falls short of it. This makes the estimate for $n = 2$ larger. **Q2.** (a) $n = 1$: lower 4, upper 68. $n = 2$: lower 12, upper 44. $n = 4$: lower 18, upper 34. (b) For equal intervals: (i) factor out the interval length; (ii) obtain the upper sum from the lower one by subtracting the first term and adding a new last term.

§2.2.4

Q1. (a) $\frac{1}{3}$ (b) The first term of the sequence differs from $\frac{1}{3}$ by less than 0.1 (because $0.3 < \frac{1}{3} < 0.4$), the second by less than 0.01, etc. It follows that any neighborhood of $\frac{1}{3}$, however small, contains all terms from some point on. **Q2.** 1 **Q3.** (a) As the figure shows, $\lambda(\xi_1) = \lambda(1/n) = 1/n$, $\lambda(\xi_2) = \lambda(2/n) = 2/n$, etc., and every $\Delta x_i = 1/n$. Hence the formula $\lambda(\xi_1)\Delta x_1 + \lambda(\xi_2)\Delta x_2 + \cdots + \lambda(\xi_n)\Delta x_n$ gives the estimate stated. (b) See Fig. A-6. (c) $\dfrac{1}{n^2}(1 + 2 + \cdots + n) = \dfrac{1}{n^2} \cdot \dfrac{1}{2} n(n + 1) = \dfrac{1}{2} \cdot \dfrac{n+1}{n}$ (d) $\frac{1}{2}$ **Q4.** The nth estimate is $\dfrac{x}{n} \cdot \dfrac{x}{n} + \dfrac{2x}{n} \cdot \dfrac{x}{n} + \dfrac{3x}{n} \cdot \dfrac{x}{n} + \cdots + \dfrac{nx}{n} \cdot \dfrac{x}{n}$,

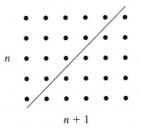

$n + 1$

Fig. A-6

which reduces to $\dfrac{x^2}{2} \cdot \dfrac{n+1}{n}$, whose limit is $\dfrac{x^2}{2}$. Hence $m(x) = M([0,x]) = \dfrac{x^2}{2}$.
Q5. If M is an endpoint, terms can be arbitrarily close to it without being in the interval.

§2.2.10

Q2. (a) The integral from 1 to 3 of x^3; the mass of the segment of wire from $x = 1$ to $x = 3$ whose line density there is $\lambda(x) = x^3$—follow the interpretation in §2.2.8 to give more detail. **Q3.** (a) $\int_5^{10} \sqrt{x}\,dx$ (b) Let the wire extend from $x = 0$ to $x = 2$, then $\lambda(x) = 2x + 3$ and the mass is $\int_0^2 (2x+3)\,dx$. **Q4.** (a) The sum of the lengths of the (finite or infinitesimal) subintervals is the length of the whole interval. When line density is unity, the number of kilograms in a segment equals the number of meters in it. (b) $c \int_a^b dx$ —argue as in (a), factoring out c.

§2.3.5

Q1. (a) $\displaystyle\int_5^{11} (4t + 2)\,dt$ (b) $\displaystyle\int_0^{10} \dfrac{t}{t^2 + 1}\,dt$ (c) (i) $\displaystyle\int_{t_0}^{t_0+4} g\,dt$ (ii) $\displaystyle\int_{t_0}^{t_0+4} mg\,dt$ (the force is mg, by §1.4.4) (iii) $\int_{x_1}^{x_2} mg\,dx$, where x_1 is the location of the particle at t_0 and x_2 its location at t_0+4. **Q2.** (a) The mass between $x = 1$ and $x = 4$ of a wire of line density $(x+1)^2$; the work done by a force $(x+1)^2$ on a particle that moves from $x = 1$ to $x = 4$; equivalently, the kinetic energy acquired by the particle. (b) Volume added; distance; velocity added; impulse; equivalently, momentum added. (c) All possibilities, since the variable of integration may be x or t.

§2.4.6

Q1. (a) $3(b - a)$ (b) $\frac{1}{2}$ (c) $\pi/2$ (the area under the semicircle $y = \sqrt{1 - x^2}$)
Q2. The area of the segment of the parabola $y = 1 - x^2$ cut off by the x-axis is $\int_{-1}^1 (1 - x^2)\,dx$. **Q3.** As along an infinitesimal segment of wire we neglect variation in line density, considering the wire uniform, so over an infinitesimal interval we neglect variation in the height of the graph, considering it horizontal. Then we apply the known facts that the mass of a uniform wire is the product of length and (constant) line density, and the rectangular area under a horizontal line is the product of base and (constant) height. **Q4.** To treat the monotone intervals separately means always to use their endpoints a_1, a_2, etc., as points of division in forming subintervals. The question then is: suppose a sequence of estimates S is formed in which these points are *not* used (Fig. A-7(a) shows an estimate that does not use a_1 and a_2); need it have the same limit as a sequence which does use them? Yes, because the points $a_1, a_2, \ldots$ can be added as new points of division so as to modify any S to a new estimate S' which differs from S only over *one* subinterval at *each*

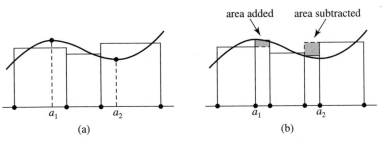

Fig. A-7

of these points (Fig. A-7(b)). Since there are only a finite number of such points, and the length of the subintervals approaches zero, the difference between sequences S and S' approaches zero; hence S' has the same limit as S. **Q5.** (a) The case in which $\lambda_1 = \lambda_2$ everywhere is trivial. In the other case, form a wire with line density $\lambda_3 = \lambda_1 - \lambda_2$; this "wire" will not exist where $\lambda_1 = \lambda_2$, but will exist on intervals where $\lambda_1 > \lambda_2$. Then wire 2 and wire 3 together have the same line density and mass as wire 1; hence $M_1 > M_2$. (b) The same argument as in the geometrical case. For example, $M \leq S_U$ because on each subinterval $[x_{i-1}, x_i]$ the line density of the wire is not greater than that of the upper estimate, hence the mass of the wire there is not greater than $f(x_i)\Delta x_i$. (c) The algebraic argument:

$$S_U - S_L = f(x_1)\Delta x_1 + f(x_2)\Delta x_2 + \cdots + f(x_n)\Delta x_n$$

$$- [f(x_0)\Delta x_1 + f(x_1)\Delta x_2 + \cdots + f(x_{n-1})\Delta x_n]$$

$$= [f(x_1) - f(x_0)]\Delta x_1 + [f(x_2) - f(x_1)]\Delta x_2 + \cdots$$

$$+ [f(x_n) - f(x_{n-1})]\Delta x_n$$

$$\leq [f(x_1) - f(x_0)]\Delta x_m + [f(x_2) - f(x_1)]\Delta x_m + \cdots$$

$$+ [f(x_n) - f(x_{n-1})]\Delta x_m$$

$$= \{[f(x_1) - f(x_0)] + [f(x_2) - f(x_1)] + \cdots + [f(x_n) - f(x_{n-1})]\}\Delta x_m$$

$$= \{f(x_n) - f(x_0)\}\Delta x_m = [f(b) - f(a)]\Delta x_m,$$

which approaches zero as Δx_m does. **Q6.** All estimates S equal $c(b-a)$. **Q7.** We could describe the area as the result of adding and subtracting areas under curves (§1.5.5.Q2), regard each curve as the graph of a function, and apply the numerical definition of the integral. The outcome would have to be proved independent of the particular decomposition of the figure. (Cf. §7.5.8.Q14.) **Q8.** Argue either as in Q4 or as follows: if S_1 is an estimate of the first integral and S_2 an estimate of the second, $S_1 + S_2$ is an estimate of the third; and the limit of $S_1 + S_2$ is the sum of the limits of its summands (see the argument in §3.1.9).

§2.4.9

Q1. (a) $\int_a^b 0\,dx = 0$ (b) $\int_a^b 2\,dx = 2(b-a)$ **Q2.** A curve will have this property for every value of b if its slope is x, which increases uniformly from $x = 0$. Its vertical position can be arbitrary. See Fig. A-8. (The curve is an arc of a parabola.) **Q3.** See Fig. A-9. The slope of the lower polygonal line over $[x_{i-1}, x_i]$ is the slope of the curve at x_{i-1}; the slope of the upper polygonal line, the slope of the curve at x_i.

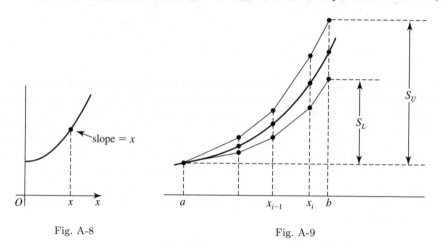

Fig. A-8   Fig. A-9

§2.5.3

Q1. $m(x) = \int_{x_0}^x \lambda(x)\,dx$, $u(t) = \int_{t_0}^t \gamma(t)\,dt$, etc.   **Q2.** See Fig. A-10.   **Q3.** 8

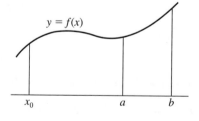

$y = f(x)$

Fig. A-10

§2.6.4

Q1. (a) It follows from §2.4.6.Q5(a) that if $x_1 < 2$, the mass of the given wire over $[x_1, 2]$ is less than the mass of a uniform wire of line density $\lambda(2)$ over the same interval; i.e., $m(2) - m(x_1) < \lambda(2)(2 - x_1)$, or $\dfrac{m(2) - m(x_1)}{2 - x_1} < \lambda(2)$. (b) 0.04, 0.07, 0.1141, 0.119401; 0.28, 0.19, 0.1261, 0.120601; $0.119401 \le \lambda(2) \le 0.120601$.

Q2. Yes; for example, the wire of §1.1.2.Q2.   **Q3.** (a) $\dfrac{\left(1 - \dfrac{1}{x_2}\right) - \left(1 - \dfrac{1}{x_1}\right)}{x_2 - x_1} =$

$\dfrac{\dfrac{1}{x_1} - \dfrac{1}{x_2}}{x_2 - x_1} = \dfrac{1}{x_1 x_2}$. (b) $1/x_0^2$, because as x_1 and x_2 approach x_0 their product will approach x_0^2.

§2.7.6

Q1. (a) $dx\,dx$ at 1; the line density at $x = 1$ of a wire whose cumulative mass is $m(x) = x$—follow the interpretation in §2.7.3 to give more detail. (b) $d\,dx$ of

$\dfrac{x}{x+1}$, etc.   **Q2.** (a) $\left.\dfrac{d(8x^2)}{dx}\right|_3$ (b) If we take (arbitrarily) the segment to be $[1,3]$,

$m(x) = 2x + 1$ there, so for x in the segment the line density function is $\dfrac{d}{dx}(2x+1)$.

Q3. No: changing a changes $m(x)$ by an additive constant, which does not affect the difference quotient (§2.9.2).   **Q4.** $x_0 = 10/\sqrt{3} \approx 5.8$ **Q5.** Calculation as in §2.7.5, using dx in place of Δx, yields $dm/dx = c \cdot (3x_0^2 + 3x_0\,dx + (dx)^2)$; neglecting the infinitesimals, this is $dm/dx = c \cdot 3x_0^2$.   **Q6.** If $x \neq 0$, the last two terms of the second expression for dm are negligible in comparison to the first term (see §2.4.8). The following argument applies even if $x = 0$: differentiability of m means that dm/dx equals some *number* (see the remark at the end of §2.9.9); the first expression for dm gives that number as $c \cdot 3x^2$; the second expression is consistent with this, for it implies that the number $\dfrac{dm}{dx} - c \cdot 3x^2$ is negligible in comparison to non-zero numbers, and the only such number is 0. (Discomfort with this kind of reasoning is only natural.) **Q7.** Suppose $c \cdot (3x_0 + \Delta x)\Delta x$ is to be made smaller in absolute value than 0.01, say; that is, we want $|(3x_0 + \Delta x)\Delta x| < 1$. Let A be any number $> |3x_0|$, and take Δx so small that (i) $|3x_0 + \Delta x| < A$ (this must be possible because $|3x_0| < A$) and (ii) $|\Delta x| < 1/A$. Then $|(3x_0 + \Delta x)\Delta x| < A \cdot (1/A) = 1$. Similarly if the quantity is to be made smaller than 0.001 or any other number.   **Q8.** (a) $\dfrac{\Delta m}{\Delta x} = \dfrac{x_0 + \Delta x - x_0}{\Delta x} =$

$\dfrac{\Delta x}{\Delta x} = 1$, so its limit is $dm/dx = 1$. The wire is uniform, of line density 1. (b) Calculation gives $dm/dx = \lim\limits_{\Delta x \to 0} (2x_0 + \Delta x) = 2x_0$.   **Q9.** $\dfrac{m(x_2) - m(x_1)}{x_2 - x_1} =$

$\dfrac{(x_0 + \varepsilon)^2 - (x_0 - \varepsilon)^2}{(x_0 + \varepsilon) - (x_0 - \varepsilon)} = \dfrac{4x_0\varepsilon}{2\varepsilon} = 2x_0$ —independent of ε.

§2.8.5

Q1. (a) $\dfrac{d}{dt}(t + 1)^4$ (b) (i) $\left.\dfrac{d}{dt}(t^3 + 5t)\right|_1$ (ii) $\left.\dfrac{d(ct)}{dt}\right|_{60}$ (c = the constant of proportionality) (c) (i) $\dfrac{d(2t)}{dt}$ (ii) $k(x) = \frac{1}{2}mv^2$, hence $f = \dfrac{d}{dx}(10x^2)$.   **Q2.** (a) The line density of a wire whose cumulative mass from 0 to x is x^2; the force on a particle whose kinetic energy at any x is x^2; equivalently, the force that has done work x^2 on the particle. (b) The growth rate at $t = 3$ if the volume present at any time is t; the velocity at $t = 3$ if the location at any time is t; the acceleration, etc.; the force at $t = 3$ if the momentum at any time is t.   **Q4.** If the infinitesimal increment of v is dv, the infinitesimal increment of mv is $m\,dv$; i.e. $d(mv) = m\,dv$, hence $\dfrac{d(mv)}{dt} = m\dfrac{dv}{dt}$. (A proof by limits will be given in §3.2.5.Q4.)

§2.9.3

Q1. (a) 0 (b) $w^2 - 1$ **Q2.** (a) By §2.9.2(2), the same as the derivative of cx^3, namely $c \cdot 3x^2$. (b) No.   **Q3.** (a) If $f(x)$ is a constant function c, $f(x + \Delta x) - f(x) = c - c = 0$, so the difference quotient is 0. (b) Yes, *if* it is known that the derivative of a sum of functions is the sum of their derivatives (§3.2.4).

§2.9.5

Q1. The tangent line has slope 6 and passes through $(3, 9)$, hence has equation $y - 9 = 6(x - 3)$, or $y = 6x - 9$.   **Q2.** $y - y_0 = f'(x_0)(x - x_0)$.   **Q3.** (a) acceleration $a(t)$ (b) location $x(t)$ **Q4.** Yes.   **Q5.** (1) A horizontal line is level. (2) Shifting a graph vertically up or down does not affect its slope.   **Q6.** dy changes sign together with dx.   **Q7.** No, because the rise must be 0 at $x = a$, but there need not exist a such that $f(a) = 0$; i.e., the graph may fail to cross the axis. An example is $f(x) = x^2 + 1$, for $x \geq 0$.

§2.9.8

Q1. Yes: on each side of P the picture is like Fig. 2-30.   **Q2.** Yes, in the sense that the secant approaches the common tangent to the circles. Furthermore, the curve keeps bending toward the left as you pass through P in the direction shown. But the curve is not smooth there: its direction changes abruptly by $180°$.

§2.9.11

Q1. For $x = 0$, $dy/dx = 1$ if $dx > 0$ and $= -1$ if $dx < 0$. In the example at the end of §2.9.9, forming the quotient of differentials yielded *no* number; here it yields *two* numbers. For differentiability *one* number is needed.   **Q2.** Let the true line density function be $\lambda(x)$, whose graph looks like part of a polygon or something worse. We can (probably) draw a smooth curve, the graph of a differentiable function $f(x)$, so close to the graph of $\lambda(x)$ that $f(x) - \lambda(x)$ is smaller in magnitude than the least variation in line density we are able to measure; then $\lambda(x)$ is indistinguishable from $f(x)$, hence can be represented by it.

§2.9.13

Q1. "Height is the rate of change of cumulative area," etc.

§2.10.9

Q1. (a) 0 (b) -1.5 coulomb   **Q2.** $mv = -10$ kg m/sec   **Q3.** There are several cases; e.g. if $a < c < b$, §2.4.4 gives $\int_a^c + \int_c^b = \int_a^b$; but $\int_b^c = -\int_c^b$, therefore $\int_a^c = \int_a^b + \int_b^c$.   **Q4.** (a) Assuming that the spring is perfectly elastic within the range of its compression, the net work is zero. The force of the spring at each point is the same during expansion as during compression, but the direction of motion is reversed; hence the positive work integral for the expansion cancels the negative work integral for the compression. (b) The same speed with which it first touched the spring. Since zero work was done, there was zero change in kinetic energy, hence zero change in speed. Velocity, however, changed sign.   **Q5.** $M([x_1, x_2]) = \int_{x_1}^{x_2} \lambda(x)\,dx$, $m(x_2) - m(x_1) = \int_a^{x_2} \lambda(x)\,dx - \int_a^{x_1} \lambda(x)\,dx$; apply Q3. The other relation follows as in §§2.6.2, 2.7.1.   **Q6.** The metal has "positive weight," the wood "negative weight." Whether the rod tends to rise or fall is determined by the predominance of one substance over the other, not the quantity of both that is present. At a point where the rod is mostly metal, "weight density" is positive; where it is mostly wood, negative.   **Q7.** (a) See Fig. A-11, a qualitative picture with arbitrary vertical scales. The instant of greatest compression is t_2. (b) Yes, because $m > 0$. (c) No, as (a) shows.   **Q8.** Among other things: $v(t)$ changes sign an odd number of times; each time it does, $a(t) = 0$.   **Q9.** Shrinkage is the negative of growth. For a model, let the water tank have an outlet as well as an inlet.   **Q10.** The graph of the function $y = |f(x)|$ is obtained from that of $f(x)$ by replacing portions below the x-axis by their mirror images above it (Fig. A-12; cf. Fig. 2-39), hence $\int_a^b |f(x)|\,dx$ gives total area.   **Q11.** x^3; $-x^3$; x^2 **Q12.** (a) $-f(x)$ (b) The part of the spring other than B exerts on B a force slightly greater than $f(x)$. Taking it for granted that forces in opposite directions on the same

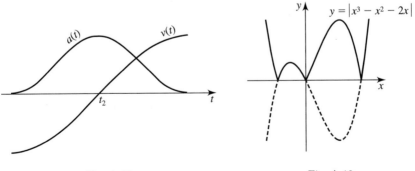

Fig. A-11   Fig. A-12

particle cancel one another, we can say that there is a small positive net force on B, sufficient to accelerate its tiny mass towards the right—that is, to decelerate its leftward motion. (c) If x_2 is the point of maximum compression, the work A does is $\int_{x_1}^{x_2} -f(x)\,dx$. To reduce the initial kinetic energy of A to 0, B does negative work $\int_{x_1}^{x_2} f(x)\,dx$; then the kinetic energy lost is $-\int_{x_1}^{x_2} f(x)\,dx$.

CHAPTER 3

§3.1.4

Q1. The decimal representation begins 1.4, so $\sqrt{2}$ cannot exceed $1.5 \,(= 1.4999\ldots)$; it begins 1.41, so $\sqrt{2}$ cannot exceed 1.42; etc. **Q2.** (a) 16; $(x+2)^2$ will be as close as we please to 16, if x is close enough to 2; show that if any arbitrary small number is specified, $(x+2)^2$ will differ from 16 by less than that number for all values of x within some neighborhood of 2. (b) 1; as in (a). **Q3.** (a) (i) $x \geq 1000$ (ii) $|x| < 1/\sqrt{1000} = 0.0316\ldots$ (iii) $x < -10$ (b) (i) If any arbitrary (small) number is specified, $1 + \dfrac{1}{x}$ will differ from 1 by less than that number for all values of x greater than some (large) number. (ii) If any arbitrary (large) number is specified, $1/x^2$ will be greater than that number for all values of x near enough to 0. (iii) If any arbitrary (large negative) number is specified, x^3 will be less than that number for all values of x less than some (large negative) number. **Q4.** No, $1/x$ has no limit as $x \to 0$ because its sign is not fixed (see Fig. 3-2). **Q5.** $\lim\limits_{x\to\infty} \dfrac{1}{x}$ refers to the behavior of the

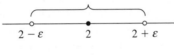

$$2 - \varepsilon \qquad 2 \qquad 2 + \varepsilon$$

Fig. A-13

continuously varying quantity $\dfrac{1}{x}$, $\lim\limits_{n\to\infty} \dfrac{1}{n}$ only to those of its values corresponding to positive integral x. Of course, both equal 0. **Q6.** (a) (i) is equivalent to $|x^2 - 2| < \varepsilon$, which is equivalent to (ii); Fig. A-13 displays the interval of numbers that differ from 2 by less than ε (the endpoints are excluded). (b) $\sqrt{2-\varepsilon} \leq x \leq \sqrt{2+\varepsilon}$ (we know that this is a neighborhood, i.e. $\sqrt{2-\varepsilon} < \sqrt{2} < \sqrt{2+\varepsilon}$, because $2 - \varepsilon < 2 < 2 + \varepsilon$). (c) Let δ be the lesser of $\sqrt{2+\varepsilon} - \sqrt{2}$ and $\sqrt{2} - \sqrt{2-\varepsilon}$. **Q7.** $f(x)$ will be in any neighborhood of l we choose, provided only that x is in a small enough neighborhood of a; in other words, all values $f(x)$ corresponding to x-values that lie within a small enough neighborhood of a are found within any arbitrary neighborhood of l. **Q8.** $\lim\limits_{x\to\infty} f(x) = l$: for any $\varepsilon > 0$ there exists $M > 0$ such that if $x > M$ then $|f(x) - l| < \varepsilon$. $\lim\limits_{x\to a} f(x) = \infty\,[-\infty]$: for any $M > 0$ there exists $\delta > 0$ such that if $|x - a| < \delta$ then $f(x) > M\,[< -M]$. Similarly for the other five cases. **Q9.** $x_n \leq \sqrt{2}$, hence $x_n^2 \leq 2$; and $2 - x_n^2 = (\sqrt{2} + x_n)(\sqrt{2} - x_n) \leq 2\sqrt{2}\,(\sqrt{2} - x_n) < 3 \cdot 10^{-n}$. (We know $2\sqrt{2} < 3$ because $8 < 9$.)

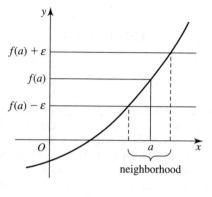

Fig. A-14

§3.1.6

Q1. Its graph is a semicircular arc.   **Q2.** It is discontinuous at $x = 1$; nevertheless $\lim_{x \to 1} \dfrac{1}{(x-1)^2}$ does exist—it equals ∞.   **Q3.** Suppose we want $f(x)$ to be within ε of $f(a)$. Draw the horizontal lines $y = f(a) + \varepsilon$ and $y = f(a) - \varepsilon$ (Fig. A-14). Because the graph is a continuous curve, there must be a neighborhood of a such that when x is in that neighborhood, the curve lies between the horizontal lines. Since this is the case for any ε, $\lim_{x \to a} f(x) = f(a)$. (See Q6 for detail about neighborhoods and continuity.)   **Q4.** The difference between $|f(x)|$ and $|f(a)|$ cannot exceed that between $f(x)$ and $f(a)$, because of the universal inequality $|\,|u| - |v|\,| \le |u - v|$. Hence as $f(x) \to f(a)$, $|f(x)| \to |f(a)|$.   **Q5.** Like $\sqrt{2}$ (§3.1.2), every number x is the limit of its partial decimal representations x_n, and these are rational (e.g. $1.41 = 141/100$). Then $f(x_n) = g(x_n)$ for all n, so by continuity $f(x) = \lim f(x_n) = \lim g(x_n) = g(x)$: the two functions are identical.   **Q6.** (a) Call the condition in the question N. (i) Suppose $f(x)$ continuous at a, and let V be a neighborhood of $f(a)$. Since $f(a)$ is in the interior of V (not an endpoint), there is a number $\varepsilon > 0$ such that the interval $[f(a) - \varepsilon, f(a) + \varepsilon]$ is contained in V. By §3.1.3 there is $\delta > 0$ such that if $|x - a| < \delta$ then $|f(x) - f(a)| < \varepsilon$; the last inequality implies that $f(x)$ is in V. Now let U be any neighborhood of a contained in the interior of $[a - \delta, a + \delta]$, e.g. $U = \left[a - \dfrac{\delta}{2}, a + \dfrac{\delta}{2}\right]$. Then if x is in U, $|x - a| < \delta$, hence $f(x)$ is in V. Thus given V we have found U; this proves N. (ii) Suppose N holds, and let ε be any positive number. Let V be a neighborhood of $f(a)$ contained in the interior of $[f(a) - \varepsilon, f(a) + \varepsilon]$. Then if y is in V, $|y - f(a)| < \varepsilon$. By N there is a neighborhood U of a such that if x is in U, $f(x)$ is in V. Since a is in the interior of U, there is a number $\delta > 0$ such that the interval $[a - \delta, a + \delta]$ is contained in U. Then if $|x - a| < \delta$, x is in U, hence $f(x)$ is in V and $|f(x) - f(a)| < \varepsilon$. Thus given ε we have found δ; this proves continuity. (b) The same as (a), except that each neighborhood of a is replaced by that part of it on which f is defined (and the restriction $|x - a| < \delta$ is supplemented by the condition that $f(x)$ be defined).

§3.1.8

Q1. Not necessarily, e.g. $y = |x|$ at $x = 0$ (§2.9.10).

§3.1.11

Q1. (a) $\frac{1}{2}$ (b) -1 (c) $\dfrac{1}{a^2 - 1}$ if $a \neq 1$ or -1 (d) $\lim\limits_{x \to 5} 0 = 0$ or $\lim\limits_{x \to 5} x - \lim\limits_{x \to 5} x = 5 - 5 = 0$ (e) $\lim\limits_{n \to \infty} \dfrac{1}{1 + \dfrac{1}{n}} = \dfrac{\lim 1}{\lim\left(1 + \dfrac{1}{n}\right)} = \dfrac{1}{1} = 1$ **Q2.** Yes, where h is non-zero.

Q3. See §3.1.6.Q1. **Q4.** (a) If $m = 0$, $l = 0$ too; otherwise the magnitude of $f(x)/g(x)$ would approach infinity as $x \to a$. If $m \neq 0$, the rule for the limit of a quotient gives $l/m = 1$. (b) Let $a = 0$, $f(x) = x^2$, $g(x) = x$. **Q5.** If $l - \dfrac{\varepsilon}{2} \leq f(x) \leq l + \dfrac{\varepsilon}{2}$ and $m - \dfrac{\varepsilon}{2} \leq g(x) \leq m + \dfrac{\varepsilon}{2}$ then by addition, $l + m - \varepsilon \leq f(x) + g(x) \leq l + m + \varepsilon$.

Q6. If $l > m$, let $p = (l + m)/2$, the number midway between l and m. By taking x near enough to a we can make $f(x) > p$ and $g(x) < p$, hence $f(x) > g(x)$, a contradiction. **Q7.** E.g.: if $f(x) \to \infty$ and $g(x) \to \infty$, $f(x) + g(x) \to \infty$, but no conclusion can be drawn about $f(x) - g(x)$. **Q8.** Since $g'(a) \neq 0$, the difference quotient of g at $x = a$ is non-zero for x near a. Then

$$\frac{f'(a)}{g'(a)} = \frac{\lim\limits_{x \to a} \dfrac{f(x) - f(a)}{x - a}}{\lim\limits_{x \to a} \dfrac{g(x) - g(a)}{x - a}} = \lim\limits_{x \to a} \frac{\dfrac{f(x)}{x - a}}{\dfrac{g(x)}{x - a}} = \lim\limits_{x \to a} \frac{f(x)}{g(x)}.$$

§3.2.3

Q1. (a) 0 (b) 0 (c) 1 (d) $4x^3$ (e) $11t^{10}$ (f) $3x^2$ (see §2.9.2(2)) **Q2.** (a) $5(2)^4 = 80$ (b) 20 **Q3.** 6 **Q4.** (a) The function is clearly differentiable for $x \neq 0$. At $x = 0$ the left-hand derivative of $y = 0$ and the right-hand derivative of $y = x^2$ are both equal to 0, so y is differentiable there. Its derivative y' is 0 for $x \leq 0$ and $2x$ for $x \geq 0$. (b) y' is differentiable everywhere except at $x = 0$, where its graph has a corner. **Q5.** Taking $a = 1$ and $b = x$ in the expansion of $(a + b)^n$ gives $(1 + x)^n = 1 + nx + $ (terms in x^2 and higher powers of x); since the quantity in parentheses is positive, the inequality follows. **Q6.** (a) For given i, each summand $a^{n-i}b^i$ is obtained as the result of selecting i of the n factors $a + b$ to contribute b's (the others then contribute a's). Hence the coefficient of $a^{n-i}b^i$, which is the number of these summands, is the number of different selections of i things (factors $a + b$) from among n things. Here the n things are the *first* factor $a + b$, the *second* factor $a + b$, etc. (b) Let us first consider selecting i things and labeling them no. 1, no. 2,

..., no. i as we do so. There are n possibilities for no. 1; once it has been chosen there are $n-1$ choices left for no. 2; thus we can choose nos. 1 *and* 2 in $n(n-1)$ ways. Then there are $n-2$ choices for no. 3, etc.; the result is that the number of ways of selecting *and labeling* the i things is the numerator N of the given fraction. But a given *unlabeled* set of i things can be labeled in D different ways, where D is the denominator of the fraction. For there are i possibilities for no. 1, then $i-1$ for no. 2, etc. It follows that the number of *different unlabeled* sets of i things is N/D.

§3.2.5

Q1. (a) $2x + 2$ (b) $12t^2 - 2\sqrt{2}\,t$ (c) $8 - 32x^3 + x^2$ **Q2.** (a) Yes: $\dfrac{d}{dx}(x+1)^3 =$

$\dfrac{d}{dx}(x^3 + 3x^2 + 3x + 1) = 3x^2 + 6x + 3 = 3(x^2 + 2x + 1) = 3(x+1)^2$. (b) No: $\dfrac{d}{dx}(x+x)^3 =$

$\dfrac{d}{dx}(2x)^3 = \dfrac{d}{dx}8x^3 = 24x^2$, while $3(x+x)^2 = 3(2x)^2 = 12x^2$.   **Q3.** $2h(x) + k(x)$

Q4. Since $p = mv$, $\dfrac{dp}{dt} = m\dfrac{dv}{dt} = ma$.   **Q5.** (a) $x(0) = 4$ (b) The velocity is $v(t) =$

$dx/dt = -gt + 3$; at $t = 0$ this is 3. (c) $a(t) = dv/dt = -g$ at all times.

Q6. $\dfrac{d}{dx}(-f(x)) = \dfrac{d}{dx}[(-1)f(x)] = (-1)\dfrac{d}{dx}f(x) = -\dfrac{d}{dx}f(x)$.

§3.3.3

Q1. (a) x^{-1} (b) t^{-1} (c) $x^{1+\alpha}$ (d) t^p **Q2.** (a) $23x^{22}$ (b) $2x - 3x^{-4}$ (c) $-1/x^2$

(d) $\dfrac{3}{2\sqrt{t}}$ (e) $2\alpha t^{2\alpha-1}$ (f) $-(5n/2)x^{-(n/2)-1}$ **Q3.** (a) $\dfrac{d}{dx}x^{1/5} = \dfrac{1}{5x^{4/5}} > 0$,

because $x^{4/5} = (x^{1/5})^4$ is always positive; hence the function is increasing.

(b) $\dfrac{d}{dx}(x^{-5}) = -\dfrac{5}{x^6} < 0$; decreasing.   **Q4.** For all $a \neq 1$, x^α is the derivative of

$\dfrac{1}{\alpha+1}x^{\alpha+1}$.   **Q5.** (a) To show that x^α becomes and remains larger than any given

number M, let $x_0 = M^{1/\alpha}$; then $x_0^\alpha = M$, and since x^α is increasing, $x^\alpha > M$ for $x > x_0$. (b) Since $x^\alpha \to 0$ if and only if $|x^\alpha| \to 0$, and $|x^\alpha| = |x|^\alpha$, we must prove $|x|^\alpha \to 0$. We have $1/|x|^\alpha = (1/|x|)^\alpha$, and as $x \to 0$, $1/|x| \to \infty$, so by (a) $1/|x|^\alpha \to \infty$. It follows that $|x|^\alpha = \dfrac{1}{1/|x|^\alpha} \to 0$.   **Q6.** (a) Let $x = 1 + u$, $u > 0$;

then $x^n = (1 + u)^n > 1 + nu$, which clearly $\to \infty$. (b) $x^n = 1^n = 1$. (c) The case $x = 0$ is trivial, since $0^n = 0$, so assume $x \neq 0$. Since $x^n \to 0$ if and only if $|x^n| \to 0$,

and $|x^n| = |x|^n$, we must prove $|x|^n \to 0$. We know $|x| < 1$, hence $1/|x| > 1$ and

(a) implies $1/|x|^n \to \infty$. It follows that $|x|^n = \dfrac{1}{1/|x|^n} \to 0$. (d) As n increases x^n

remains ≥ 1 in magnitude, while its sign alternates. **Q7.** (a) If $\sqrt{2}$ is rational, write it as a fraction in lowest terms. (b) Square both sides of the equation in (a) and multiply by q^2. (c) By (b), p^2 is even, which is impossible if p is odd. (d) Substitute $2r$ for p in (b) and divide by 2. (e) As in (c). The contradiction is to the hypothesis that p and q have no common factor.

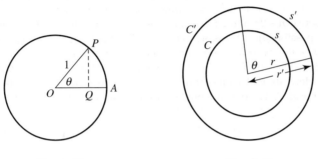

Fig. A-15 Fig. A-16

§3.3.6

Q1. $s = r\theta = \sqrt{2} \cdot \sqrt{2} = 2$ **Q2.** 0, $\pi/6$, $\pi/4$, $\pi/3$, $\pi/2$, π, $3\pi/2$, 2π **Q3.** Drop $PQ \perp OA$ (Fig. A-15); the hypotenuse of $\triangle OQP$ equals 1, hence $OQ = \cos\theta$, $QP = \sin\theta$. **Q4.** Sines: 0, $1/2$, $1/\sqrt{2}$, $\sqrt{3}/2$, 1, 0, -1, 0. Cosines: 1, $\sqrt{3}/2$, $1/\sqrt{2}$, $1/2$, 0, -1, 0, 1. **Q5.** $(r\cos x, r\sin x)$ **Q6.** See Fig. A-16. By elementary geometry (theory of ratios), $\dfrac{\text{arc } s}{\text{circumference } C} = \dfrac{\text{arc } s'}{\text{circumference } C'}$, i.e. $\dfrac{s}{2\pi r} = \dfrac{s'}{2\pi r'}$, or $\dfrac{s}{r} = \dfrac{s'}{r'}$.

Q7. $\dfrac{\text{area of sector}}{\text{area of circle}} = \dfrac{\text{arc of sector}}{\text{circumference of circle}}$, i.e. $\dfrac{A}{\pi r^2} = \dfrac{r\theta}{2\pi r}$.

§3.3.9

Q2. (a) $\sin\dfrac{5\pi}{2} = \sin\left(\dfrac{\pi}{2} + 2\pi\right) = \sin\dfrac{\pi}{2} = 1$; $\cos\dfrac{5\pi}{2} = 0$ (b) -1; 0

(c) $\sin\dfrac{2\pi}{3} = 2\sin\dfrac{\pi}{3}\cos\dfrac{\pi}{3} = \dfrac{\sqrt{3}}{2}$; $-\dfrac{1}{2}$ (d) $-\dfrac{1}{2}$; $-\dfrac{\sqrt{3}}{2}$ (e) $-\dfrac{1}{\sqrt{2}}$; $\dfrac{1}{\sqrt{2}}$ (f) $\sin\dfrac{\pi}{12} =$

$\sin\left(\dfrac{\pi}{3} - \dfrac{\pi}{4}\right)$, and an addition formula gives $\dfrac{\sqrt{3}-1}{2\sqrt{2}}$; $\dfrac{\sqrt{3}+1}{2\sqrt{2}}$ (g) $\cos x$; $-\sin x$

Q3. $\sin\left(x + \dfrac{\pi}{2}\right) = \cos x$: the graph of $y = f(x+a)$ is the graph of $y = f(x)$ shifted left by a.   **Q4.** Changing x to $-x$ reverses the direction of the angle around the circle from A, hence changes the sign of the vertical but not the horizontal coordinate of P. The third identity states that $OP^2 = 1$.   **Q5.** (a) $2x + \cos x$ (b) $3\sec^2 t - 4\sin t$ (c) $\sin^2(x/2) = \frac{1}{2}(1 - \cos x) = \frac{1}{2} - \frac{1}{2}\cos x$ has derivative $\frac{1}{2}\sin x$.   **Q6.** (a) At any x the ordinate $\cos x$ is the slope of $y = \sin x$. E.g., at $x = 0$ the slope is $\cos 0 = 1$. Where the graph of $y = \cos x$ lies above the x-axis, the graph of $y = \sin x$ slopes upward, etc.   **Q7.** (a) (i) The coordinates of P in Fig. 3-6 determine the angle at the center. (ii) $\cos(u - v) = \cos^2 u + \sin^2 u = 1$. Since $-2\pi < u - v < 2\pi$, $u - v = 0$; for the only number x between -2π and 2π at which $\cos x = 1$ is $x = 0$. (b) They differ by a multiple of 2π (possibly 0). This is obvious from Fig. 3-6, in light of (a)(i). Alternatively, write $u = u' + m \cdot 2\pi$, $v = v' + n \cdot 2\pi$, where m and n are integers and $0 \le u' < 2\pi$, $0 \le v' < 2\pi$. Then by (a) $u' = v'$, hence $u - v = (m - n) \cdot 2\pi$.   **Q8.** By §2.8.4, force is the derivative of momentum with respect to time, $f(t) = -10\sin t$ newton. **Q9.** (a) See Fig. A-17. If P lies on the graph of an odd function $f(x)$, i.e. $y = f(x)$,

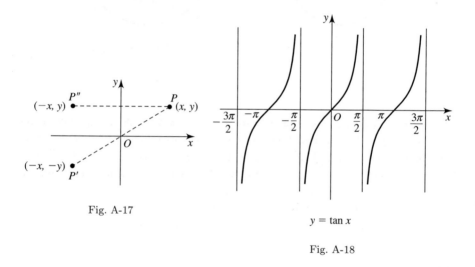

Fig. A-17

$y = \tan x$

Fig. A-18

so does P', because $-y = f(-x)$. Similarly for an even function, with P'' in place of P'. (b) Let $Q_1 = \dfrac{f(x + \Delta x) - f(x)}{\Delta x}$, $Q_2 = \dfrac{f(-x - \Delta x) - f(-x)}{-\Delta x}$. As $\Delta x \to 0$, $Q_1 \to f'(x)$ and $Q_2 \to f'(-x)$. If f is odd, $Q_1 = Q_2$; if f is even, $Q_1 = -Q_2$. The result follows.   **Q10.** (a) The slope of line OP is $\dfrac{\sin x}{\cos x} = \tan x$, hence the ordinate to the line at A is $\tan x$. (b) The period is π. (c) See Fig. A-18. The vertical lines $y = \pi/2$, etc., are asymptotes (for the definition see §3.9.8). (d) See Fig. A-19.   **Q11.** See Fig. A-20, in which $QR \perp OA$. Both QR and

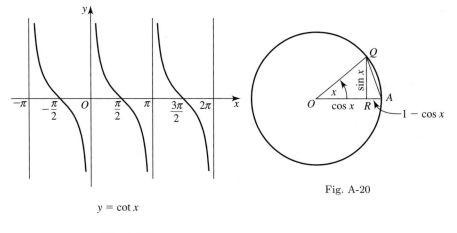

$y = \cot x$

Fig. A-19

Fig. A-20

AR are $< AQ$, which $\to 0$ as $Q \to A$, i.e. as $x \to 0$. Hence $\sin x \to 0$ and $\cos x \to 1$. **Q12.** Assume $(\sin x)' = \cos x$ (the opposite case is similar). Then

$$\frac{\cos(x + \Delta x) - \cos x}{\Delta x} = \frac{\sin\left(x + \frac{\pi}{2} + \Delta x\right) - \sin\left(x + \frac{\pi}{2}\right)}{\Delta x} \to \cos\left(x + \frac{\pi}{2}\right) =$$

$-\sin x$, i.e. $(\cos x)' = -\sin x$. **Q13.** $c^2 = (a\cos\theta - b)^2 + (a\sin\theta)^2 = a^2(\sin^2\theta + \cos^2\theta) - 2ab\cos\theta + b^2 = a^2 + b^2 - 2ab\cos\theta$. **Q14.** By the law of cosines $PQ^2 = 2 - 2\cos(u - v)$; by the distance formula $PQ^2 = (\cos u - \cos v)^2 + (\sin u - \sin v)^2 = 2 - 2\cos u\cos v - 2\sin u\sin v$; equate these. (b) First, the formula holds if $u < v$ too, since neither side changes if u and v are interchanged. Thus it holds for all u and v between 0 and π. Now given *any* u and v, write $u = u' + m\pi$, $v = v' + n\pi$, u' and v' between 0 and π, m and n integers. Then the formula can be written $\cos[(u' - v') + (m - n)\pi] = \cos(u' + m\pi)\cos(v' + n\pi) + \sin(u' + m\pi)\sin(v' + n\pi)$. This follows from $\cos(u' - v') = \cos u'\cos v' + \sin u'\sin v'$ because $\cos(x + k\pi) = \cos x$ if k is even or $-\cos x$ if k is odd, and $\sin(x + k\pi) = \sin x$ if k is even or $-\sin x$ if k is odd. (c) E.g.: first deduce $\cos\left(\frac{\pi}{2} - x\right) = \sin x$, from which follows $\cos x = \sin\left(\frac{\pi}{2} - x\right)$; then $\sin(u + v) = \cos\left(\frac{\pi}{2} - (u + v)\right) = \cos\left(\left(\frac{\pi}{2} - u\right) - v\right) = \cos\left(\frac{\pi}{2} - u\right)\cos v + \sin\left(\frac{\pi}{2} - u\right)\sin v = \sin u\cos v + \cos u\sin v$. (d) As in (c), $\cos\left(\frac{\pi}{2} - x\right)$ and the like are found by direct expansion. Other identities are found by expanding the sine or cosine of $-x = 0 - x$, $0 = x - x$, and $2x = x + x$. The half-angle formulas follow from the double-angle formulas for $2\left(\frac{x}{2}\right) = x$ by substituting for $\sin^2 \frac{x}{2}$ or $\cos^2 \frac{x}{2}$ from $\sin^2 \frac{x}{2} + \cos^2 \frac{x}{2} = 1$. **Q15.** The derivative of a polynomial $f(x)$ of degree n is a polynomial of degree $n-1$, which has at most $n-1$ distinct roots, hence can change sign at most $n-1$ times. This means that the graph of $f(x)$ can change from rising to falling or from falling to rising at most $n-1$ times; that is, it cannot have more than a total of $n-1$ hills and valleys together. **Q16.** See Fig. A-21, in which PR is an arc of a circle with center O: $OP = r = f(\theta)$, $OQ = r + dr = f(\theta + d\theta)$. Consider arc $PR = rd\theta$. Assuming that PQR can be regarded as a rectilinear triangle, the angle ζ at P equals $\angle PQR$, and since $\angle QRP$ is right, $\cot \zeta = \dfrac{dr}{rd\theta}$.

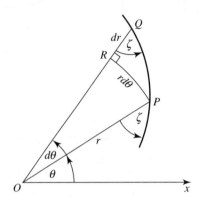

Fig. A-21

§3.3.13

Q1. (a) The function decreases as $x \to \infty$, and its value is halved with each unit decrease in x; hence $y \to 0$ and the graph approaches the axis. (In fact it is asymptotic to it: see §3.9.8.) (b) They are symmetric about the y-axis, because $(\frac{1}{2})^x = 2^{-x}$ is obtained from 2^x by changing the sign of the independent variable. **Q2.** (a) $2x$ (b) $(e^{\log x})^2 = x^2$ **Q3.** (a) The statement is equivalent to $1/x = e^{-\log x}$, which is true because $e^{-\log x} = 1/e^{\log x} = 1/x$. (b) Similarly, $e^{\alpha \log x} = (e^{\log x})^\alpha = x^\alpha$. **Q4.** (a) e^x (b) $2e^x + \cos x$ (c) $(\log 2)2^x$ (d) $e^x + \dfrac{1}{x}$ (e) $(\log a)a^x + (\log b)b^x$ (f) $\dfrac{2}{x \log 2} + 16x^3$ **Q5.** ce^x (c any non-zero constant) **Q6.** By the differentiation formula, $\dfrac{d}{dx}(e^\alpha)^x = (\log e^\alpha)(e^\alpha)^x = \alpha e^{\alpha x}$. **Q7.** (a) 2 (b) If $4 = 2e^{3t}$, $2 = e^{3t}$ and hence $\log 2 = 3t$, so $t = \frac{1}{3}\log 2$. **Q8.** Since $a > 1$, $\sqrt{a} = 1 + h$ with $h > 0$. Then $a^n = ((\sqrt{a})^2)^n = ((\sqrt{a})^n)^2 > (1 + nh)^2 > n^2 h^2$, hence $\dfrac{a^n}{n} > \dfrac{n^2 h^2}{n} = nh^2 \to \infty$. **Q9.** (a) Immediate consequence of the cited question. (b) $r = r_0 e^{\theta \cot \zeta}$

§3.4.4

Q1. (a) $\cos t + \sin t$ (b) $\log x$ (c) $(x+1)e^x$ (d) $e^x(\tan x + \sec^2 x)$ (e) $-\dfrac{2}{x(\log x)^2}$

(f) $8(\sin x + x\cos x)$ (g) $\dfrac{2^t((t-1)\log 2 - 1)}{(t-1)^2}$ (h) $e^x[(2x^2 + 4x - 1)\cos x -$

$(2x^2 - 1)\sin x]$ **Q2.** E.g., by the quotient rule

$$(\tan x)' = \frac{(\sin x)'\cos x - (\sin x)(\cos x)'}{(\cos x)^2} = \frac{\cos^2 x + \sin^2 x}{\cos^2 x} = \frac{1}{\cos^2 x} = \sec^2 x.$$

Q3. (a) Since $V = \dfrac{C}{P}$, $V' = -\dfrac{CP'}{P^2}$. (b) As a logical consequence of Boyle's law it contains no new information; yet it does relate quantities not mentioned in Boyle's law. **Q4.** $(fgh)' = ((fg)h)' = (fg)'h + (fg)h' = (f'g + fg')h + fgh' = f'gh + fg'h + fgh'$. **Q5.** $x^n = xx\cdots x$ (n factors x), so $(x^n)' = x'xx\cdots x + xx'x\cdots x + \cdots + xxx\cdots x' = 1xx\cdots x + x1x\cdots x + \cdots + xxx\cdots 1 = nx^{n-1}$. **Q6.** (a) The quantity $g(x + \Delta x)$ is non-zero for Δx near 0, i.e. for $x + \Delta x$ near x, because $g(x + \Delta x) \to g(x) \neq 0$. Hence for small Δx we can write

$$\frac{\frac{1}{g(x+\Delta x)} - \frac{1}{g(x)}}{\Delta x} = -\frac{1}{g(x)g(x+\Delta x)} \cdot \frac{g(x+\Delta x) - g(x)}{\Delta x} \text{ and then pass to the limit.}$$

(b) $\left(\dfrac{f}{g}\right)' = \left(f \cdot \dfrac{1}{g}\right)' = f'\dfrac{1}{g} + f\left(-\dfrac{g'}{g^2}\right) = \dfrac{f'g - fg'}{g^2}$. **Q7.** (a) $(\cos 0)/1 = 1$

(b) $(\sin 0)/1 = 0$ (c) 2 (d) $\log a$ (e) -1

§3.5.5

Q1. See the following sections, especially Fig. 3-17. **Q2.** E.g., in the case of mass, $F(x)$ is cumulative mass m from a to x, say (so $m(a) = 0$), and $f(x)$ is line density $\lambda(x)$. Part I: $m(x)$ is the sum of infinitesimal masses $\lambda(x)\,dx$, i.e. its differential is $dm = \lambda(x)\,dx$, so $\lambda(x) = dm/dx$. Part II: $\lambda(x) = dm/dx$ implies $dm = \lambda(x)\,dx$, and $m(x)$ is the sum of these differentials.

§3.5.9

Q1. (a) C (b) $x + C$ (c) $t^3 + C$ (d) $\cos x + C$ (e) $e^t + C$ (f) $G(x) + C$ **Q2.** (a) $u(x) = v(x) + C$ (b) $u(x) = v(x)$ **Q3.** (a) E.g., if velocity is zero, location is fixed; if slope is zero, graph is level. (b) If $F(x)$ is not constant, $F(x_1) \neq F(x_2)$ for some x_1, x_2; then the graph of F from x_1 to x_2 cannot be level. This intuitive argument will be refined in §§3.7.5–9. (c) See §3.7.9 (first paragraph). **Q4.** No: e^x itself is not, because it is everywhere non-zero, while any such integral vanishes at a point $x = a$. **Q5.** A primitive has the form $\log x + C$. Any number C can be written in the form $\log a$ for some $a > 0$ (let $a = e^C$, then $C = \log a$), so we can write $\log x + C = \log x + \log a = \log ax$ (§3.3.11).   **Q6.** Infinity begets paradox. An analogous correspondence within the set of positive integers can be defined as follows. The numbers which are *powers of primes* $(2^1, 2^2, 2^3, \ldots, 3^1, 3^2, 3^3, \ldots, 5^1, 5^2, 5^3, \ldots)$ are only a part of all positive integers, yet we can easily match each positive integer n to a distinct infinite class of powers of primes, namely the collection of powers with exponent n. Thus 1 corresponds to $2^1, 3^1, 5^1, \ldots, 2$ to $2^2, 3^2, 5^2, \ldots$, etc.

§3.6.6

Q1. (a) $\left. \dfrac{x^5}{5} \right|_0^1 = \frac{1}{5}$ (b) $\frac{2}{3}t^{3/2} + C$ (c) $\left. -\cos x \right|_0^{\pi/2} = 1$ (half the area under an arch calculated in §3.6.1) (d) $\frac{3}{2}t^2 + 5e^t + C$ (e) $\left[2\log x + \dfrac{3}{x} \right]_1^2 = 2\log 2 - \frac{3}{2}$.

(f) $-a\cos x + b\sin x + C$ **Q2.** $\displaystyle\int_{-2}^2 (4 - x^2)\, dx = \frac{32}{3}$ **Q3.** $\displaystyle\int_1^{10} \left(1 + \dfrac{1}{t^2}\right) dt = \left[t - \dfrac{1}{t} \right]_1^{10} = 9.9$ m^3 **Q4.** The work is the integral of the force with respect to x, $W = \displaystyle\int_{x_1}^{x_2} c(x_1 - x)\, dx$ (note that $x_1 > x_2$) $= \left. c\left(x_1 x - \dfrac{x^2}{2} \right) \right|_{x_1}^{x_2} = c\left[\left(x_1 x_2 - \dfrac{x_2^2}{2} \right) - \left(x_1^2 - \dfrac{x_1^2}{2} \right) \right] = -\dfrac{c}{2}(x_1^2 - 2x_1 x_2 + x_2^2) = -\dfrac{c}{2}(x_1 - x_2)^2$.

Q5. (a) $x(t) = 2\sin t + C$ (b) $C = x(0)$ **Q6.** The original graph and the shifted graph represent different primitives $F(x)$, $F(x)+c$ of the slope function $f(x)$. Changing the primitive does not affect f or its definite integrals $\int_a^b f(x)\, dx = F(b) - F(a) = [F(b) + c] - [F(a) + c]$, which represent total quantity.   **Q7.** (a) E.g., in the case of addition, with the notation introduced after the rules, we have $\int_a^b (f(x) + g(x))\, dx = (F + G)(b) - (F + G)(a) = [F(b) + G(b)] - [F(a) + G(a)] = [F(b) - F(a)] + [G(b) - G(a)] = \int_a^b f(x)\, dx + \int_a^b g(x)\, dx$. (b) For the addition rule let $S = f(\xi_1)\Delta x_1 + \cdots + f(\xi_n)\Delta x_n$ and $T = g(\xi_1)\Delta x_1 + \cdots + g(\xi_n)\Delta x_n$ be sums based on the same subintervals and evaluation points; then $S \to \int_a^b f(x)\, dx$ and $T \to \int_a^b g(x)\, dx$, hence $S + T \to \int_a^b f(x)\, dx + \int_a^b g(x)\, dx$; also $S + T = (f(\xi_1) + g(\xi_1))\Delta x_1 + \cdots + (f(\xi_n) + g(\xi_n))\Delta x_n \to \int_a^b (f(x) + g(x))\, dx$.   **Q8.** In uniformly accelerated motion: (i) velocity is a linear function of time; (ii) change in velocity is proportional to length of time interval; (iii) in equal time intervals, equal non-zero changes in velocity occur; (iv) acceleration, the rate of change of velocity, is constant (in §1.3.3 it was non-zero).

Q9. (ii) $\dfrac{\Delta y}{\Delta x} = \dfrac{x_2^2 - x_1^2}{x_2 - x_1} = x_1 + x_2$ is not constant; (iii) $\Delta y = (x_1 + x_2)\Delta x$ is not determined by Δx alone; (iv) $y' = 2x$. **Q10.** $\dfrac{g(x + \Delta x) - g(x)}{\Delta x} =$

$$\dfrac{\log(-x - \Delta x) - \log(-x)}{\Delta x} = -\dfrac{\log(-x + (-\Delta x)) - \log(-x)}{-\Delta x} \to -\dfrac{1}{-x} = \dfrac{1}{x}.$$

§3.7.3

Q1. Interval lacking an endpoint: $f(x) = x$, $0 \le x < 1$, has no maximum. Function not continuous: $f(x) = x$, $0 \le x < 1$, and $= 0$, $1 \le x \le 2$, has no maximum on $[0, 2]$. **Q2.** $g(x) \le |g(x)|$ everywhere. **Q3.** (a) Each $f(\xi_i)$ is ≥ 0, hence $S = f(\xi_1)\Delta x_1 + \cdots + f(\xi_n)\Delta x_n \ge 0$. Then the limit of sums S is also ≥ 0 (§3.1.9(1)). (b) $h(x) - g(x) \ge 0$ on $[a, b]$, hence by (a) $\int_a^b h(x)\,dx - \int_a^b g(x)\,dx = \int_a^b (h(x) - g(x))\,dx \ge 0$. **Q4.** (a) Assume that there is a point x_0 such that $g(x_0) < h(x_0)$, then by continuity $g(x) < h(x)$ for all x in some interval $[c, d]$ contained in $[a, b]$. Let m be the minimum value of $h(x) - g(x)$ on $[c, d]$, then by (1) $\int_c^d (h(x) - g(x))\,dx \ge m(d - c) > 0$. The integral of $h(x) - g(x)$ over $[a, b]$ is the sum of the integrals over $[c, d]$ and over the one or two other intervals that compose $[a, b]$. The integral over $[c, d]$ is > 0 and by (1) the other(s) are ≥ 0; the result follows. (b) For the first inequality assume that there is x_0 such that $f(x_0) > m$; similarly for the second.

§3.7.6

Q1. (a) $b^2 - a^2 = 2\xi(b - a)$ implies $\xi = (a + b)/2$—the average, or (arithmetic) mean, of a and b. (b) The tangent to $y = x^2$ parallel to a secant AB occurs at the value of x halfway between the x-coordinates of A and B. **Q2.** (a) $\dfrac{1}{b} - \dfrac{1}{a} = -\dfrac{1}{\xi^2}(b - a)$ implies $\xi = \sqrt{ab}$—the geometric mean of a and b (§6.2.4). (b) Cf. Q1(b). **Q3.** Fig. A-22 graphs the location of two particles which are together at $t = a$ and again at $t = b$, when the accelerating particle passes the uniformly moving one. At $t = \tau$, the time

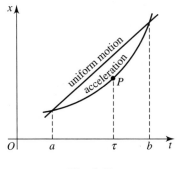

Fig. A-22

prescribed by the mean value theorem, their velocities agree.   **Q4.** No. Continuity
requires $f'(\xi) \to f'(x_0)$ regardless of how ξ approaches x_0, but $\Delta f/\Delta x = f'(\xi)$ is
true only for certain values of ξ.   **Q5.** Suppose in general that $g(x) \le f(x) \le h(x)$
and as $x \to a$, both $g(x)$ and $h(x)$ have the finite limit l. Given $\varepsilon > 0$, there is $\delta > 0$
so small that if $|x - a| < \delta$ then both (i) $|g(x) - l| < \varepsilon$ and (ii) $|h(x) - l| < \varepsilon$. (There
is δ_1 such that (i) holds, and there is δ_2 such that (ii) holds; take $\delta \le$ both δ_1 and
δ_2.) Then if $|x - a| < \delta$, $l - \varepsilon < g(x) \le f(x) \le h(x) < l + \varepsilon$, i.e. $|f(x) - l| < \varepsilon$.

§3.7.13

Q1. On the interval $|x^9| \le 1$, hence $|10x^9| \le 10$ and $|0.99^{10} - 1| = |0.99^{10} - 1^{10}| \le$
$10\,|0.99 - 1| = 0.1$.   **Q2.** $(\tan x)' = 1/\cos^2 x$. On $[44°, 45°]\cos x \ge \cos 45° = 1/\sqrt{2}$,
hence $|(\tan x)'| \le 2$ and $|\tan 44° - 1| = |\tan 44° - \tan 45°| \le 2(\pi/180) < 0.035$.
Therefore $0.965 < \tan 44° < 1$.   **Q3.** The graph has a secant of slope α; but by
the mean value theorem, any secant has slope $f'(\xi)$ for some ξ; therefore $|\alpha| \le 2$.

Q4. (a) $\left|-\dfrac{1}{x^2}\right| = \dfrac{1}{x^2}$ will be larger than M if $x < \dfrac{1}{\sqrt{M}}$. Consequently no M bounds
it. (b) Suppose it is claimed that if x_1 is within distance δ of x_0, $1/x_1$ is within ε
of $1/x_0$ (cf. §3.1.3). Take $x_0 \le \delta$; then choose $x_1 < x_0$ so small that $\dfrac{1}{x_1} > \dfrac{1}{x_0} + \varepsilon$.
Then certainly $x_0 - x_1 < \delta$, while $\dfrac{1}{x_1} - \dfrac{1}{x_0} > \varepsilon$ (Fig. A-23; there δ and ε are large,
but the illustration is still valid).   **Q5.** If the maximum width of the blocks is small
enough, uniform continuity ensures that they will *all* be $\le$ any given $\varepsilon > 0$ in height.
Their total width is $b - a$, hence their total area does not exceed $\varepsilon(b - a)$. This can
be made arbitrarily small.

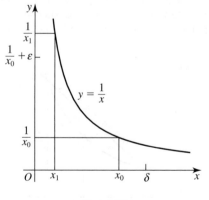

Fig. A-23

§3.7.17

Q1. (a) $\int_a^b x\,dx = \xi(b-a)$, or $\frac{1}{2}(b^2 - a^2) = \xi(b-a)$, gives $\xi = (a+b)/2$. (b) If a, $b > 0$: the area under $y = x$ from a to b equals the area of the rectangle of base $[a, b]$ and height $(a+b)/2$. Examine the other cases also, especially $a < 0 < b$, to illustrate (c). (c) The signed area between the line $y = \mu$ and the x-axis equals the net signed area between the graph of $y = f(x)$ and the x-axis (§2.10.6). **Q2.** (a) $e - 1$

(b) $\log(e - 1) \approx 0.541$ **Q3.** $\dfrac{1}{\pi}\displaystyle\int_0^\pi \sin t\,dt = \dfrac{2}{\pi}$ **Q4.** Call them I, II, and III, in the order presented (I being the intermediate value property). I $\Rightarrow$ II $\Rightarrow$ III (notation as in §3.6.5) is obvious. To prove III $\Rightarrow$ I, let $m < c < M$ and consider the function $g(x) = f(x) - c$ on the interval between x_1 and x_2, where $f(x_1) = m$ and $f(x_2) = M$. We have $g(x_1) = m - c < 0$ and $g(x_2) = M - c > 0$, hence there is x_0 between x_1 and x_2 with $g(x_0) = 0$, i.e. $f(x_0) = c$. **Q5.** (a) Since $-1 \le \alpha \le 1$ there is v such that $\cos v = \alpha$, by the intermediate value property. Then $\sin v = \pm\sqrt{1 - \alpha^2} = \pm\beta$. If $+$, let $u = v$; if $-$, let $u = -v$. (b) Put $r = \sqrt{x^2 + y^2}$ and apply (a) with $\alpha = x/r$, $\beta = y/r$. **Q6.** (a) If $f(x)$ is continuous, the Fundamental Theorem provides F such that $F' = f$ and $F(b) - F(a) = \int_a^b f(x)\,dx$; by the differential mean value theorem $F(b) - F(a) = f(\xi)(b - a)$, and the integral mean value theorem follows. (b) If $f(x)$ is continuously differentiable, $\int_a^b f'(x)\,dx = f(b) - f(a)$ by the Fundamental Theorem, while $\int_a^b f'(x)\,dx = f'(\xi)(b - a)$ by the integral mean value theorem; the differential theorem follows. **Q7.** $\dfrac{1}{\Delta x}\displaystyle\int_{x_0}^{x_0+\Delta x} f(x)\,dx = f(\xi)$; as $\Delta x \to 0$, $\xi \to x_0$ and $f(\xi) \to f(x_0)$.

§3.8.3

Q1. $y' = 10t^4 - 4t^3 - 9t^2 + 8t + 4$, $y'' = 40t^3 - 12t^2 - 18t + 8$, $y''' = 120t^2 - 24t - 18$, $y^{(4)} = 240t - 24$, $y^{(5)} = 240$; $y^{(n)} = 0$, $n \ge 6$ **Q2.** $y' = 2x \sin x + x^2 \cos x$ (product rule), $y'' = 4x \cos x - (x^2 - 2)\sin x$ **Q3.** $y' = e^x + xe^x = (x + 1)e^x$, $y'' = e^x + (x+1)e^x = (x+2)e^x$, etc.—at each stage e^x is added. **Q4.** (a) $n!$ ("n factorial," or $1 \cdot 2 \cdot 3 \cdots n$) (b) $\alpha(\alpha-1)(\alpha-2)\cdots(\alpha-n+1)x^{\alpha-n}$ **Q5.** The first set provides three linear equations in the unknowns c_1, c_2, c_3: $c_3 = 1$, $\dfrac{c_1}{2} + c_2 + c_3 = 0$, $\dfrac{c_1}{2} - c_2 + c_3 = 4$. Solving these yields the result. The values in the second set are applied successively, in reverse order, to determine c_1, c_2, and c_3 in that order. **Q6.** By integration, $y' = 2x + c_1$. Since $y'(0) = 1$, $c_1 = 1$ and $y' = 2x + 1$. Integrating again gives $y = x^2 + x + c_2$, and $y(0) = 0$ implies $c_2 = 0$, so $y = x^2 + x$. **Q7.** $p^{(n+1)}(x) = 0$, so $p^{(n)}(x) = c_0$ for some constant c_0. Then by integration, (i) $p^{(n-1)}(x) = c_0 x + c_1$, (ii) $p^{(n-2)}(x) = \dfrac{c_0}{2}x^2 + c_1 x + c_2$, etc. We have $c_0 = p^{(n)}(a)$; then (i) $p^{(n-1)}(a) = c_0 a + c_1$ determines c_1, (ii) $p^{(n-2)}(a) = \dfrac{c_0}{2}a^2 + c_1 a + c_2$ determines c_2, etc.; finally all the c_i are determined, so p is determined. **Q8.** $y = \cos x$ **Q9.** If f is odd [even], (a) $f^{(n)}$ is odd if n is even [odd], and even if n is odd [even]; hence (b) $f^{(n)}(0) = 0$ if n is even [odd]. **Q10.** Differentiation of fg produces two summands, $f'g$ and fg'; differentiation of each of them produces two summands, e.g. $(f'g)' = f''g + f'g'$; and so on. A summand is produced from $f^{(j)}g^{(k)}$ by the choice to differentiate either the f factor $f^{(j)}$ or the g factor $g^{(k)}$. Thus each of the 2^n summands of $(fg)^{(n)}$ is the result of n choices of f or g. The summand $f^{(n-i)}g^{(i)}$ results from $n-i$ choices of f and i of g, hence occurs in the sum as many times as $a^{n-i}b^i$ in the expansion of $(a+b)^n$.

§3.8.5

Q1. The first of the two given geometrical interpretations applies if we graph $v(t)$, the second if we graph $x(t)$. **Q2.** The slope $y' = \cos x$ increases from 0 at $x = -\pi/2$ to 1 at $x = 0$, then falls back to 0 at $x = \pi/2$. That is, the graph steepens as far as $x = 0$, then declines in steepness. The *rate* of steepening $y'' = -\sin x$ is accordingly positive at first, negative later; it continuously declines from 1 at $x = -\pi/2$ to -1 at $x = \pi/2$. **Q3.** (a) $x(t) = c_1 t + c_2$ for some constants c_1 and c_2, either or both of which may be 0. If $c_1 = 0$ the particle is at rest at $x = c_2$; if $c_1 \neq 0$ it is in uniform motion at velocity c_1, its location at $t = 0$ being c_2. (b) $y'' = 0$ (assuming y twice differentiable) **Q4.** $\dfrac{d^3 x}{dt^3} = \dfrac{da}{dt}$; a jerk is a noticeable sudden change in acceleration, a Δa occurring in a short enough Δt that $\Delta a/\Delta t$ is appreciable. If, say, a is constant, then changes abruptly to another constant value, da/dt is briefly non-zero. (But non-zero $x'''(t)$ need not mean jerky motion.)

§3.8.9

Q1. $f(x) = x^{10}$, $f'(x) = 10x^9$, $f''(x) = 90x^8$. Let $a = 1$, $x = 0.99$. Then $(0.99)^{10} \approx 1^{10} + 10(1^9)(0.99 - 1) = 0.9$, and since $\xi \leq 1$, $|R_1| \leq \frac{1}{2} \cdot 90(1^8)(0.99 - 1)^2 = 0.0045$.

Q2. $y = \tan\dfrac{\pi}{4} + \left(\sec^2\dfrac{\pi}{4}\right)\left(x - \dfrac{\pi}{4}\right) = 1 + 2\left(x - \dfrac{\pi}{4}\right)$, or (approximately) $y = 2x - 0.57$. **Q3.** (a) $f(r) = 4\pi r^2$, so $f'(r) = 8\pi r$. Therefore $\Delta f \approx 8\pi a \Delta r$, and $\dfrac{\Delta f}{f(a)} \approx 2\dfrac{\Delta r}{a}$, $100\dfrac{\Delta f}{f(a)} \approx 2\left(100\dfrac{\Delta r}{\Delta a}\right)$. Absolute error is multiplied by $8\pi a = 1.6 \times 10^8$, e.g. an error of 100 m results in an error of 1.6×10^{10} m^2 (the whole area being about $f(a) = 5.1 \times 10^{14}$ m^2); relative and percentage error are doubled. (b) $f(r) = \frac{4}{3}\pi r^3$. Similar reasoning gives: absolute error is multiplied by 5.1×10^{14}; relative and percentage errors are tripled.

§3.9.3

Q1. If $f' \geq 0$ throughout an interval, f is increasing in the weaker sense there, and conversely; similarly for $f' \leq 0$. There is a (weak) minimum at x_0 if $f(x) \geq f(x_0)$

for x nearby, as happens when f' changes from non-positive to non-negative at x_0; similarly for a (weak) maximum.

§3.9.9

Q1. (a) Like x^4, Fig. 3-44. (b) $f' = 6(x-1)(x-2)$, $f'' = 12(x - \frac{3}{2})$. Increases: $x \leq 1$, $x \geq 2$; decreases: $1 \leq x \leq 2$. Concave up: $x \geq \frac{3}{2}$; down: $x \leq \frac{3}{2}$. Local max $(x, y) = (1, 2)$; local min $(2, 1)$. Inflection point $(\frac{3}{2}, \frac{3}{2})$. $f(0) = -3$. Approaches ∞ as $x \to \infty$, $-\infty$ as $x \to -\infty$ (either because of the concavity, or because the x^3 term comes to dominate the others). See Fig. A-24. (c) $f' = x^3 - 3x + 2$. Since $f'(1) = 0$, $x - 1$ is a factor and we find $f' = (x + 2)(x - 1)^2$. $f'' = 3(x + 1)(x - 1)$. Increases: $x \geq -2$; decreases: $x \leq -2$. Concave up: $x \leq -1$, $x \geq 1$; down: $-1 \leq x \leq 1$. Local min $(-2, -6)$. Inflection points $(-1, -\frac{13}{4})$, $(1, \frac{3}{4})$ (horizontal tangent). $f(0) = 0$, $f'(0) = 2$. Approaches ∞ as $x \to \infty$ or $-\infty$. See Fig. A-25. (d) f is periodic with period π (§3.3.9.Q10(b)); it is sufficient to consider it for $-\pi/2 < x < \pi/2$. $f' = \sec^2 x > 0$: f increases. $f'' = 2\sec^2 x \tan x$. Concave up: $x \geq 0$; down: $x \leq 0$. Inflection point $(0, 0)$. Approaches ∞ as $x \to \pi/2$, $-\infty$ as $x \to -\pi/2$; asymptotes $x = \pi/2$, $x = -\pi/2$. Graph has odd-function symmetry. See Fig. A-18. (e) Even function, positive. $f' = -\dfrac{2x}{(x^2 + 1)^2}$, $f'' = \dfrac{2(\sqrt{3}\,x + 1)(\sqrt{3}\,x - 1)}{(x^2 + 1)^3}$. Increases: $x \leq 0$;

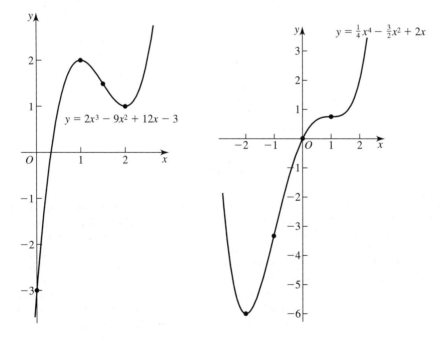

Fig. A-24   Fig. A-25

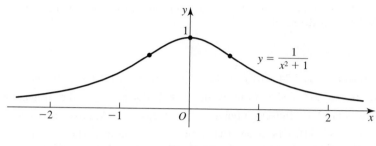

Fig. A-26

decreases: $x \geq 0$. Concave up: $x \leq -1/\sqrt{3}$, $x \geq 1/\sqrt{3}$; down: $-1/\sqrt{3} \leq x \leq 1/\sqrt{3}$. Max $(0,1)$. Inflection points $(-1/\sqrt{3}, \frac{3}{4})$, $(1/\sqrt{3}, \frac{3}{4})$. Asymptotic to x-axis in both directions. See Fig. A-26. (f) Odd function, positive for $x > 0$. $f' = -\dfrac{(x+1)(x-1)}{(x^2+1)^2}$, $f'' = \dfrac{2x(x+\sqrt{3})(x-\sqrt{3})}{(x^2+1)^3}$. Increases: $-1 \leq x \leq 1$; decreases: $x \leq -1$, $x \geq 1$. Concave up: $-\sqrt{3} \leq x \leq 0$, $x \geq \sqrt{3}$; down: $x \leq -\sqrt{3}$, $0 \leq x \leq \sqrt{3}$. Max $(1, \frac{1}{2})$; min $(-1, -\frac{1}{2})$. Inflection points $(0,0)$, $(-\sqrt{3}, -\sqrt{3}/4)$, $(\sqrt{3}, \sqrt{3}/4)$. $f(0) = 0$, $f'(0) = 1$. Asymptotic to x-axis in both directions. See Fig. A-27.

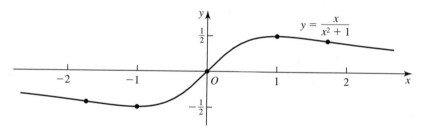

Fig. A-27

Q2. Let $f(x) = e^x - x$, then $f'(x) = e^x - 1$, $f''(x) = e^x > 0$. Concave up, so at most one critical point; there is one at $x = 0$: $f'(0) = e^0 - 1 = 0$. Hence f takes its minimum value at 0, viz. $f(0) = e^0 - 0 = 1$, so $e^x - x > 0$. (The other argument that $e^x \to \infty$: by §3.3.3.Q6(a), $e^n \to \infty$; since e^x is an increasing function, this implies $e^x \to \infty$.) **Q3.** $f(0) = 0$; $f > 0$ for $x > 0$, $f < 0$ for $x < 0$. $f' = -(x-1)e^{-x}$, $f'' = (x-2)e^{-x}$. Increasing: $x \leq 1$; decreasing: $x \geq 1$. Concave up: $x \geq 2$; down: $x \leq 2$. Max $(1, e^{-1})$. Inflection point $(2, 2e^{-2})$. As $x \to \infty$, $x/e^x \to 0$ since $e^x/x \to \infty$ (§3.3.13.Q8 shows this for $x = n$, and e^x/x is increasing for $x \geq 1$). As $x \to \infty$, $f \to -\infty$ because $x \to -\infty$ and $e^{-x} \to \infty$. See Fig. A-28. **Q4.** At $x = 1$, where $m' = \sin x$ takes its greatest value on $[0, 1]$. **Q5.** Let the area be A and one

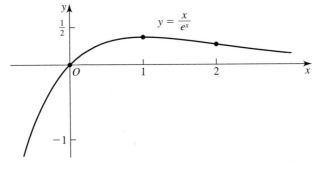

Fig. A-28

side x, then an adjacent side is A/x and the perimeter is $p = 2\left(x + \dfrac{A}{x}\right)$. Then

$p' = 0$ at $x = \pm\sqrt{A}$, but we must have $x > 0$, so $x = \sqrt{A}$: the figure is a square.
Q6. $a = v' = -t(t-2)e^{-t}$, so the maximum v is at $t = 2$. The acceleration a becomes negative then; i.e., deceleration begins.   **Q7.** Volume $= \pi r^2 h = 1$, so $h = 1/\pi r^2$.
Area $A = 2\pi r h + 2\pi r^2 = 2\left(\dfrac{1}{r} + \pi r^2\right)$. $A' = 2\left(-\dfrac{1}{r^2} + 2\pi r\right) = 0$ when $r = \dfrac{1}{(2\pi)^{1/3}}$,

which gives the minimum of A. Then $h = \dfrac{2}{2\pi r^2} = \dfrac{2}{(2\pi)^{1/3}} = 2r$—the height equals

the diameter.   **Q8.** The critical points are $t = 1$ and, if n is odd, $t = -1$. $f''(1) > 0$, so there is a minimum at $t = 1$, and $f(1) = 0$; then for n even $f(t) > 0$ for all $t < 1$, so $(1 + x)^n > 1 + nx$ for all $x < 0$. For n odd, $f''(-1) < 0$, so there is a maximum at $t = -1$, and $f(-1) > 0$. To the left of $t = -1$, as t *decreases* $f(t)$ decreases and at some point becomes negative. Then $(1 + x)^n > 1 + nx$ for $u < x < 0$, where u (which depends on n) is some number < -2.   **Q9.** $f'(x) - f'(x_0) = f''(\xi)(x - x_0)$, and $f''(\xi) > 0$; hence f' changes sign at x_0 from negative to positive.   **Q10.** The tangent line at a is $y = f(a) + f'(a)(x - a)$, so the difference between $f(x)$ and the height of the tangent at x is $f(x) - f(a) - f'(a)(x - a) = \frac{1}{2}f''(\xi)(x - a)^2$, by the second-order mean value theorem. This quantity is positive except at $x = a$, where it vanishes.   **Q11.** As in Q10, the second-order mean value theorem shows that if f'' changes sign at x_0 the graph is above the tangent at x_0 on one side of the point and below it on the other.   **Q12.** Suppose there is a point C on the graph between A and B which lies on or above AB. Then at least one of the points A and B lies on or below the tangent at C, in contradiction to Q10 (Fig. A-29).   **Q13.** If

$f''(x_0) > 0$, $\dfrac{f'(x) - f'(x_0)}{x - x_0} > 0$ near x_0; then $f'(x) - f'(x_0)$ has the same sign as

$x - x_0$.   **Q14.** (a) Let $0 < x \le b$. By the mean value theorem there is ξ, $0 < \xi < x$, such that $f(x) - f(0) = f'(\xi)(x - 0)$, i.e. $f(x) = f'(\xi)x > 0$. (b) Apply (a) to $x - \sin x$ with $b = 1$. This proves $x > \sin x$ for $0 < x \le 1$. For $x > 1$ we know $x > \sin x$,

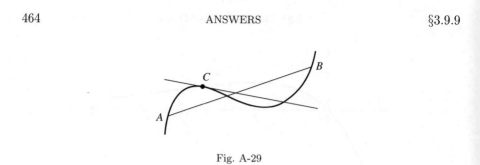

Fig. A-29

because $\sin x \leq 1$.   **Q15.** It asks for a whole *function* (whose graph is the desired curve) rather than a *value* of a function.

CHAPTER 4

§4.1.2

Q1. (a) $y = \tan \log x$ (b) $y = e^{x^2}$ (c) $y = 2\cos^3 x$ (d) $y = \cos 2x^3$ (e) $y = e^{\sqrt{\sec x}}$ (f) $y = \dfrac{1}{\tan^2 x + 3\tan x}$ **Q2.** (Only one form is given.) (a) $2^{(\ )} \circ \dfrac{1}{(\ )}$ (b) $| \ \ | \circ \log$ (c) $\dfrac{1}{(\ )} \circ (\ )^4 \circ \tan$ (d) $\sin \circ e^{(\ )}$ (e) $\cos \circ \sqrt{\ } \circ ((\ )+a^2)\circ(\ )^2$ (f) $\log \circ \log$

Q3. (a) x; defined for $x \geq 0$. Thus the given function is not the same as the function x, which is defined everywhere. (b) $|x|$; defined for all x.   **Q4.** (a) If $\lim_{x\to\infty} g(x) = l$ and f is continuous then $\lim_{x\to\infty}(f\circ g)(x) = f(l)$. (b) Let $g(x) = \dfrac{1}{x^2}+1$, $f(u) = \sqrt{u}$, then as $x \to \infty$, $g(x) \to 1$ and $f(g(x)) \to f(1) = 1$. (c) $\dfrac{x}{\sqrt{1+x^2}} = \dfrac{1}{\sqrt{\dfrac{1}{x^2}+1}} \to 1$.

§4.1.5

Q1. (a) $-3\cos^2 x \sin x$ (b) $8(2x+1)^3$ (c) $-e^{-x}$ (d) $-\dfrac{1}{3(1-x)^{2/3}}$ (e) $-\dfrac{1}{(1+x)^2}$ (f) $\sin\dfrac{1}{t} - \dfrac{1}{t}\cos\dfrac{1}{t}$ (g) $\dfrac{1}{x\log x}$ (h) $12(1+4\sin^2 t)^{1/2}\sin t \cos t = 6(1+4\sin^2 t)^{1/2}\sin 2t$ (i) $-\dfrac{xe^{\sqrt{1-x^2}}}{\sqrt{1-x^2}}$ (j) $\dfrac{2}{(1-t)^2}\sec^2\dfrac{1+t}{1-t}$ **Q2.** (a) $-\dfrac{1}{2\sqrt{x}}$ (b) -1 **Q3.** $1/g = g^{-1}$, hence $(1/g)' = -g^{-2}g'$.   **Q4.** $a^x = e^{x\log a}$; differentiate.   **Q5.** $f'(x)/f(x)$ **Q6.** $f''(g(x))g'(x)^2 + f'(g(x))g''(x)$ **Q7.** (a) $\sinh(x+y) = \sinh x \cosh y + \cosh x \sinh y$, $\cosh(x+y) = \cosh x \cosh y + \sinh x \sinh y$; and the same formulas with "+" replaced everywhere by "−". (b) $(\sinh x)' = \cosh x$, $(\cosh x)' = \sinh x$ (c) $1/\cosh^2 x$ (d) (i) $f(x) = \sinh x$ is an odd function, positive for $x > 0$. $f' = \cosh x > 0$ for all x, $f'' = \sinh x$. Increases. Concave up: $x \geq 0$; down: $x \leq 0$. Inflection point $(0,0)$. $f'(0) = 1$. Asymptotic to $y = g(x) = \frac{1}{2}e^x$ as $x \to \infty$, in the sense that $f - g$ and $f' - g' \to 0$; asymptotic to $y = -\frac{1}{2}e^{-x}$ as $x \to -\infty$. See Fig. A-30. (ii) $f(x) = \cosh x$ is an even function, ≥ 1 for all x. $f' = \sinh x$, $f'' = \cosh x$. Increases: $x \geq 0$; decreases: $x \leq 0$. Concave up. Min $(0,1)$. Asymptotic to $y = \frac{1}{2}e^x$ as $x \to \infty$, to $y = \frac{1}{2}e^{-x}$ as $x \to -\infty$. See Fig. A-30. (iii) $f(x) = \tanh x$ is an odd function, positive for $x > 0$. $f' = \dfrac{1}{\cosh^2 x}$, $f'' = -\dfrac{2\tanh x}{\cosh^2 x}$. Increases.

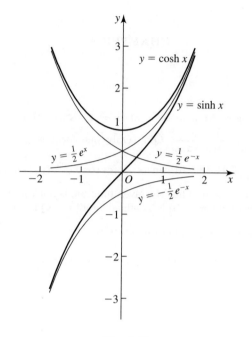

Fig. A-30

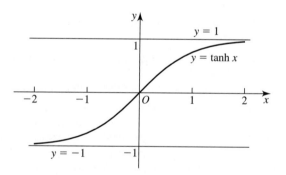

Fig. A-31

Concave up: $x \leq 0$; down: $x \geq 0$. Inflection point $(0,0)$. $f'(0) = 1$. Asymptotic to $y = 1$ as $x \to \infty$, to $y = -1$ as $x \to -\infty$. See Fig. A-31.

§4.2.2

Q1. If m is measured from $x = a$, $m(x(t))$ is the mass from a to the location of the particle at time t. **Q2.** Since $v = \alpha t$, $x = \int_0^r v \, dt = \frac{1}{2}\alpha t^2$. Consequently

$dm/dt = (2 + \sin(\frac{1}{2}\alpha t^2))\alpha t$ kg/sec. **Q3.** The current is $f(t) = te^{-t^2}$ kg/sec. We have $f'(t) = (1 - 2t^2)e^{-t^2}$, so the maximum is at $t = 1/\sqrt{2}$.

§4.2.5

Q1. (a) $y = +\sqrt{4 - x^2}$, hence by (2) $y' = -\dfrac{x}{\sqrt{4 - x^2}}$. (b) By the chain rule $y' = $

$\dfrac{1}{2\sqrt{4 - x^2}}(-2x) = -\dfrac{x}{\sqrt{4 - x^2}}$. **Q2.** Differentiating implicitly gives $\dfrac{x}{2} + \dfrac{2yy'}{9} = 0$,

or $y' = -\dfrac{9}{4}\dfrac{x}{y}$. At $x = 1$ we have $\dfrac{1}{4} + \dfrac{y^2}{9} = 1$, or $y = \pm\dfrac{3\sqrt{3}}{2}$. Then at $\left(1, \dfrac{3\sqrt{3}}{2}\right)$ $y' = $

$-\dfrac{\sqrt{3}}{2}$ and the tangent line is $y - \dfrac{3\sqrt{3}}{2} = -\dfrac{\sqrt{3}}{2}(x - 1)$; at $\left(1, -\dfrac{3\sqrt{3}}{2}\right)$ $y' = \dfrac{\sqrt{3}}{2}$ and

the line is $y + \dfrac{3\sqrt{3}}{2} = \dfrac{\sqrt{3}}{2}(x - 1)$. **Q3.** $\dfrac{dP}{dt}V + P\dfrac{dV}{dt} = 0$, therefore (denoting

differentiation with respect to t by the prime) $V' = -\dfrac{VP'}{P}$. Agreement with $V' = $

$-\dfrac{CP'}{P^2}$ is obtained by using $C = PV$. **Q4.** $V = \frac{4}{3}\pi r^3$, hence $4\pi r^2 \dfrac{dr}{dt} = \dfrac{dV}{dt} = $

-0.2. When $V = 36$, $r^3 = 27/\pi$, hence $r^2 = 9/\pi^{2/3}$ and $\dfrac{dr}{dt} = -\dfrac{0.2}{4\pi r^2} = -\dfrac{1}{180\pi^{1/3}} \approx$

-0.004 m/sec.

§4.2.7

Q1. $f''(x) = \dfrac{a^2}{(x^2 + a^2)^{3/2}} + \dfrac{b^2}{((c - x)^2 + b^2)^{3/2}} > 0$ **Q2.** To solve $f'(x) = 0$, put

the terms on opposite sides, square, and simplify to get $b^2 x^2 = a^2(c - x)^2$. Since

all quantities are non-negative this implies $bx = a(c - x)$. Thus $x_0 = \dfrac{ac}{a + b}$ and

$c - x_0 = \dfrac{bc}{a + b}$ —line CD is divided in the ratio $a{:}b$. **Q3.** Within either medium

(from A or B to the boundary) the path of shortest time is the shortest path, which

is straight. **Q4.** As before, show that $\dfrac{\sin\alpha}{v_1} - \dfrac{\sin\gamma}{v_2}$ is increasing. **Q5.** (a) $v_1 > $

$v_2 \Rightarrow \sin\alpha > \sin\gamma \Rightarrow \alpha > \gamma$ because the sine is increasing for acute angles. (b) Since

light travels faster in medium 1, least time requires $AP > PB$, which implies $\alpha > \gamma$

(if $\alpha = \gamma$, $AP = PB$). **Q6.** (a) See Fig. A-32: the light ray reaching A comes from

the direction of B'. (The path of least time from B to A is the same as from A to B.)

(b) $BB' = QB' - QB = b\tan\alpha - b\tan\gamma = b(\tan\alpha - \tan\gamma)$ (c) $\sin\gamma = \dfrac{\sin\alpha}{4/3} = \dfrac{3\sqrt{3}}{8}$,

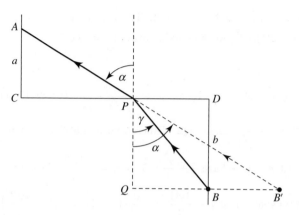

Fig. A-32

hence $\cos \gamma = \sqrt{1 - \left(\dfrac{3\sqrt{3}}{8}\right)^2} = \dfrac{\sqrt{37}}{8}$ and $\tan \gamma = \dfrac{3\sqrt{3}}{\sqrt{37}}$. The displacement is

then $1\left(\sqrt{3} - \dfrac{3\sqrt{3}}{\sqrt{37}}\right) \approx 0.88$ m. **Q7.** See Fig. A-33. $AP + PB = A'P + PB$ is

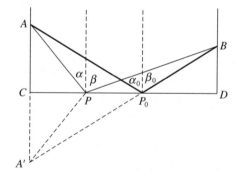

Fig. A-33

shortest when $A'PB$ is straight, as happens when P is at P_0; and it is easy to see that $\alpha_0 = \beta_0$. **Q8.** The function whose graph is a path APB is completely determined by specifying the single number x. **Q9.** Let the line have equation $y - y_0 = \mu(x - x_0)$. The x-intercept is then at $x = x_0 - \dfrac{y_0}{\mu}$, the y-intercept at $y = y_0 - \mu x_0$. Therefore

the length of the line is $\sqrt{\left(x_0 - \dfrac{y_0}{\mu}\right)^2 + (y_0 - \mu x_0)^2}$. To minimize this we could differentiate it with respect to μ, but it is easier to differentiate its square $f(\mu) = \left(x_0 - \dfrac{y_0}{\mu}\right)^2 + (y_0 - \mu x_0)^2$, which is a minimum when the length is. Let $\alpha = \dfrac{y_0}{x_0}$, then $f(\mu) = x_0^2 \left[\left(1 - \dfrac{\alpha}{\mu}\right)^2 + (\alpha - \mu)^2\right]$ and the derivative can be put in the form

$$f'(\mu) = \frac{2x_0^2}{\mu^3}(\mu^3 + \alpha)(\mu - \alpha).$$ Since μ must be negative, the critical point is at $\mu = -\alpha^{1/3}$. This is a minimum, since f' changes sign there. **Q10.** Let A be at the origin, let B be at $(1,1)$, and let $P(x,0)$ be the point where the wire leaves the road, which is the x-axis (Fig. A-34). Let the cost of stringing wire along the road be 1,

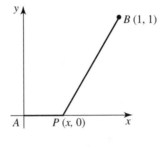

Fig. A-34

then the cost overland is 2. The cost along APB is $f(x) = x + 2\sqrt{(x-1)^2 + 1}$. Then $f'(x) = 1 + \dfrac{2x - 2}{\sqrt{x^2 - 2x + 2}}$. The only solution of $f'(x) = 0$ is $x = 1 - \dfrac{1}{\sqrt{3}}$, which gives the minimum. The cost for this route is about 2.73, for the straight route 2.83, and for the right-angle route 3. **Q11.** As in Q9 it is easier to deal with the square of the distance, $\phi(x) = PQ^2 = (x - x_0)^2 + (y - y_0)^2$, where $y = f(x)$. Where the extremum occurs, $\phi'(x) = 2(x - x_0) + 2(y - y_0)f'(x) = 0$. There are two possibilities. If $x = x_0$, so that PQ is vertical, then $y \neq y_0$ (otherwise Q would be P), and therefore the equation shows $f'(x) = 0$—the graph is horizontal at Q. If $x \neq x_0$, the equation shows $f'(x) \neq 0$, and $\dfrac{y - y_0}{x - x_0} = -\dfrac{1}{f'(x)}$; i.e., the slope of PQ is the negative reciprocal of the slope of the graph. **Q12.** $\cos\alpha = x/OQ$, hence the work is $W(x) = f \cdot OQ = \dfrac{x^2}{OQ} = \dfrac{x^2}{\sqrt{x^2 + (1 - 3x^2)^2}} = \dfrac{x^2}{\sqrt{9x^4 - 5x^2 + 1}}$. Maximizing this (as in Q9 and Q11, maximize the square) gives $x = \sqrt{\tfrac{2}{5}}$ (the critical point 0 is a minimum, and the maximum $-\sqrt{\tfrac{2}{5}}$ is outside the interval). Hence the maximum occurs at $Q = (\sqrt{\tfrac{2}{5}}, -\tfrac{1}{5})$ (thus $\alpha < 0$).

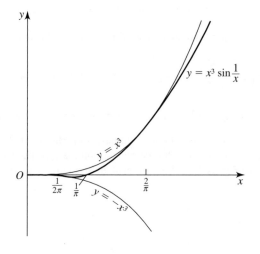

Fig. A-35

§4.2.10

Q1. For the derivatives the product rule and the chain rule are used. **Q2.** At $x = 1/n\pi$, n odd, $f'(x) = +1$. At $x = 1/n\pi$, n even, $f'(x) = -1$. **Q3.** See Fig. A-35. **Q4.** At points $x = 1/n\pi$ where $\sin(1/x) = 0$, f' is negative if n is even, positive if n is odd. Then between these points there are critical points (see the remark after the intermediate value property, §3.7.14). Thus no neighborhood of 0 is free of critical points; in other words, 0 is not an isolated critical point. Also, every neighborhood of 0 contains points $2/n\pi$, n odd, at some of which f is positive and at others negative. Hence $f(0) = 0$ cannot be a local extremum; nor is it an inflection point, since on neither side of $x = 0$ is f' monotone (it keeps changing sign as $x \to 0$). **Q5.** Since x^α is defined for all α only for $x \geq 0$, we ask whether $x^\alpha \sin(1/x)$, augmented by the value 0 at $x = 0$, is continuous as $x \to 0$ from the right and whether it has a right-hand derivative at 0, and likewise for its derivatives. With respect to these properties, for $n - 1 < \alpha < n$ the function agrees with $x^n \sin(1/x)$. E.g., $\sqrt{x} \sin(1/x)$ is continuous because $\sqrt{x} \to 0$ as $x \to 0$.

§4.3.3

Q1. x^q is non-negative and increasing for $x \geq 0$, because if $0 \leq x_1 < x_2$ then $0 \leq x_1 x_1 \cdots x_1 < x_2 x_2 \cdots x_2$ (q factors). Furthermore, $x^q \to \infty$ as $x \to \infty$, because given any number $M \geq 1$ we have $M^q \geq M$. Then x^q takes all the non-negative numbers as values (by the intermediate value property). If q is odd, it follows easily from the fact that x^q is an odd function that x^q is increasing for all x and takes all numbers as values. Thus $y = x^q$ is invertible as claimed, and its inverse $y = x^{1/q}$ is defined everywhere if q is odd and for $x \geq 0$ if q is even. **Q2.** (a) $\pi/2$ (b) $-\pi/6$ (c) $\pi/3$ **Q3.** No, only for $-\pi/2 \le x \le \pi/2$. E.g. $\arcsin \sin \pi = 0$. **Q4.** To show that $\arcsin(-x) = -\arcsin x$, let $u = \arcsin x$. Then $\sin u = x$, hence $\sin(-u) = -\sin u = -x$, i.e. $\arcsin(-x) = -u$ (because $-\pi/2 \le -u \le \pi/2$) $= -\arcsin x$. **Q5.** Reflection in the line $y = x$ takes a point (u, v) to (v, u); i.e., it interchanges coordinates. By definition, the graph of f consists of the points such that 2nd coordinate $= f(\text{1st coordinate})$. Likewise, the graph of f^{-1} consists of the points such that 2nd coordinate $= f^{-1}(\text{1st coordinate})$, which is equivalent to 1st coordinate $= f(\text{2nd coordinate})$. This latter condition describes the reflected graph of f. **Q6.** Let $f(x)$ be increasing. (i) To prove f invertible we must show that different inputs do not yield the same output. Let $x_1 \ne x_2$, say $x_1 < x_2$; then $f(x_1) < f(x_2)$ because f is increasing. Thus $f(x_1) \ne f(x_2)$. (ii) To prove f^{-1} increasing, let $x_1 < x_2$ and let $y_1 = f^{-1}(x_1)$, $y_2 = f^{-1}(x_2)$; we must show $y_1 < y_2$. We have $f(y_1) = x_1$, $f(y_2) = x_2$. If $y_1 = y_2$, $x_1 = x_2$, which is not the case; if $y_1 > y_2$, $x_1 > x_2$, because f is increasing—again, not the case. Thus $y_1 < y_2$. **Q7.** Each function is x, the "identity" function (it makes each number correspond to itself); i.e. $f^{-1}(f(x)) = x$, $f(f^{-1}(x)) = x$. However, if f or f^{-1} is not defined everywhere, this function is restricted. E.g., $\arcsin \sin x = x$ for $-\pi/2 \le x \le \pi/2$; $\sin \arcsin x = x$ for $-1 \le x \le 1$. (Q3 shows that $\arcsin \sin x$ can differ from x, but only if "$\sin x$" denotes a function defined on some values not in the domain of the restricted sine function whose inverse is the arc sine.) **Q8.** (a) $f \circ g$ takes x to y via u, $x \to u$ (by g) $\to y$ (by f). (Here the arrow means "goes to.") Let us for the moment *not* interchange variables when taking inverses; then clearly $g^{-1} \circ f^{-1}$ takes y back to x via u, $x \leftarrow u$ (by g^{-1}) $\leftarrow y$ (by f^{-1}). The condition of forming the composite $g^{-1} \circ f^{-1}$, that g^{-1} be defined on all values of f^{-1}, is satisfied, as we see as follows. Let $u = f^{-1}(y)$; then $y = f(u)$, so by hypothesis u is a value of g, which means that $g^{-1}(u)$ is defined. (b) $x^{3/5}$ is $(\ )^3 \circ (\ )^{1/5}$; its inverse is $(\ )^5 \circ (\ )^{1/3}$, or $x^{5/3}$. (c) $x^{2/3}$ is $(\ )^2 \circ (\ )^{1/3}$. It is invertible if restricted to $x \ge 0$, where $(\ )^2$ is invertible. There its inverse is $x^{3/2}$. **Q9.** (a) We are to show that as x grows, $f^{-1}(x)$ becomes and remains $\ge$ any given number M. Since f^{-1} is increasing, it is sufficient to find x_0 such that $f^{-1}(x_0) = M$; then for $x \ge x_0$ we will have $f^{-1}(x_0) \ge M$. Let $x_0 = f(M)$. (b) By §3.3.10 a^x satisfies the hypothesis; hence $\lim_{x \to \infty} \log_a x = \infty$. **Q10.** The correspondence pictured in §3.5.8 between the class A of continuous functions and the class B of groups of smooth functions (the members of a group differing by constants) is given by two

functions, $\int$ and $\dfrac{d}{dx}$, which are inverses of each other. Of course, these are not

numerical-valued functions of numbers: $\int$ goes from A to B, $\dfrac{d}{dx}$ from B to A.

§4.3.6

Q1. (a) $f'(x) = 2x + 2 > 0$ for $x > -1$, so f is increasing there. (b) Solving $f(x) = 3$, we find the only solution > -1 to be $x_0 = 0$. Since $f'(0) = 2$, $(f^{-1})'(3) = \frac{1}{2}$. **Q2.** $\dfrac{dy}{dx} = \frac{3}{2}x^{1/2}$, therefore $\dfrac{dx}{dy} = \dfrac{1}{\frac{3}{2}x^{1/2}} = \frac{2}{3}x^{-1/2}$. Substituting $x = y^{2/3}$ gives

$\frac{2}{3}y^{-1/3}$, and interchanging variables then yields $\dfrac{d}{dx}(x^{2/3}) = \frac{2}{3}x^{-1/3}$. **Q3.** $\dfrac{dx}{dy} =$

$\dfrac{1}{(\log a)a^x} = \dfrac{1}{(\log a)y}$, etc. **Q4.** That is what $(f^{-1})'(y) = \dfrac{1}{f'(x(y))}$ means.

Q5. f^{-1} is continuous, and by hypothesis f' is continuous; therefore $f' \circ f^{-1}$ is continuous (§4.1.1), and then so is its reciprocal.

§4.4.3

Q1. $\mathrm{SIN}\,x = \sin\dfrac{\pi}{180}x$ and $\mathrm{COS}\,x = \cos\dfrac{\pi}{180}x$, so $(\mathrm{SIN}\,x)' = \dfrac{\pi}{180}\cos\dfrac{\pi}{180}x =$

$\dfrac{\pi}{180}\mathrm{COS}\,x$. **Q2.** $\dfrac{\sin\Delta x}{\Delta x} = \dfrac{\sin\Delta x - \sin 0}{\Delta x}$ is the difference quotient of $\sin x$ at $x =$

0, which $\to \cos 0 = 1$. **Q3.** Example 1. As $x \to \pm\infty$, $1/x \to 0$, so $y \to \sin 0 = 0$.

Example 2. As $x \to \pm\infty$, $y = \dfrac{\sin(1/x)}{1/x} \to 1$. Example 3. As $x \to \infty$, $y =$

$x \cdot \dfrac{\sin(1/x)}{1/x} \to \infty$; as $x \to -\infty$, $y \to -\infty$. Example 4. As $x \to \pm\infty$, $y \to \infty$.

§4.4.6

Q1. For a given central angle h the arcs of circles are in the ratio of the radii (principle of radian measure); by similar triangles the same is true for chords and tangents. Hence the ratios whose limits are taken are independent of the radius. **Q2.** (a) In Fig. 4-21 drop a perpendicular PN from P to OQ; $PN = QS$, and the inequalities are equivalent to $PN < \mathrm{arc}\,PQ < PR$. (b) The inequalities are equivalent to $\sin h < h < \tan h$. The first is (b) of the cited question; the second is proved by applying (a) of that question to $\tan x - x$. (c) No, since the deduction requires knowledge of derivatives.

§4.4.9

Q1. (a) $\pi/3$ (b) $2\pi/3$ (c) $\pi/4$ (d) $-\pi/4$ (e) 0 (f) 0 **Q2.** (a) $\dfrac{1}{\sqrt{x - x^2}}$

(b) $-\dfrac{2x}{\sqrt{1 - x^4}}$ (c) $\dfrac{5}{2\sqrt{x}\,(1 + x)}$ **Q3.** (a) $\arctan x\Big|_0^1 = \dfrac{\pi}{4}$ (b) $\arctan x\Big|_{-1}^1 =$

$\dfrac{\pi}{2}$ (c) $\arcsin x\Big|_{1/\sqrt{2}}^{\sqrt{3}/2} = \dfrac{\pi}{12}$ **Q4.** (a) $\arctan x = \displaystyle\int_0^x \dfrac{1}{1 + x^2}\,dx$ (b) $\arcsin x =$

$\displaystyle\int_0^x \dfrac{1}{\sqrt{1 - x^2}}\,dx$ for $-1 < x < 1$, but the integrand is discontinuous at $x = \pm 1$; we would like to write $\arccos x = \int_{1}^{x} -\dfrac{1}{\sqrt{1-x^2}}\,dx$, but the integrand is discontinuous at the lower limit. See §4.7.3.Q8.   **Q5.** The (one-sided) tangent is vertical there.

Q6. From $y = \cos x$, $0 \le x \le \pi$, we obtain $\dfrac{dx}{dy} = -\dfrac{1}{\sin x}$, $0 < x < \pi$. On this interval $\sin x$ is positive, hence $\sin x = \sqrt{1-\cos^2 x} = \sqrt{1-y^2}$ and the result follows. **Q7.** The two graphs are reflections of one another in the line $y = \pi/4$, so $\arcsin x + \arccos x = 2(\pi/4) = \pi/2$.   **Q8.** (a) It is decreasing there (cf. §3.3.9.Q10(d)). (b) The same argument as was used in §4.4.7 to show $\arccos x = \dfrac{\pi}{2} - \arcsin x$. (c) $-\dfrac{1}{1+x^2}$ (d) Reflect part of Fig. A-19 in $y = x$. (e) (i) Let $y = \operatorname{arccot} x$, $x \ne 0$. If $x > 0$, $0 < y < \pi/2$ and $\tan y = 1/\cot y = 1/x$, so $\arctan(1/x) = y$. If $x < 0$, $\pi/2 < y < \pi$, or $-\pi/2 < y - \pi < 0$, and $\tan(y - \pi) = \tan y = 1/\cot y = 1/x$, so $\arctan(1/x) = y - \pi$. (ii) For $x \ne 0$, $\operatorname{arccot} x$ and $\arctan(1/x)$ have the same derivative. Therefore over *each* of the two intervals $x > 0$ and $x < 0$, $\operatorname{arccot} x = \arctan(1/x) + C$; but the constant need not be the same on the two intervals. For $x > 0$ we substitute $x = 1$ and get $\dfrac{\pi}{4} = \dfrac{\pi}{4} + C$, or $C = 0$; for $x < 0$ we substitute $x = -1$ and get $\dfrac{3\pi}{4} = -\dfrac{\pi}{4} + C$, or $C = \pi$.   **Q9.** (a) By §3.3.7(1), $\alpha(x) = \arctan f'(x)$. (b) α is $\arctan \circ f'$; the composite of continuous functions is continuous. Conversely, f' is $\tan \circ \alpha$. (c) As in (b); $\alpha' = \dfrac{f''}{1+f'^2}$. (d) Concave up and down correspond respectively to increasing and decreasing α.   **Q10.** (a) Use §4.1.5.Q7(d). (b) The first two follow the pattern for the arc sine, and the third follows that for the arc tangent. All three use §4.1.5.Q7(a). (c) For $\operatorname{arsinh} x$ solve $2x = e^y - \dfrac{1}{e^y}$ for e^y (which must be > 0), then take the logarithm. Similarly for $\operatorname{arcosh} x$ (with $e^y \ge 1$) and $\operatorname{artanh} x$.

§4.5.3

Q1. $4x$; $-y^2$ **Q2.** (i) A point (x, y, z) lies in the plane containing the x- and z-axes if and only if $y = 0$. When $y = 0$ the given equation is $z = x^2$. Hence the parabola $z = x^2$ in that plane is the cross-section of the graph by that plane. (ii) A horizontal cross-section is specified by assigning a value c to z. In the plane $z = c$ the equation is $x^2 + y^2 = c$, which represents a circle about the z-axis. Thus every horizontal cross-section is such a circle. (iii) Putting together (i) and (ii), the graph is obtained by rotating $z = x^2$ about the z-axis.   **Q3.** The case $y_0 = 0$ was discussed in the preceding answer. The case $y_0 = 1$ is the parabola $z = x^2 + 1$, which if drawn in the xz-plane passing through $y = 1$ (and perpendicular to the y-axis) is the cross-section of the paraboloid by that plane (Fig. A-36). In general, $z = f(x, y_0)$ is the parabola $z = x^2 + y_0^2$ in an xz-plane through $y = y_0$, which is the cross-section of the paraboloid by that plane.   **Q4.** As $((x-x_0)^2 + (y-y_0)^2)^{1/2} \to 0$, $|z - z_0| \to 0$.

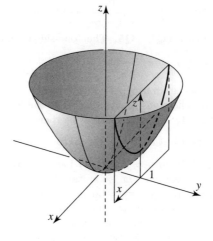

Fig. A-36

§4.5.6

Q1. (a) $\dfrac{\partial z}{\partial x} = 3(x + 3y^2)^2$, $\dfrac{\partial z}{\partial y} = 18y(x + 3y^2)^2$ (b) $\dfrac{\partial z}{\partial x} = \dfrac{1}{y^2}$, $\dfrac{\partial z}{\partial y} = -\dfrac{2x}{y^3}$

(c) $\dfrac{\partial z}{\partial x} = \sin xy + xy\cos xy$, $\dfrac{\partial z}{\partial y} = x^2\cos xy$ **Q2.** $\dfrac{\partial z}{\partial x} = 2xe^{x^2-y}$, $\dfrac{\partial z}{\partial y} = -e^{x^2-y}$;

$\dfrac{\partial^2 z}{\partial x^2} = 2e^{x^2-y}(2x^2 + 1)$, $\dfrac{\partial^2 z}{\partial x\partial y} = -2xe^{x^2-y}$, $\dfrac{\partial^2 z}{\partial y^2} = e^{x^2-y}$ **Q3.** At most 9; at

most (and ordinarily) 6, if mixed partials are equal. **Q4.** (a) $\dfrac{\partial P}{\partial T} = \dfrac{\alpha}{V}\left(=\dfrac{P}{T}\right)$

(b) $\dfrac{\partial P}{\partial V} = -\dfrac{\alpha T}{V^2}\left(=-\dfrac{P}{V}\right)$ **Q5.** $\left.\dfrac{\partial f}{\partial x}\right|_{y=y_0} = \lim\limits_{\Delta x\to 0}\dfrac{f(x + \Delta x, y_0) - f(x, y_0)}{\Delta x}$,

$\left.\dfrac{\partial f}{\partial y}\right|_{x=x_0} = \lim\limits_{\Delta y\to 0}\dfrac{f(x_0, y + \Delta y) - f(x_0, y)}{\Delta y}$ **Q6.** $\left.\dfrac{\partial f}{\partial x}\right|_{y=y_0}$ at any x is the slope at

that x of the curve which is the intersection of the plane $y = y_0$ with the graph of $f(x,y)$. E.g., in Fig. A-36 the slope at any x of the parabola $z = x^2 + 1$ in the xz-plane

at $y = 1$ is $\left.\dfrac{\partial}{\partial x}(x^2 + y^2)\right|_{(x,1)} = 2x$. **Q7.** (a) $g'(x) = 3x^2 - 2xy_0 = 3x(x - \frac{2}{3}y_0)$.

If $y_0 > 0$ there is a maximum at 0 and a minimum at $\frac{2}{3}y_0$; if $y_0 = 0$ there is no extremum; if $y_0 < 0$ there is a maximum at $\frac{2}{3}y_0$ and a minimum at 0.

(b) $g'(x) = \left.\dfrac{\partial}{\partial x}(x^3 - x^2y + 3)\right|_{y=y_0}$: the curves we are examining are cross-sections

of a curved surface. **Q8.** (a) Differentiate. (b) If y is held constant, $f_x = \alpha$ is an equation of functions of x. Integrating gives $f(x,y) = \alpha x + \text{constant}$, where *the constant may depend on the value of* y. Thus $f(x,y) = \alpha x + g(y)$ for some function g. Now $\beta = f_y = g'(y)$ implies $g(y) = \beta y + \gamma$ for some constant γ. Hence $f(x,y) = \alpha x + \beta y + \gamma$. **Q9.** See Fig. A-37. We have $\Delta z = \alpha(x + \Delta x) + \beta(y + \Delta y) + \gamma - [\alpha x + \beta y + \gamma] = \alpha \Delta x + \beta \Delta y = \alpha r \cos\theta + \beta r \sin\theta = (\alpha\cos\theta + \beta\sin\theta)r$. **Q10.** $f(x,y)$ has a local minimum at (x_0, y_0) if for all $x \neq x_0$ in some neighborhood of x_0 *and* all $y \neq y_0$ in some neighborhood of y_0, $f(x_0, y_0) < f(x,y)$. Then $f(x, y_0)$ has a minimum at x_0, so its derivative vanishes there, i.e. $f_x(x_0, y_0) = 0$. Similarly for f_y, and for a maximum. **Q11.** $Q(a,b)$ is the sum of the e_i^2, i.e. of the $(y_i - a - bx_i)^2$, so $\partial Q/\partial a$ is the sum of the $2(y_i - a - bx_i)(-1)$, or $-2(T - na - S_1 b)$, and $\partial Q/\partial b$ is the sum of the $2(y_i - a - bx_i)(-x_i)$, or $-2(U - aS_1 - bS_2)$.

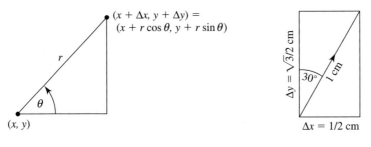

Fig. A-37   Fig. A-38

§4.5.8

Q1. (a) dx (b) $a\,dx + b\,dy$ (c) $y\,dx + x\,dy$ (d) $-\dfrac{y}{x^2}\,dx + \dfrac{1}{x}\,dy = \dfrac{-y\,dx + x\,dy}{x^2}$

(e) $\dfrac{-y\,dx + x\,dy}{x^2 + y^2}$ **Q2.** Given that $\dfrac{\partial T}{\partial x} = 5°/\text{m}$ and $\dfrac{\partial T}{\partial y} = -2°/\text{m}$ we find

(Fig. A-38) $\Delta T \approx \dfrac{\partial T}{\partial x}\,\Delta x + \dfrac{\partial T}{\partial y}\,\Delta y = (5)\left(\dfrac{1}{2} \times 10^{-2}\right) + (-2)\left(\dfrac{\sqrt{3}}{2} \times 10^{-2}\right) \approx$

$0.008°$. **Q3.** $f(x,y) = x^2 + y^2$, so $f_x(x_0, y_0) = 2x_0$ and $f_y(x_0, y_0) = 2y_0$. The equation is then $z = 2x_0(x - x_0) + 2y_0(y - y_0) + z_0$. **Q4.** $z = x + y - 1$

§4.5.11

Q1. At $t = 1$, $x = e^0 = 1$ and $y = 2$. We have $f_x = \dfrac{y}{2\sqrt{xy}} = \dfrac{1}{\sqrt{2}}$, $f_y = \dfrac{x}{2\sqrt{xy}} = \dfrac{1}{2\sqrt{2}}$, $x'(t) = e^{t-1} = 1$, $y'(t) = 3t^2 + 1 = 4$. Then $F'(1) = \dfrac{1}{\sqrt{2}} \cdot 1 + \dfrac{1}{2\sqrt{2}} \cdot 4 = \dfrac{3}{\sqrt{2}}$.

Q2. $f_x = \dfrac{x}{\sqrt{x^2 + y^2}}$, $f_y = \dfrac{y}{\sqrt{x^2 + y^2}}$, hence $F'(t_0) = \dfrac{x_0}{\sqrt{x_0^2 + y_0^2}} \cdot 1 + \dfrac{y_0}{\sqrt{x_0^2 + y_0^2}} \cdot 2 =$

$\dfrac{x_0 + 2y_0}{\sqrt{x_0^2 + y_0^2}}$. **Q3.** $T_x = 2x$, $T_y = 2y$, and $y' = 2x$. Let $F(x) = T(x, y(x))$, then

$F'(x) = 2x + 2y \cdot 2x = 2x + 4xy$. **Q4.** $F_r = f_x \cos\theta + f_y \sin\theta = \dfrac{1}{\sqrt{x^2 + y^2}}(xf_x + yf_y)$,

$F_\theta = -f_x r \sin\theta + f_y r \cos\theta = -yf_x + xf_y$

§4.6.5

Q1. (a) Let $u = 2x - 1$. The integral is $\dfrac{(2x - 1)^{11}}{22} + C$. (b) Let $u = \sin x$. The

integral is $\dfrac{u^3}{3}\Big|_0^{1/2} = \frac{1}{24}$. (c) Let $u = 5t + 4$. The integral is $\frac{1}{5}\log(5t + 4) + C$.

(d) $\frac{1}{3}e^{t^3} + C$ (e) Let $u = x/2$, or $x = 2u$, then the integral is $\displaystyle\int \dfrac{1}{4u^2 + 4}(2\,du) =$

$\dfrac{1}{2}\displaystyle\int \dfrac{1}{u^2 + 1}\,du = \dfrac{1}{2}\arctan u + C = \dfrac{1}{2}\arctan\dfrac{x}{2} + C$. (f) Completing the square

gives $\displaystyle\int \dfrac{1}{(x - 1)^2 + 4}\,dx$. Let $u = \dfrac{x - 1}{2}$, or $x - 1 = 2u$, and proceed as in (e).

Or: let $u = x - 1$, then the integral becomes $\displaystyle\int \dfrac{1}{u^2 + 4}\,du$, the same as (e):

$\dfrac{1}{2}\arctan\dfrac{u}{2} + C = \dfrac{1}{2}\arctan\dfrac{x - 1}{2} + C$. **Q2.** $du = (\sec x \tan x + \sec^2 x)\,dx =$

$(\tan x + \sec x)\sec x\,dx = u \sec x\,dx$, hence $\displaystyle\int \sec x\,dx = \int \dfrac{1}{u}\,du = \log|u| + C =$

$\log|\sec x + \tan x| + C$. **Q3.** $\displaystyle\int g(u(x))\,dx = \int g(u(x))\left[\dfrac{dx}{du}\dfrac{du}{dx}\right]dx \left(\text{because } \dfrac{dx}{du}\dfrac{du}{dx}\right.$

$\left. = 1\right) = \displaystyle\int \left[g(u(x))\dfrac{dx}{du}\right]\dfrac{du}{dx}\,dx = \int g(u)\dfrac{dx}{du}\,du$ (by §4.6.2(1)). **Q4.** In $F(-x_0) =$

$\int_0^{-x_0} f(x)\,dx$ let $u = -x$; then if f is odd [even], the integral is $\int_0^{x_0} -f(-u)\,du =$

$[-]\int_0^{x_0} f(u)\,du = [-]F(x_0)$. **Q5.** (a) Let $x = \cosh u$, i.e. $u = \operatorname{arcosh} x$, $x \geq 1$;

then $dx = \sinh u\,du$ and $\sqrt{x^2 - 1} = \sinh u$, so (using §4.1.5.Q7 repeatedly)

$\displaystyle\int \sqrt{x^2 - 1}\,dx = \int \sinh^2 u\,du = \int \dfrac{1}{2}(\cosh 2u - 1)\,du = \dfrac{1}{2}\left(\dfrac{\sinh 2u}{2} - u\right) + C =$

$\frac{1}{2}(\sinh u \cosh u - u) + C = \frac{1}{2}(x\sqrt{x^2 - 1} - \operatorname{arcosh} x) + C$. By §4.4.9.Q10(c) an

alternate form of the solution is $\frac{1}{2}[x\sqrt{x^2 - 1} - \log(x + \sqrt{x^2 - 1})] + C$. (b) Let

$x = \sinh u$ and use $\cosh^2 u = \frac{1}{2}(\cosh 2u + 1)$. The result is $\frac{1}{2}(x\sqrt{x^2 + 1} +$

$\operatorname{arsinh} x) + C = \frac{1}{2}[x\sqrt{x^2 + 1} + \log(x + \sqrt{x^2 + 1})] + C$. **Q6.** (a) Let $u = \sqrt{1 + x}$,

then $du = \dfrac{dx}{2u}$ and $x = u^2 - 1$, so $\displaystyle\int \sqrt{\dfrac{x}{1 + x}}\,dx = 2\int \sqrt{u^2 - 1}\,du =$

$u\sqrt{u^2-1}\ -\ \text{arcosh}\,u\ +\ C$ (by Q5) $=\ \sqrt{x(1+x)}\ -\ \text{arcosh}\,\sqrt{1+x}\ +\ C\ =$

$\sqrt{x(1+x)}-\log\left(\sqrt{x}+\sqrt{1+x}\,\right)+C.$ (b) $a\left[\sqrt{\dfrac{x}{a}\left(1+\dfrac{x}{a}\right)}-\text{arcosh}\,\sqrt{1+\dfrac{x}{a}}\,\right]+C=$

$a\left[\sqrt{\dfrac{x}{a}\left(1+\dfrac{x}{a}\right)}-\log\left(\sqrt{\dfrac{x}{a}}+\sqrt{1+\dfrac{x}{a}}\,\right)\right]+C.$

§4.6.8

Q1. (a) $\int x\sin x\,dx\ =\ x(-\cos x)\ -\ \int 1(-\cos x)\,dx\ =\ -x\cos x\ +\ \sin x\ +\ C.$
(b) $\int x^2 e^x\,dx = x^2 e^x - \int 2xe^x\,dx = x^2 e^x - 2\int xe^x\,dx$; by Example 1 this is $x^2 e^x -$
$2(xe^x - e^x)+C = (x^2-2x+2)e^x+C.$ (c) $\int e^x\cos x\,dx = e^x\cos x - \int e^x(-\sin x)\,dx =$
$e^x\cos x + \int e^x\sin x\,dx$; by Example 3 this is $e^x\cos x + \dfrac{e^x}{2}(\sin x - \cos x) + C =$
$\dfrac{e^x}{2}(\sin x + \cos x) + C.$ **Q2.** (a) Example 4, §4.6.7, gives $\frac{1}{2}\sin x\cos x + \frac{1}{2}x + C.$
(b) $\int \cos^6 x\,dx\ =\ \frac{1}{6}\sin x\cos^5 x + \frac{5}{6}\int \cos^4 x\,dx,$ $\int \cos^4 x\,dx\ =\ \frac{1}{4}\sin x\cos^3 x +$
$\frac{3}{4}\int \cos^2 x\,dx,$ $\int \cos^2 x\,dx = \frac{1}{2}\sin x\cos x + \frac{1}{2}\int 1\,dx = \frac{1}{2}\sin x\cos x + \frac{1}{2}x + C.$ Putting
these together gives $\frac{1}{6}\sin x\cos^5 x\ +\ \frac{5}{24}\sin x\cos^3 x\ +\ \frac{5}{16}\sin x\cos x\ +\ \frac{5}{16}x\ +\ C.$

Q3. $\displaystyle\int \arcsin x\,dx = x\arcsin x - \int \dfrac{x}{\sqrt{1-x^2}}\,dx.$ The new integral is $-\sqrt{1-x^2}+C,$
so the answer is $x\arcsin x+\sqrt{1-x^2}+C.$ **Q4.** (a) $\int \sec^3 x\,dx = \int \sec^2 x\sec x\,dx =$
$\tan x\sec x\ -\ \int \tan^2 x\sec x\,dx\ =\ \tan x\sec x\ -\ \int(\sec^2 x\ -\ 1)\sec x\,dx\ =\ \tan x\sec x-$
$\int \sec^3 x\,dx + \int \sec x\,dx$; therefore $\int \sec^3 x\,dx = \frac{1}{2}\,(\sec x\tan x + \log|\sec x + \tan x|) +$
$C.$ (b) Let $x = \tan u,\ -\pi/2 < u < \pi/2.$ Then $\sec u$ is positive, so $\sec u = \sqrt{x^2+1}\,.$
Also $dx\ =\ \sec^2 u\,du,$ so the integral becomes $\int \sec^3 u\,du,$ which by (a) equals
$\frac{1}{2}\,[x\sqrt{x^2+1} + \log(x + \sqrt{x^2+1}\,)] + C$ (the quantity $x + \sqrt{x^2+1}$ is positive for
all x).

§4.6.10

Q1. $\dfrac{1}{(x-2)(x+3)}\ =\ \dfrac{1}{5}\left[\dfrac{1}{x-2}-\dfrac{1}{x+3}\right],$ so $\displaystyle\int \dfrac{1}{(x-2)(x+3)}\,dx\ =$

$\dfrac{1}{5}\left[\displaystyle\int \dfrac{1}{(x-2)}\,dx - \int \dfrac{1}{(x+3)}\,dx\right] = \dfrac{1}{5}\,[\log|x-2|-\log|x+3|]+C = \dfrac{1}{5}\log\left|\dfrac{x-2}{x+3}\right| +$

$C.$ **Q2.** Multiplication by $x^2(x+1)$ gives $x+2\ =\ Ax(x^2+1) + B(x^2+1)+$
$(Cx+D)x^2 = (A+C)x^3 + (B+D)x^2 + Ax + B.$ Then $A+C=0,\ B+D=0,\ A=1,$

and $B = 2$, so also $C = -1$ and $D = -2$.   **Q3.** Let $\dfrac{x^3}{(x-1)(x+1)(x^2+2)} =$
$\dfrac{A}{x-1} + \dfrac{B}{x+1} + \dfrac{Cx+D}{x^2+2}$. Clearing denominators and equating coefficients of equal
powers of x as in Q2 results in the four equations $A + B + C = 1$, $A - B + D = 0$, $2A + 2B - C = 0$, $2A - 2B - D = 0$, which have the solution $A = \frac{1}{6}$, $B = \frac{1}{6}$, $C = \frac{2}{3}$, $D = 0$. The integrand is then $\dfrac{1}{6}\left[\dfrac{1}{x-1} + \dfrac{1}{x+1} + \dfrac{4x}{x^2+2}\right]$, and
the integral is $\frac{1}{6}\left[\log|x-1| + \log|x+1| + 2\log(x^2+2)\right] + C = \frac{1}{6}\log\left[|x^2-1|(x^2+2)^2\right] + C = \frac{1}{6}\log|x^6 + 3x^4 - 4| + C$.

§4.7.3

Q1. (a) $\frac{1}{2}$ (b) Does not exist. (c) 6   **Q2.** $\displaystyle\int_a^b \dfrac{1}{x^\alpha}\,dx = \left.-\dfrac{1}{(\alpha-1)x^{\alpha-1}}\right|_a^b =$
$\dfrac{1}{\alpha-1}\left[\dfrac{1}{a^{\alpha-1}} - \dfrac{1}{b^{\alpha-1}}\right] \to \dfrac{1}{(\alpha-1)a^{\alpha-1}}$ as $b \to \infty$, because $b^{\alpha-1} \to \infty$ (§3.3.3.Q5(a)).

Q3. For $\alpha < 1$, as in Q2 we obtain $\displaystyle\int_a^b \dfrac{1}{x^\alpha}\,dx = \dfrac{1}{1-\alpha}(b^{1-\alpha} - a^{1-\alpha}) \to \infty$
as $b \to \infty$. For $\alpha = 1$ the integral is $\log(b/a) \to \infty$ (§4.3.3.Q9(b)).   **Q4.** As in

Q3, $\displaystyle\int_a^b \dfrac{1}{x^\alpha}\,dx = \dfrac{1}{1-\alpha}(b^{1-\alpha} - a^{1-\alpha}) \to \dfrac{b^{1-\alpha}}{1-\alpha}$ as $a \to 0$, because $a^{1-\alpha} \to 0$

(§3.3.3.Q5(b)).   **Q5.** For $\alpha > 1$, $\displaystyle\int_a^b \dfrac{1}{x^\alpha}\,dx = \dfrac{1}{\alpha-1}\left[\dfrac{1}{a^{\alpha-1}} - \dfrac{1}{b^{\alpha-1}}\right] \to \infty$ as $a \to 0$.

For $\alpha = 1$ the integral is $\log(b/a) \to \infty$.   **Q6.** (a) $\displaystyle\int_{-1}^2 x^{-1/3}\,dx = \int_{-1}^0 x^{-1/3}\,dx +$
$\displaystyle\int_0^2 x^{-1/3}\,dx = \frac{3}{2}(2^{2/3} - 1)$ (b) $\frac{1}{2}$ (c) Does not exist (at each endpoint there
is a vanishing quantity to the first power in the denominator). (d) Does not exist.
Q7. As $x \to a$, $x/a \to 1$, so the right-hand side of §4.6.4(3) approaches C. Then the
value of the integral is $-a\left[\sqrt{\dfrac{b}{a}\left(1 - \dfrac{b}{a}\right)} + \arcsin\sqrt{1 - \dfrac{b}{a}}\right]$.   **Q8.** Yes: $\arcsin x =$
$\displaystyle\int_0^x \dfrac{1}{\sqrt{1-x^2}}\,dx$, improper if $x = \pm 1$—see Example 4 of §4.7.2; similarly $\arccos x =$
$\displaystyle\int_1^x -\dfrac{1}{\sqrt{1-x^2}}\,dx$, improper for all x.

§4.7.5

Q1. It suffices to prove continuity for $-\pi/2 < x \le \pi/2$, because the function on other intervals $-\dfrac{\pi}{2} + n\pi < x \le \dfrac{\pi}{2} + n\pi$ is identical (except for sign, if n is odd). We know that $\sin x$ is continuous for $-\pi/2 < x < \pi/2$, and that its left-hand limit as $x \to \pi/2$ is $\sin(\pi/2) = 1$. Then the only question is whether its right-hand limit at $\pi/2$ is also 1. That limit is the negative of the right-hand limit as $x \to -\pi/2$, because $\sin(x + \pi) = -\sin x$; and the formula by which $\sin x$ was defined shows the latter limit to be -1, because $\tan x \to -\infty$ as $x \to -\pi/2$.

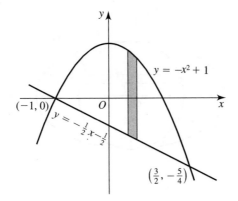

Fig. A-39

§4.8.5

Q1. See Fig. A-39: $\displaystyle\int_{-1}^{3/2} [(-x^2 + 1) - (-\tfrac{1}{2}x - \tfrac{1}{2})] \, dx = \tfrac{125}{48}$. **Q2.** See Fig. A-40: $\int_{-3\pi/4}^{\pi/4} (\cos x - \sin x) \, dx = 2\sqrt{2}$. **Q3.** If $f(x) \ge g(x)$ for $a \le x \le b$, the area

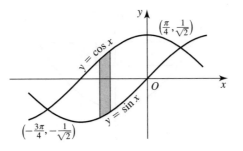

Fig. A-40

between the graphs from a to b is $\int_a^b (f(x) - g(x))\,dx$. If in addition $f(a) = g(a)$
and $f(b) = g(b)$, the area is entirely enclosed by the graphs. If $f(x)$ is not necessarily
$\geq g(x)$ this integral gives net signed area; the total area is $\int_a^b |f(x) - g(x)|\,dx$
(cf. §2.10.9.Q10).   **Q4.** (a) If $x = (b/a)\,x'$ and $y = y'$ are substituted in the equation
of the circle, there results the equation of the ellipse with x' and y' in place of x and
y. Conversely, if $x' = (a/b)\,x$ and $y' = y$ are substituted for x and y in the equation
of the ellipse, the equation of the circle results. Thus if P is on the circle, P' is
on the ellipse, and conversely. (b) See Fig. A-41: $dA':dA = TP':TP = a:b$, so

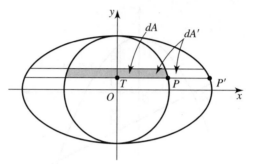

Fig. A-41

$$\int dA' = \int \frac{a}{b}\,dA = \frac{a}{b}\int dA = \frac{a}{b}\cdot(\text{area of circle}) = \frac{a}{b}\,(\pi b^2) = \pi ab.\quad \textbf{Q5.}\ \text{The}$$

length of the strip shown in Fig. A-42 is twice the x-coordinate of P, or $2\sqrt{y+1}$;
hence its area is $2\sqrt{y+1}\,dy$. Let the pressure at any level y be $-\gamma y$, where $\gamma > 0$
(recall that $y \leq 0$). The force on the strip is then $-2\gamma y\sqrt{y+1}\,dy$, and the total

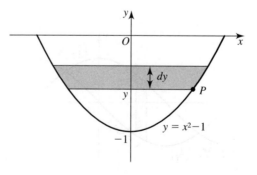

Fig. A-42

force on the dam is $\int_{-1}^{0} -2\gamma y \sqrt{y+1}\, dy = -2\gamma \int_{-1}^{0} y\sqrt{y+1}\, dy$. The integral is readily evaluated with the aid of the substitution $u = y + 1$, and the result is $8\gamma/15$ (newton).   **Q6.** Let the rate at radius r be αr. Using ring elements as in Example 4 we find the total rate to be $\int_{0}^{0.2} \alpha r \cdot 2\pi r\, dr \approx 0.017\alpha$ kg/sec.   **Q7.** The volume under a surface (§4.5.1).   **Q8.** Represent one of the figures (F_1) as composed of infinitesimal rectilinear elements, then change the scale so that it becomes the size of the other figure (F_2). The altered elements are still infinitesimal, they represent F_2, and they stand in the desired ratio to their originals. (This argument is easily made rigorous by using approximation by finite elements.)   **Q9.** (a) Just as the area under a curve $y = f(x)$ from x to $x + dx$ can be regarded as a rectangle of height $f(x)$, this area can be regarded as a circular sector of radius $f(\theta)$; the formula follows.

(b) $\displaystyle\int_{\theta_1}^{\theta_2} \tfrac{1}{2} r_0^2 e^{2\alpha\theta}\, d\theta = \frac{r_0^2}{4\alpha}\left(e^{2\alpha\theta_2} - e^{2\alpha\theta_1}\right).$

§4.8.7

Q1. (a) $\displaystyle\int_{0}^{H} \pi(\alpha x)^2\, dx = \tfrac{1}{3}\pi\alpha^2 H^3$.   Since the radius at $x = H$ is αH, $B = \pi(\alpha H)^2$ and the formulas agree. (b) $\displaystyle\int_{0}^{H} \pi(\alpha x)^2(1 + 2x)\, dx = \pi\alpha^2 H^3\left(\frac{1}{3} + \frac{H}{2}\right).$

Q2. (a) $\displaystyle\int_{-\rho}^{\rho} \pi(\rho^2 - x^2)(x - 1)\, dx = -\tfrac{4}{3}\pi\rho^3$ coulomb. (b) By symmetry the variable term in $\delta(x)$ contributes zero net charge.   **Q3.** (a) $\displaystyle\int_{1}^{b} \frac{\pi}{x^2}\, dx = \pi\left(1 - \frac{1}{b}\right)$ (b) $V \to$ π, the "volume" of the infinite horn.   **Q4.** As elements we use spherical shells of volume $dV = 4\pi r^2\, dr$.   Then $M = \displaystyle\int dm = \int_{0}^{\rho} 4\pi r^2 \delta_0\left(1 - \frac{r}{\rho}\right)dr = \tfrac{1}{3}\pi\delta_0\rho^3.$

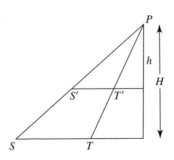

Fig. A-43

Q5. $\displaystyle\int_{-a}^{a} \pi b^2\left(1 - \frac{x^2}{a^2}\right) dx = \frac{4}{3}\pi ab^2.$ **Q6.** (a) We can regard the base as composed of rectilinear elements (cf. §4.8.4), hence can regard the cylinder as composed of cylinders based on these elements, i.e. prisms. (b) Because it is thin, we can treat the slice as a right cylinder as far as volume is concerned, even though its sides may slant.   **Q7.** The figures differ only in scale, because if S, T are any points in the base and S', T' the corresponding points at level h (Fig. A-43), we have $\dfrac{S'T'}{ST} = \dfrac{h}{H}$.

CHAPTER 5

§5.1.3

Q1. $x'(t) = -gt + v_0$, $x''(t) = -g$. Substitution gives $x(0) = x_0$, $x'(0) = v_0$.
Q2. $x(t) = -\frac{1}{2}gt^2 + 20$ (m), $v(t) = -gt$, $t_1 = 2.0$ sec, $|v| = 20$ m/sec. **Q3.** Since the distance of fall is proportional to the square of the time, it falls $4p$ m in $2q$ sec, hence $3p$ m in the second q sec. **Q4.** (a) $\sqrt{2gx_0} = 5.4$ m/sec, so $x_0 = 1.5$ m. Or use the calculation in §5.1.2: the ratio of the heights ($\propto t^2$) is the square of the ratio of the speeds ($\propto t$), so $\dfrac{x_0}{3.0} = \left(\dfrac{12}{17}\right)^2 = 0.50$. (b) $\dfrac{v_0^2}{2g} = 1.5$ m—the same height. See Q7. **Q5.** (a) $x_0 =$ initial position, $v_0 t =$ change of position due to initial velocity v_0, $-\frac{1}{2}gt^2 =$ change of position due to acceleration $-g$. (b) $x(t) = v_0 t + x_0$, motion at constant velocity v_0 —uniform motion. **Q6.** (a) By the quadratic formula

$$t_1 = \frac{v_0 + \sqrt{2gx_0 + v_0^2}}{g}.$$ (b) The time of fall is $\sqrt{\dfrac{2\left(x_0 + \dfrac{v_0^2}{2g}\right)}{g}} = \dfrac{\sqrt{2gx_0 + v_0^2}}{g}$, so

the total time is $\dfrac{v_0}{g} + \dfrac{\sqrt{2gx_0 + v_0^2}}{g}$. **Q7.** By Q6 with $x_0 = 0$, $t_1 = \dfrac{2v_0}{g}$. Then $v = -gt_1 + v_0 = -v_0$. **Q8.** Consider an observer moving horizontally in a straight line at uniform velocity. In the vertical plane that is parallel to his motion and contains the x-axis he draws horizontal and vertical coordinate axes, fixed with respect to him, and marks the location of the particle at each instant on this coordinate plane. A graph like Fig. 5-3 results (the observer's motion is in the negative t-direction). **Q9.** (a) $f = ma$ becomes $-mg - \rho v = ma$. As the speed $|v| = -v$ increases, a increases toward 0; a would vanish at $v = -mg/\rho$, the terminal velocity. (b) Taking $m = 75$ kg gives $|v_T| \approx 57$ m/sec ≈ 130 mph.

§5.2.5

Q1. (a) π sec (b) 6 m/sec **Q2.** $4g/\pi^2$ m ≈ 4.0 m ≈ 13 ft **Q3.** Once around the circle is 2π radians, or 1 cycle. Thus $\dfrac{\omega \text{ rad}}{1 \text{ sec}} = \dfrac{2\pi\omega \text{ rad}}{2\pi \text{ sec}} = \dfrac{\omega \text{ cycle}}{2\pi \text{ sec}}$.
Q4. A m/rad $\times \omega$ rad/sec $= A\omega$ m/sec $=$ maximum speed of particle (attained at O, when P is moving parallel to the particle). **Q5.** (a) v: $-$ for leftward motion, $+$ for rightward. a: $-$ to right of O, $+$ to left of O. (b) See Fig. A-44. Each arrow reverses direction when its length passes zero. **Q6.** Since the two functions are the same, they must have the same maximum values, i.e. $A' = A$. Then $\cos(\omega t + \delta') = \cos(\omega t + \delta)$ for all t. Taking $t = 0$, $\cos\delta' = \cos\delta$, and

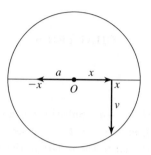

Fig. A-44

taking $t = -\dfrac{\pi}{2\omega}$, $\cos\left(\delta' - \dfrac{\pi}{2}\right) = \cos\left(\delta - \dfrac{\pi}{2}\right)$, i.e. $\sin\delta' = \sin\delta$. The result then follows from the reference cited. **Q7.** (a) We must have $x_0^2 + (v_0/\omega)^2 = A^2$, so let $A = \sqrt{x_0^2 + (v_0^2/\omega^2)}$. Now we are to find δ such that $\cos\delta = x_0/A$, $\sin\delta = -v_0/A\omega$. Since $(x_0/A)^2 + (v_0/A\omega)^2 = 1$, δ exists and is determined as claimed (§3.7.17.Q5). (b) The resultant amplitude must be ≤ 0.3, so $x_0^2 + (v_0^2/\omega^2) \le 0.09$. Thus, e.g., if the particle is released from rest at x_0, $|x_0| \le 0.3$; if the particle is launched from $x_0 = 0.2$ at v_0, $|v_0| \le 0.22\,\omega$ (approximately). **Q8.** (a) Differentiation verifies it. (b) Yes, $A\cos(\omega t + \delta) = (A\cos\delta)\cos\omega t + (-A\sin\delta)\sin\omega t$. (c) Yes: we can find A and δ such that $A\cos\delta = \alpha$, $-A\sin\delta = \beta$. Let $A = \sqrt{\alpha^2 + \beta^2}$; if $A = 0$, δ can be chosen arbitrarily; if $A > 0$ there is δ such that $\cos\delta = \alpha/A$, $\sin\delta = -\beta/A$, because $(\alpha/A)^2 + (-\beta/A)^2 = 1$; cf. Q7(a). **Q9.** $p' = mv' = m(-A\omega\sin\omega t)' = -m\omega^2 A\cos\omega t = -m\omega^2 x = -m(c/m)x = -cx = f(x)$. **Q10.** The quarter-cycle goes from $t = 0$ to $t = \pi/2\omega$, so the mean velocity is $\dfrac{1}{\pi/2\omega}\displaystyle\int_0^{\pi/2\omega} -A\omega^2\cos\omega t\,dt = -\dfrac{2A\omega^2}{\pi}$. **Q11.** (a) The frequency is proportional to the square root of the acceleration of gravity, which is smaller on the moon (about $\frac{1}{6}g$). (b) The system of suspended mass + earth (or just earth's gravitational attraction). **Q12.** (a) Measure the period by dividing the duration of a series of n swings to and fro by n. (b) $\tau \propto \sqrt{l}$. **Q13.** The pendulum will keep time, but will soon stop because of friction if it is not supplied with energy.

§5.3.8

Q1. The decrease in potential energy is $mgx_0 - mgx = 98m$ joule. The kinetic energy gains the same amount; since it was initially 0, at $x = 10$ we have $\frac{1}{2}mv^2 = 98m$, so $v = 14$ m/sec. **Q2.** No. The initial kinetic energy, which is the total energy (taking $\phi(0) = 0$), is $\frac{1}{2}(0.4)(2)^2 = 0.8$. Since $c = \frac{4}{3}$, if the particle reached $x = 1.1$

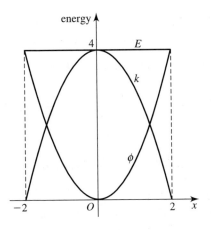

Fig. A-45

its potential energy would be $\frac{1}{2}(\frac{4}{3})(1.1)^2 > 0.8$, which is impossible. **Q3.** We have $\omega^2 = 2$ and $m = 1$, hence $c = m\omega^2 = 2$ and $\phi(x) = \frac{1}{2}cx^2 = x^2$. The total energy is $E = \frac{1}{2}cA^2 = 4$, so $k(x) = 4 - x^2$ and the graphs are as in Fig. A-45. **Q4.** Taking $b \to a$ as the positive direction, the force on the charge has the constant value $-q\alpha$, so a potential energy function is $\phi(x) = q\alpha x$ and $\phi(a) - \phi(b) = q\alpha(a - b)$, whose sign is that of q. **Q5.** (a) For a pendulum $c = mg/l$ (§5.2.4), so (with $\phi(0) = 0$)

$$E = \tfrac{1}{2}cA^2 = \left(\frac{mg}{2}\right)\frac{A^2}{l} \text{—it varies as the square of the amplitude and inversely as}$$

the length. (b) The maximum speed is attained at $x = 0$, where $\frac{1}{2}mv^2 = E$, hence $v \propto A/\sqrt{l}$. **Q6.** (a) According to §4.2.8 $dk/dx = dp/dt$, while by definition $d\phi/dx = -f$. Hence differentiating $k + \phi = E$ with respect to x gives the result. (b) $dk/dx = dp/dt = f$, hence by integrating with respect to x we get $k = -\phi + E$ for some constant E. **Q7.** (a) For $x < x_0$, $dx/dt < 0$, hence $x(t)$ is invertible and

its derivative is $1\Big/\dfrac{dx}{dt}$. (b) For clarity let the upper limit be x_1:

$$\int_{x_0}^{x_1} -\frac{1}{\sqrt{2g(x_0 - x)}}\,dx = \frac{1}{\sqrt{2g}}\int_{x_0}^{x_1} -\frac{1}{\sqrt{x_0 - x}}\,dx. \text{ The substitution } u = x_0 - x$$

transforms the integral into $\displaystyle\int_0^{x_0 - x_1} \frac{1}{\sqrt{u}}\,du$, which as in the example cited is found

to have the value $2\sqrt{x_0 - x_1}$. Consequently the original integral is $\dfrac{2\sqrt{x_0 - x_1}}{\sqrt{2g}} =$

$\sqrt{\dfrac{2(x_0 - x_1)}{g}}$. **Q8.** Conservation of energy gives $\frac{1}{2}mv_0^2 + mgx_0 = \frac{1}{2}mv^2 + mgx$, or $v^2 = 2g(x_1 - x)$, where $x_1 = x_0 + \dfrac{v_0^2}{2g}$. Then $v = \pm\sqrt{2g(x_1 - x)}$. If $v_0 \leq 0$ the particle falls and the sign is negative; if $v_0 > 0$ it rises until $x = x_1$, then falls, so the sign changes from plus to minus. In the first case $t = \displaystyle\int_{x_0}^{x} -\dfrac{1}{\sqrt{2g(x_1 - x)}}\,dx =$

$\sqrt{\dfrac{2(x_1 - x)}{g}} - \sqrt{\dfrac{2(x_1 - x_0)}{g}} = \sqrt{\dfrac{2(x_1 - x)}{g}} + \dfrac{v_0}{g}$ (the last plus sign appears because $\sqrt{v_0^2} = -v_0$). In the second case, while $v \geq 0$ we have $t =$

$\displaystyle\int_{x_0}^{x} \dfrac{1}{\sqrt{2g(x_1 - x)}}\,dx = -\sqrt{\dfrac{2(x_1 - x)}{g}} + \dfrac{v_0}{g}$. When $x = x_1$ this gives $t = \dfrac{v_0}{g}$. Then

for $t \geq \dfrac{v_0}{g}$, $t = \dfrac{v_0}{g} + \displaystyle\int_{x_1}^{x} -\dfrac{1}{\sqrt{2g(x_1 - x)}}\,dx = \sqrt{\dfrac{2(x_1 - x)}{g}} + \dfrac{v_0}{g}$. When solved for x

each expression yields $x = -\frac{1}{2}gt^2 + v_0 t + x_0$. **Q9.** (a) As in §5.3.7, $\phi(x) = \frac{1}{2}cx^2$; at $t = 0$, $E = k + \phi = \phi(A) = \frac{1}{2}cA^2$. (b) Because the force is negative at the beginning of motion, so is v. Hence $t = \displaystyle\int_{A}^{x} \dfrac{1}{-\sqrt{\dfrac{2}{m}\left(\dfrac{1}{2}cA^2 - \dfrac{1}{2}cx^2\right)}}\,dx =$

$\displaystyle\int_{A}^{x} -\dfrac{1}{\omega\sqrt{A^2 - x^2}}\,dx = \dfrac{1}{\omega}\arccos\dfrac{x}{A}$ (cf. Example 7, §4.6.4, and Example 4, §4.7.2), so $x = A\cos\omega t$. Then $v = -A\omega\sin\omega t$ remains negative until $t = \pi/\omega = \tau/2$, when $v = 0$ and $x = -A$. At that point v turns positive, so on the next interval

$t = \dfrac{\pi}{\omega} + \displaystyle\int_{-A}^{x} \dfrac{1}{\omega\sqrt{A^2 - x^2}}\,dx = \dfrac{1}{\omega}\left(\pi - \arccos\dfrac{x}{A} + \pi\right) = \dfrac{1}{\omega}\left(2\pi - \arccos\dfrac{x}{A}\right)$, and

again $x = A\cos\omega t$, this time until $t = 2\pi/\omega = \tau$. The motion then repeats. It is simpler to deduce the second half-period from the first: the particle is released from $-x = A$ at $t - \dfrac{\pi}{\omega} = 0$, hence $-x = A\cos\omega\left(t - \dfrac{\pi}{\omega}\right) = A\cos\left(\omega t - \pi\right)$, or $x = A\cos\omega t$.

Q10. (a) The particle is at a point where $\phi(x) \leq E$. (b) See Fig. A-46: ϕ has a local

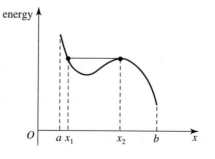

Fig. A-46

maximum at x_2, and $\phi(x_1) = \phi(x_2)$. The particle will reach b if $x_0 > x_2$, because $\phi(x)$ is decreasing, so that $f(x) > 0$ (the force is non-zero and positively directed). Similarly, if $x_0 < x_1$ it will reach x_1 with $v(x_1) > 0$; then since $\phi(x) \leq \phi(x_1)$ for $x \geq x_1$, the kinetic energy will become no smaller, so that v will remain $\geq v(x_1)$ and the particle will reach b. For $x_1 \leq x_0 \leq x_2$ the condition for it to reach b is $E = \frac{1}{2}mv_0^2 + \phi(x_0) > \phi(x_2)$, because this means that v does not fall to zero: for $x \geq x_0$, $\frac{1}{2}mv^2 = E - \phi(x) \geq E - \phi(x_2) > 0$.

§5.4.5

Q1. $\dfrac{GM}{(4002)^2} \Big/ \dfrac{GM}{(4000)^2} = \left(\dfrac{4000}{4002}\right)^2 \approx 0.999$ **Q2.** $x_0 = 2R$, $x = 4100$ miles $=$

6.6×10^6 m; $|v| = \sqrt{2GM\left(\dfrac{1}{x} - \dfrac{1}{2R}\right)} = 7.7 \times 10^3$ m/sec $= 17{,}000$ mph.

Q3. (a) $x_0 = 2R$, $x = R$. In the formula for t recall that $\arcsin(1/\sqrt{2}) = \pi/4$, then

$t = \dfrac{R^{3/2}}{\sqrt{GM}}\left(1 + \dfrac{\pi}{2}\right) = 2.1 \times 10^3$ sec $= 35$ min. (b) $x_0 = R$, $x = 0$ (as a limiting

location). Then $t = \dfrac{\pi R^{3/2}}{2\sqrt{2GM}} = 900$ sec $= 15$ min. From §5.4.2(3) with $v_0 = 0$ and

$x_0 = R$, $v = c$ when $c^2 = 2GM\left(\dfrac{1}{x} - \dfrac{1}{R}\right)$, or $x = 1\Big/\left(\dfrac{c^2}{2GM} + \dfrac{1}{R}\right) = 0.89$ cm.

Q4. We use §5.4.4(6). Here $x_0 = R$, $x = 3.8 \times 10^8$ m. Then $t = \sqrt{\dfrac{2}{9GM}} \times$

$(x^{3/2} - x_0^{3/2}) \approx \sqrt{\dfrac{2}{9GM}}\, x^{3/2}$ (since x_0 is much smaller than x) $= 1.7 \times 10^5$ sec $=$

47 hours. **Q5.** Eschewing experiment, we read that such an elephant weighs 5400–7200 kg. The midpoint of the range is 6300 kg; thus the potential energy difference for a moderate elephant is $mgh = (6.3 \times 10^3)(9.8)(1) = 6.2 \times 10^4$ joule. **Q6.** The

work is $\displaystyle\int_R^\infty -f(x)\,dx$ (cf. §2.10.9.Q12) $= \displaystyle\int_R^\infty \dfrac{GMm}{x^2}\,dx = \lim_{b\to\infty}\displaystyle\int_R^b \dfrac{GMm}{x^2}\,dx =$

$\displaystyle\lim_{b\to\infty} -\dfrac{GMm}{x}\Big|_R^b = \dfrac{GMm}{R}$. **Q7.** (a) 0. Since the potential energy is 0 at ∞, the

kinetic energy $k(x)$ at any x exactly equals the potential energy difference $-\phi(x)$

between ∞ and x. (b) Since $\phi(x) = -\dfrac{GMm}{x}$, $\dfrac{2}{m}(E - \phi(x)) = \dfrac{2GM}{x}$ and the

integral is $\displaystyle\int_{x_0}^x \sqrt{\dfrac{x}{2GM}}\,dx$. **Q8.** Not for a *projectile*, if other forces are ruled out. If the gravity of other bodies is admitted, a projectile can rise to some height and then be captured by another body, e.g. the moon. If a rocket engine is permitted, you can go as far as you want as slowly as you want, while there is fuel. **Q9.** (a) By

§5.4.3(4), $v = 0$ at $\bar{x} = \dfrac{2GM}{v_{x_0}^2 - v_0^2} = \left[\dfrac{1}{1 - (v_0/v_{x_0})^2}\right] x_0$. As $v_0 \to v_{x_0}$ this $\to \infty$.

(b) By the same equation, as $x \to \infty$, $v^2 \to v_0^2 - v_{x_0}^2$, so the limiting velocity is
$\sqrt{v_0^2 - v_{x_0}^2} = [1 - (v_{x_0}/v_0)^2]^{1/2} v_0$. If $v_0 = v_{x_0}$ this is 0. **Q10.** (a) By §5.4.3(4),
$v^2 = 2GM/x$. Let $\alpha = 2GM$, then $v^2 = \alpha/x$, so $v = \pm\sqrt{\alpha/x}$; the sign is the same
as that of v_0. The graph of $x = \alpha/v^2$ (on xv- rather than vx-coordinates) is the
two curves in Fig. A-47, including the dashed lines; only the solid portions represent
motion which begins at x_0. For $v_0 > 0$, $x \to \infty$ and $v \to 0$; for $v_0 < 0$, $x \to 0$ and

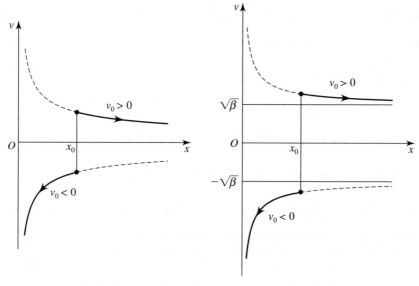

Fig. A-47   Fig. A-48

$v \to -\infty$. (b) Let $\beta = v_0^2 - v_{x_0}^2 > 0$, then $v^2 = \dfrac{\alpha}{x} + \beta$, so $v = \pm\sqrt{\dfrac{\alpha}{x} + \beta}$; the

sign is that of v_0. Figure A-48 shows the part of the graph of $x = \dfrac{\alpha}{v^2 - \beta}$ for

which $x > 0$. The motion is similar to that in (a), except that for $v_0 > 0$, $v \to$

$\sqrt{\beta}$ rather than 0. (c) Let $\gamma = v_{x_0}^2 - v_0^2 > 0$, then $v^2 = \dfrac{\alpha}{x} - \gamma$, or $x = \dfrac{\alpha}{v^2 + \gamma}$

(cf. §3.9.9.Q1(e))—see Fig. A-49. The maximum x is α/γ, which is the $\bar{x}$ of Q9(a).

Q11. $\dfrac{x(t)}{\alpha t^{2/3}} = \left[\dfrac{\alpha^{3/2} t + x_0^{3/2}}{\alpha^{3/2} t}\right]^{2/3} = \left[1 + \left(\dfrac{x_0}{\alpha}\right)^{3/2} \dfrac{1}{t}\right]^{2/3} \to 1^{2/3} = 1.$

Q12. (a) Let $GM = h$, then the equation is $\alpha\beta(\beta - 1)t^{\beta - 2} = -\dfrac{h}{\alpha^2 t^{2\beta}}$. The powers

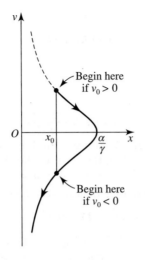

Fig. A-49

of t must agree, so $\beta - 2 = -2\beta$ or $\beta = \frac{2}{3}$. It follows that $\alpha = (9h/2)^{1/3}$. Then a solution is $x = \alpha t^{2/3}$, where $\alpha = (9GM/2)^{1/3}$. (b) It is valid for $t \neq 0$. For $t < 0$ it represents a particle approaching the center, and for $t > 0$ one leaving it. (c) As in (a) we find $\beta = \frac{2}{3}$, and then $\gamma^2 = 9h/2 = \alpha^3$. Taking $\gamma > 0$ gives $\gamma = \alpha^{3/2}$; setting $t = 0$ gives $\delta = x_0^{3/2}$. **Q13.** (i) If $v_0^2 < v_{x_0}^2$, following Q9(a) let $\bar{x} = \dfrac{2GM}{v_{x_0}^2 - v_0^2} > 0$.

Then $v^2 = \dfrac{2GM}{\bar{x}} \left(\dfrac{\bar{x} - x}{x} \right) = v_{\bar{x}}^2 \left(\dfrac{\bar{x} - x}{x} \right)$ (where $v_{\bar{x}}$ denotes the escape velocity for height $\bar{x}$), exactly as in case 2 of §5.4.4 with $\bar{x}$ replacing x_0, except that integration begins at x_0 instead of $\bar{x}$ and if $v_0 > 0$ the sign of v will be positive until $x = \bar{x}$. (ii) If $v_0^2 > v_{x_0}^2$, let $\bar{x} = \dfrac{2GM}{v_0^2 - v_{x_0}^2} > 0$. Then $v^2 = v_{\bar{x}}^2 \left(\dfrac{\bar{x} + x}{x} \right)$. The integration is as before, except that the result of §4.6.5.Q6(b) is needed this time.

CHAPTER 6

§6.2.10

Q1. The ratio of increase from t to $t+T$ is $\dfrac{q(t+T)}{q(t)} = \dfrac{e^{\alpha(t+T)}}{e^{\alpha t}} = e^{\alpha t + \alpha T}e^{-\alpha t} = e^{\alpha T}$,

independent of t.　**Q2.** (a) It triples every third of a minute, so in two minutes it will increase by the factor $3^6 = 729$; the mass will be about 7.3 g. (b) Measuring t in minutes, we have $0.03 = 0.01e^{\alpha(1/3)}$, so $\alpha = 3\log 3 \approx 3.3$. Then $q(2.5) = 0.01e^{\alpha(2.5)} \approx 0.01e^{8.25} \approx 38$ g.　**Q3.** Measuring time in years, we have $0.99957 = e^{-\frac{\log 2}{\tau}\cdot 1}$, so $\tau = -\dfrac{\log 2}{\log 0.99957} = 1600$ years.　**Q4.** (i) $a^{b+c} = e^{(b+c)\log a} = e^{b\log a + c\log a} = e^{b\log a}e^{c\log a}$ (by §6.1.4(6)) $= a^b a^c$ (ii) $\log a^b = \log e^{b\log a} = b\log a$ (iii) $(a^b)^c = e^{c\log a^b} = e^{cb\log a}$ (by (ii)) $= a^{bc}$　**Q5.** $2q_0 = q_0 a^T$; divide by q_0 and apply $\log_a$.　**Q6.** Let $y = 10^h$, where h is the (signed) height above the x-axis (e^h or any other a^h, $a > 1$, would work as well). Then $h = \log_{10} b + (c\log_{10} a)x$ is a linear function (Fig. A-50).　**Q7.** (a) $\log x$ and e^x are increasing and $\log e = 1$. (b) Clearly $\log 2 <$ the large square of area 1 in Fig. A-51. As for $\log 3$, divide $[1,3]$ into 8 equal parts and use right endpoints as shown in the figure. The estimate is then $\log 3 > \dfrac{1}{4}\left(\dfrac{4}{5} + \dfrac{4}{6} + \dfrac{4}{7} + \cdots + \dfrac{4}{12}\right) > 1$.　**Q8.** As of 2002, about $e^{-\frac{0.693}{5750}\times 315} \approx 0.96$, or 96%.　**Q9.** $(qe^{-\alpha t})' = 0$, hence $qe^{-\alpha t} = c$.　**Q10.** If the logarithm takes all numbers as values, then given $M > 0$ there are t_1 and t_2 such that $\log t_1 = M$ and $\log t_2 = -M$; then $\log t > M$ for $t > t_1$, $\log t < -M$ for $t < t_2$. Con-

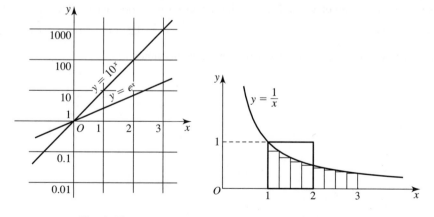

Fig. A-50　　　　　　　　　　　　Fig. A-51

versely, given any L, there are t_1 and t_2 such that $\log t_1 > L$ and $\log t_2 < L$; then by the intermediate value property there is t with $\log t = L$. **Q11.** We have $e = p(1) > p(0) = 1$, so $p(n) = e^n \to \infty$ as $n \to \infty$ (by the cited question). As an increasing function $p(t)$ grows with $p(n)$, so has limit ∞ as $t \to \infty$. The second limit follows from the first because $p(-t) = 1/p(t)$. **Q12.** Using reasoning similar to that in §6.2.2 we have $1 \geq 1/t$ on $[1, 2]$, $1/2 \geq 1/t$ on $[2, 3]$, and in general for

$1 \leq i \leq n$, $1/i \geq 1/t$ on $[i, i+1]$, hence $\dfrac{1}{i} = \displaystyle\int_i^{i+1} \dfrac{1}{i}\, dt \geq \int_i^{i+1} \dfrac{1}{t}\, dt$; it follows

that $s_n \geq \displaystyle\int_1^{n+1} \dfrac{1}{t}\, dt = \log(n+1) \to \infty$. A direct proof can also be given:

$$s_{2^n} = 1 + \frac{1}{2} + \left(\frac{1}{3} + \frac{1}{4}\right) + \left(\frac{1}{5} + \frac{1}{6} + \frac{1}{7} + \frac{1}{8}\right) + \cdots + \left(\frac{1}{1+2^{n-1}} + \cdots + \frac{1}{2^n}\right)$$

$$\geq 1 + \frac{1}{2} + 2 \cdot \frac{1}{4} + 4 \cdot \frac{1}{8} + \cdots + 2^{n-1} \cdot \frac{1}{2^n} = 1 + \frac{n}{2} \to \infty.$$

Q13. (a) The derivatives are equal, hence $\log uv = \log u + C$; put $u = 1$. (b) $\log 2^2 = \log 2 + \log 2 = 2\log 2$, $\log 2^3 = \log 2^2 + \log 2 = 3\log 2, \ldots, \log 2^n = n\log 2$. Since $\log$ is increasing, $\log 2 > \log 1 = 0$, so $n \log 2 \to \infty$. As in §6.2.2, this establishes the first limit. The second follows as in the same section, because $\log t + \log \dfrac{1}{t} = \log\left(t \cdot \dfrac{1}{t}\right) = \log 1 = 0$ implies $\log t = -\log(1/t)$. **Q14.** $\log a_n \to \log a$ (because $\log$ is continuous), hence $b_n \log a_n \to b \log a$, hence $e^{b_n \log a_n} \to e^{b \log a}$ (because e^x is continuous). **Q15.** We know that for integral m, $a^m/m \to \infty$ as $m \to \infty$. Furthermore, a^x/x is increasing for $x > 1/\log a$, because its derivative is $\dfrac{(\log a)a^x x - a^x}{x^2} = \dfrac{a^x \log a}{x^2} \times$

$\left(x - \dfrac{1}{\log a}\right) > 0$. Therefore $a^x/x \to \infty$ as $x \to \infty$. Now let $b = a^{1/n}$, then $a^x/x^n = b^{nx}/x^n = (b^x/x)^n \to \infty$ because $b^x/x \to \infty$. **Q16.** Since $q' = u'$, $q' = \alpha q$ and therefore $q = q_0 e^{\alpha t}$. But $u = q + \dfrac{c}{\alpha}$ and $q_0 = u_0 - \dfrac{c}{\alpha}$, so $u = \left(u_0 - \dfrac{c}{\alpha}\right)e^{\alpha t} + \dfrac{c}{\alpha} = u_0 e^{\alpha t} + \dfrac{c}{\alpha}(1 - e^{\alpha t})$. **Q17.** (a) Since $a = v'$ the equation can be written $v' = -\dfrac{\rho}{m}v - g$. Then by Q16 $v = -\dfrac{mg}{\rho}(1 - e^{-\rho t/m}) = v_T(1 - e^{-\rho t/m})$. (b) As $t \to \infty$, $e^{-\rho t/m} \to 0$. (c) $v = v_0 e^{-\rho t/m} + v_T(1 - e^{-\rho t/m}) \to v_T$ as $t \to \infty$. **Q18.** $I' = -\dfrac{R}{L}I + \dfrac{V}{L}$ gives (since $I_0 = 0$) $I = \dfrac{V}{R}(1 - e^{-Rt/L}) \to \dfrac{V}{R}$. **Q19.** In time dt fraction dt/T of the cells divide, each producing p additional cells. Then $dN = p \cdot \dfrac{dt}{T} \cdot N$, so $\dfrac{dN}{dt} = \dfrac{p}{T}N$ and $\alpha = \dfrac{p}{T}$. **Q20.** (a) $e^{-(0.20)(40)} = 0.00034$ atmosphere, 1000 times thinner than on Mt. Everest. (b) Since the pressure (in atmospheres) is $e^{-\beta x}$, where

$\beta \approx 0.2$, the density is $ce^{-\beta x}$ for some constant c. Then the mass of a 1 m^2 column from 0 to x is $\int_0^x ce^{-\beta x}\,dx = c\cdot\frac{1}{\beta}(1-e^{-\beta x})$, and from 0 to ∞ is $\int_0^\infty ce^{-\beta x}\,dx = c\cdot\frac{1}{\beta}$.
The ratio is $1 - e^{-\beta x}$. For $x = 40$ and $\beta = 0.2$ this is over 99.9%.   **Q21.** Draw (as accurately as possible) parallel tangents to the inner and outer arcs, at S and P, say; that is, draw a tangent at an arbitrary point S on one arc, and then a parallel tangent to the other arc, to determine P. The center lies on SP (extended), because the angle ζ is constant. This is sufficient to proceed with measurement of PS/ST, for which we only need a line on which O lies. To *locate* O, repeat the construction elsewhere.

Q22. (a) At the end of each nth part of a year A becomes $A + \frac{\alpha}{n}A = \left(1 + \frac{\alpha}{n}\right)A$;

there are nt parts. (b) Since by §6.2.4 $\left(1 + \frac{\alpha}{n}\right)^n \to e^\alpha$, $\left(1 + \frac{\alpha}{n}\right)^{nt} \to e^{\alpha t}$. (c) Let

A_0 be 1 dollar, then compounding 12 times a year gives $\left(1 + \frac{0.05}{12}\right)^{120} = 1.6470$, and

compounding continuously gives $e^{0.05(10)} = 1.6487$. The advantage is 17 hundredths

of a cent. (d) If $t = m/n$ there are m periods in t years, so $A = A_0\left(1 + \frac{\alpha}{n}\right)^m =$

$A_0\left(1 + \frac{\alpha}{n}\right)^{nt}$. Now let t be arbitrary, and for each n let $\frac{m}{n} \leq t < \frac{m+1}{n}$. The value

of the investment after t years, when interest is compounded n times a year, is the

same as the value after $\frac{m}{n}$ years, because no further interest is added until $\frac{m+1}{n}$

years. The value after $\frac{m}{n}$ years is $A_0\left(1 + \frac{\alpha}{n}\right)^{n(m/n)}$, and clearly $\frac{m}{n} \to t$ as $n \to \infty$.

Then (Q14) $A_0\left(1 + \frac{\alpha}{n}\right)^{n(m/n)} \to A_0 e^{\alpha t}$. This is then the value after t years when

interest is compounded continuously.   **Q23.** For $t > 0$ reason as before, but use
$\log(1 + tx)$ (t fixed) in place of $\log(1 + x)$. For $t < 0$, $-t > 0$, so e^{-t} is the limit

of increasing $\left(1 - \frac{t}{n}\right)^n$ and decreasing $\left(1 + \frac{t}{n}\right)^{-n}$, hence $1/e^{-t} = e^t$ is the limit of

decreasing $\left(1 - \frac{t}{n}\right)^{-n}$ and increasing $\left(1 + \frac{t}{n}\right)^n$.

CHAPTER 7

§7.2.6

Q1. (a) $\int_0^x \sqrt{1+c^2}\, dx = x\sqrt{1+c^2}$; Pythagoras gives $\sqrt{x^2 + (cx)^2}$, which is the
same. (b) $\sqrt{1+y'^2} = \sqrt{1+\tan^2 \alpha} = \sec \alpha$ (c) $\dfrac{\kappa}{y''} = \dfrac{1}{(1+y'^2)^{3/2}} = \cos^3 \alpha$

Q2. (a) $\displaystyle\int_0^x \sqrt{1+x^2}\, dx = \frac{1}{2}\left(x\sqrt{x^2+1} + \operatorname{arsinh} x\right) = \frac{1}{2}\left[x\sqrt{1+x^2} + \right.$

$\left. \log\left(x + \sqrt{1+x^2}\,\right)\right]$ (b) In $\displaystyle\int_0^x \sqrt{1+e^{2x}}\, dx$ let $u = \sqrt{1+e^{2x}}$ and the integral

becomes $\displaystyle\int \frac{u^2}{u^2-1}\, du = \int \left[1 + \frac{1}{2}\left(\frac{1}{u-1} - \frac{1}{u+1}\right)\right] du = u + \frac{1}{2}\log\frac{u-1}{u+1} + C.$ Thus

$\displaystyle\int_0^x \sqrt{1+e^{2x}}\, dx = \sqrt{1+e^{2x}} + \frac{1}{2}\log\frac{\sqrt{1+e^{2x}}-1}{\sqrt{1+e^{2x}}+1} - \sqrt{2} - \frac{1}{2}\log\frac{\sqrt{2}-1}{\sqrt{2}+1}.$ (c) $\sinh x$

Q3. (a) $1/R$, where R is the radius of the earth. The numerical value depends on the unit in which R is measured. If the mile is used, $R = 4000$ and $|\kappa| = 1/4000 = 0.00025$. (b) The mast, being perpendicular to the tangent, turns at the same rate as the tangent, which is $|\kappa| = 1/R$ (this can also be seen directly) $= 0.00025$ radian/mile $= 0.014°$/mile. **Q4.** (a) In Fig. A-52, since angles A and B are right, angles AOB and ACB are supplementary, so the angle DCB by which direction changes equals $\angle AOB$. (b) The ratio of central angle (in radians) to arc is $1/r$. **Q5.** $L = \int_{-\pi/4}^{\pi/4} \sqrt{1+\sin^2 x}\, dx$. The minimum value of the integrand on the

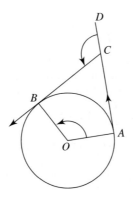

Fig. A-52

interval is 1, its maximum value $\sqrt{1 + (1/\sqrt{2})^2} = \sqrt{\frac{3}{2}}$. Since the length of the interval is $\pi/2$, the result follows. Alternatively, the inequalities are those of Proposition 3. **Q6.** (a) $|\kappa| = \dfrac{\sin(\pi/6)}{[1 + \cos^2(\pi/6)]^{3/2}} = \dfrac{4}{7^{3/2}} \approx 0.216$. (b) As in §3.8.8,

$\Delta\alpha \approx \dfrac{d\alpha}{ds}\Delta s$, so $\dfrac{\pi}{1800} \approx \dfrac{4}{7^{3/2}}\Delta s$ or $\Delta s \approx 0.0081$. **Q7.** (a) $\kappa(x) = \dfrac{e^x}{(1 + e^{2x})^{3/2}}$

(b) $\kappa'(x) = \dfrac{e^x(1 - 2e^{2x})}{(1 + e^{2x})^{5/2}} = 0$ at $x_0 = -\frac{1}{2}\log 2 \approx -0.35$. Clearly $\kappa'(x)$ is positive for $x < x_0$ and negative for $x > x_0$. $\kappa(x_0) = 2/3^{3/2} \approx 0.38$. **Q8.** Assuming $y \geq 0$, $y = \dfrac{b}{a}\sqrt{x^2 - a^2}$ and we find $\kappa = -\dfrac{ab}{\left[\left(1 + \dfrac{b^2}{a^2}\right)x^2 - a^2\right]^{3/2}}$. Thus $|\kappa| = \dfrac{b/a^2}{(l^2z^2 - 1)^{3/2}}$;

here $|z| \geq 1$. The maximum occurs at $z^2 = 1$ ($x = \pm a$) and is a/b^2. $|\kappa|$ takes the value b/a^2 where $l^2z^2 - 1 = 1$, i.e. at $z^2 = 2/l^2$ or $x = \pm a^2\sqrt{\dfrac{2}{a^2 + b^2}}$. **Q9.** (a) From §7.2.5, Example 6, we know that $|\kappa| = \dfrac{b/a^2}{(1 - k^2z^2)^{3/2}}$ has maxima at $z = \pm 1$ and minima at $z = 0$. Since $1 - k^2z^2$ increases as z^2 decreases from 1 to 0, $|\kappa|$ decreases and there are no other extrema. (b) No—try to draw such a curve. This "four-vertex theorem" is proved in books on differential geometry. **Q10.** Yes, if $\angle PRQ \leq \angle PQR$. With QR vertical let PQ be the graph of $y(x)$ and let PR have slope μ. The hypothesis ensures that $|y'| < \mu$ everywhere but at P and Q, so by §3.7.3.Q4(a) $PQ = \int \sqrt{1 + y'^2}\,dx < \int \sqrt{1 + \mu^2}\,dx = PR$. **Q11.** It usually does cross the curve. The reason is that if κ is increasing or decreasing at P, the curve turns less sharply than the circle on one side of P and more sharply on the other, so it crosses the circle at P; and κ is usually increasing or decreasing. (Cf. §8.2.10.Q19.) **Q12.** $\displaystyle\int_{s_1}^{s_2}\kappa\,ds =$

$\displaystyle\int_{x_1}^{x_2}\dfrac{y''}{(1 + y'^2)^{3/2}}\sqrt{1 + y'^2}\,dx = \int_{x_1}^{x_2}\dfrac{y''}{1 + y'^2}\,dx = \arctan y'\Big|_{x_1}^{x_2} = \alpha\Big|_{x_1}^{x_2} = \alpha(x_2) -$

$\alpha(x_1) = \Psi$. **Q13.** I will show that the arc of the curve with $0 \leq x \leq 2/\pi$ is not finite in length. It suffices to show that a polygon of arbitrarily great length can be inscribed in the arc. Let $h = 2/\pi$ and for $n \geq 1$ let P_n be the inscribed polygon with

vertices (h, h), $\left(\dfrac{h}{3}, -\dfrac{h}{3}\right)$, $\left(\dfrac{h}{5}, \dfrac{h}{5}\right)$, $\left(\dfrac{h}{7}, -\dfrac{h}{7}\right)$, ..., $\left(\dfrac{h}{2n+1}, (-1)^n\dfrac{h}{2n+1}\right)$, $(0,0)$

(Fig. A-53 shows P_3). Clearly (see figure) the length of $P_n > \dfrac{h}{3} + \dfrac{h}{3} + \dfrac{h}{5} + \dfrac{h}{5} +$

$\dfrac{h}{7} + \dfrac{h}{7} + \cdots + \dfrac{h}{2n+1} + \dfrac{h}{2n+1} > h\left[\dfrac{1}{3} + \dfrac{1}{4} + \dfrac{1}{5} + \dfrac{1}{6} + \cdots + \dfrac{1}{2n+1} + \dfrac{1}{2n+2}\right]$. But

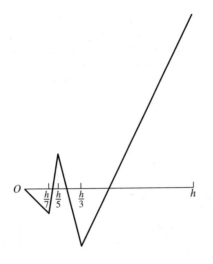

Fig. A-53

as $n \to \infty$ the term in brackets $\to \infty$, by the cited question. Thus the length of $P_n \to \infty$.

§7.3.6

Q1. The same as the given argument, except that the inequalities are reversed because the curve is concave down. **Q2.** See Fig. A-54. The distance from C to the curve has a local maximum or minimum at an interior point Q' of arc PQ, by an argument like that used for Rolle's theorem (§3.7.8); then the cited question applies. (Alternatively, the mean value theorem applied to the distance shows that it has a

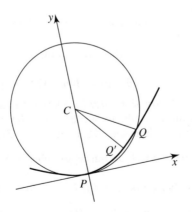

Fig. A-54

critical point Q'; the argument used for the cited question only required a critical point, not an extremum.) Now as $Q \to P$, $Q' \to P$, so by Example 3 C approaches the center of curvature. **Q3.** See Fig. A-55. By Example 3, C_1 and C_2 approach the center of curvature. The slopes of $Q_1' C$ and $Q_2' C$ eventually ensure that CC_1 and CC_2 are smaller than $C_1 C_2$; then C must approach the center of curvature also.

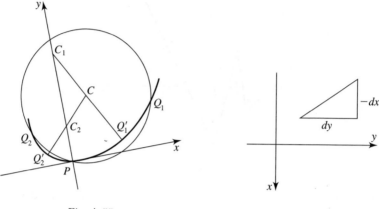

Fig. A-55   Fig. A-56

§7.4.8

Q1. The equations for p and q have the solutions $p = 1/\sqrt{5}$, $q = 2/\sqrt{5}$ and $p = -1/\sqrt{5}$, $q = -2/\sqrt{5}$. Taking (x_0, y_0) to be $(1, -1)$, we obtain either $x = 1 + (s/\sqrt{5})$, $y = -1 + (2s/\sqrt{5})$, or $x = 1 - (s/\sqrt{5})$, $y = -1 - (2s/\sqrt{5})$. In the former case the direction is towards the first quadrant, in the latter towards the third. **Q2.** (a) $\dot{x} = -\sin t$, $\dot{y} = \cos t$, therefore $\dot{s} = \sqrt{\dot{x}^2 + \dot{y}^2} = 1$. (b) $\dot{x} = 1$, $\dot{y} = -\dfrac{t}{\sqrt{1 - t^2}}$, therefore $\dot{s} = \sqrt{\dot{x}^2 + \dot{y}^2} = \dfrac{1}{\sqrt{1 - t^2}}$. Here $\dot{y}$ and $\dot{s}$ are not defined at $t = \pm 1$: the speed approaches ∞ at the endpoints of the path. **Q3.** $\dot{x} = -\sin \dfrac{s}{a}$, $\dot{y} = \cos \dfrac{s}{a}$, $\ddot{x} = -\dfrac{1}{a} \cos \dfrac{s}{a}$, $\ddot{y} = -\dfrac{1}{a} \sin \dfrac{s}{a}$, so $\kappa = \dot{x}\ddot{y} - \dot{y}\ddot{x} = 1/a$. **Q4.** Slope relative to the y-axis, defined analogously to ordinary slope dy/dx relative to the x-axis, is $-dx/dy$, as Fig. A-56 indicates, because the direction $90°$ counterclockwise from the positive y-direction is the *negative* x-direction. Then $\tan \beta = -dx/dy = -\dot{x}/\dot{y}$, and since $-\pi/2 < \beta < \pi/2$ this means $\beta = -\arctan(\dot{x}/\dot{y})$, whence $\dot{\beta} = \dfrac{d\beta}{dt} = -\dfrac{1}{1 + (\dot{x}/\dot{y})^2} \times \dfrac{\dot{y}\ddot{x} - \dot{x}\ddot{y}}{\dot{y}^2} = \dfrac{\dot{x}\ddot{y} - \dot{y}\ddot{x}}{\dot{x}^2 + \dot{y}^2} = \dot{\alpha}$. It follows that $d\beta/ds = d\alpha/ds$. **Q5.** (a) $\sec^2 t - \tan^2 t = 1$ (b) The interval $-\pi/2 < t < \pi/2$ will do: P traverses the right-hand branch in the upward direction. (c) $x = -\sec t$, $y = \tan t$, $-\pi/2 < t < \pi/2$ (d) The right-hand branch is $x = \cosh t$, $y = \sinh t$, $-\infty < t < \infty$. **Q6.** (a) $\dot{x}\ddot{x} + \dot{y}\ddot{y} = \dfrac{1}{2}\dfrac{d}{ds}(\dot{x}^2 + \dot{y}^2) =$

$\dfrac{1}{2}\dfrac{d}{ds}(1) = 0$ (b) $\kappa\dot{x} = (\dot{x}\ddot{y} - \dot{y}\ddot{x})\dot{x} = \dot{x}^2\ddot{y} - (\dot{x}\ddot{x})\dot{y} = \dot{x}^2\ddot{y} + (\dot{y}\ddot{y})\dot{y}$ (by (a)) $=$

$(\dot{x}^2 + \dot{y}^2)\ddot{y} = \ddot{y}$. Similarly $\kappa\dot{y} = -\ddot{x}$. **Q7.** The definition of κ, but not of $\dot{s}$, involves the choice of "counterclockwise," which is the direction of rotation through $90°$ from the positive x-axis to the positive y-axis, as the positive direction for angles. If x and y are interchanged that direction of rotation becomes the clockwise direction; then angles change sign, and so does $\kappa = d\alpha/ds$. **Q8.** Since U is $(a\cos t, a\sin t)$ and V is $(b\cos t, b\sin t)$, the intersection of the two lines is $(a\cos t, b\sin t)$. Substitution shows

that this point satisfies $\dfrac{x^2}{a^2} + \dfrac{y^2}{b^2} = 1$. Conversely, given $P(x,y)$ on the ellipse, choose t_1 in $[0, 2\pi]$ so that $x = a\cos t_1$ (i.e., so that the U corresponding to t_1 is on the vertical line through x). The equation gives $y = \pm b\sin t_1$. If $+$, let $t = t_1$; if $-$, let $t = 2\pi - t_1$. Then $x = a\cos t$, $y = b\sin t$. That $\dot{s} > 0$ everywhere follows from Q9. **Q9.** We have $\dot{s} = \sqrt{a^2\sin^2 t + b^2(1 - \sin^2 t)} = \sqrt{(a^2 - b^2)\sin^2 t + b^2}$; since $\sin^2 t$ is a minimum at $t = 0$ and a maximum at $t = \pi/2$, so is $\dot{s}$. The values are $\dot{s}(0) = b$, $\dot{s}(\pi/2) = a$ (cf. Example 2, §7.4.5). **Q10.** (a) Substitution verifies (i)–(iii); for (iv), calculate $(x_2 - x_1)^2 + (y_2 - y_1)^2 = (x_2' - x_1')^2 + (y_2' - y_1')^2$. Rotation through η carries the points with xy-coordinates $(0,0)$, $(1,0)$, and $(0,1)$ respectively into the points with the same $x'y'$-coordinates, and preserves distance (the latter fact is easily shown by congruent triangles). Since the location of any point is uniquely determined by its distances from three non-collinear points, the transformation defined by the equations can only be that rotation. (b) $\dot{x} = (\cos\eta)\dot{x}' - (\sin\eta)\dot{y}'$, etc., give $\dot{x}^2 + \dot{y}^2 = \dot{x}'^2 + \dot{y}'^2$ and $\dot{x}\ddot{y} - \dot{y}\ddot{x} = \dot{x}'\ddot{y}' - \dot{y}'\ddot{x}'$. (c) $x = x' + h$, $y = y' + k$ give $\dot{x} = \dot{x}'$, $\dot{y} = \dot{y}'$, so the

formulas are the same. **Q11.** (a) $\dfrac{dx}{du} = \dfrac{dx}{dt}\dfrac{dt}{du} = -\dfrac{dx}{dt}$, therefore $\dfrac{d^2x}{du^2} =$

$\dfrac{d}{dt}\left(\dfrac{dx}{du}\right)\dfrac{dt}{du} = \dfrac{d^2x}{dt^2}$. Similarly for derivatives with respect to y. Thus $\dot{x}$ and $\dot{y}$ change sign, but $\ddot{x}$ and $\ddot{y}$ do not. There is no change in $\dot{s}^2 = \dot{x}^2 + \dot{y}^2$. The direction of the curve is reversed, since we always make the positive direction the direction of increasing parameter. Hence $\dot{s} = \sqrt{\dot{x}^2 + \dot{y}^2}$ is unchanged, but s has a different meaning from before: it is measured in the opposite direction. Therefore $S = \Delta s$ from a point P to a point Q is reversed in sign. In particular, the sign of $s(P) = \Delta s$ from P_0 to P is reversed: $\int_{u_0}^{u} \dot{s}\, du = -\int_{t_0}^{t} \dot{s}\, dt$, where $u = -t$, $u_0 = -t_0$. The sign of $\kappa = (\dot{x}\ddot{y} - \dot{y}\ddot{x})/\dot{s}^3$ is also reversed. (b) (i) $\dot{s}(t)^2 = \left(\dfrac{dx}{dt}\right)^2 + \left(\dfrac{dy}{dt}\right)^2 = \left(\dfrac{dx}{du}\dfrac{du}{dt}\right)^2 +$

$\left(\dfrac{dy}{du}\dfrac{du}{dt}\right)^2 = \left[\left(\dfrac{dx}{du}\right)^2 + \left(\dfrac{dy}{du}\right)^2\right]\left(\dfrac{du}{dt}\right)^2$. Therefore $\dot{s}(t) = \pm\dot{s}(u)\dfrac{du}{dt}$, the sign being $+$

if $\dfrac{du}{dt} > 0$, $-$ if $\dfrac{du}{dt} < 0$, so $s(t) = \int_{t_0}^{t} \dot{s}(t)\, dt = \pm\int_{t_0}^{t} \dot{s}(u)\dfrac{du}{dt}\, dt = \pm\int_{u_0}^{u} \dot{s}(u)\, du =$

$\pm s(u)$. (ii) $\dfrac{dx}{dt} = \dfrac{dx}{du}\dfrac{du}{dt}$, so $\dfrac{d^2x}{dt^2} = \dfrac{d}{dt}\left(\dfrac{dx}{du}\right)\dfrac{du}{dt} + \dfrac{dx}{du}\dfrac{d}{dt}\left(\dfrac{du}{dt}\right) = \dfrac{d}{du}\left(\dfrac{dx}{du}\right)\left(\dfrac{du}{dt}\right)^2 +$

$\dfrac{dx}{du}\dfrac{d^2u}{dt^2} = \dfrac{d^2x}{du^2}\left(\dfrac{du}{dt}\right)^2 + \dfrac{dx}{du}\dfrac{d^2u}{dt^2}$. Likewise $\dfrac{d^2y}{dt^2} = \dfrac{d^2y}{du^2}\left(\dfrac{du}{dt}\right)^2 + \dfrac{dy}{du}\dfrac{d^2u}{dt^2}$. Then

calculation shows that $\dfrac{dx}{dt}\dfrac{d^2y}{dt^2} - \dfrac{dy}{dt}\dfrac{d^2x}{dt^2} = \left(\dfrac{du}{dt}\right)^3\left[\dfrac{dx}{du}\dfrac{d^2y}{du^2} - \dfrac{dy}{du}\dfrac{d^2x}{du^2}\right]$. By (i), $\dot{s}(t)^3 =$

$\pm\left(\dfrac{du}{dt}\right)^3 \dot{s}(u)^3$, and dividing each side of the preceding equation by the corresponding side of this one gives $\kappa(t) = \pm\kappa(u)$.   **Q12.** If $\kappa = 0$, (1) of §7.4.6 gives $\psi = 0$ and then (2) gives $x = s\cos\gamma + c_1$, $y = s\sin\gamma + c_2$, in agreement with §7.4.4, Example 0: $x_0 = c_1$, $y_0 = c_2$, $p = \cos\gamma$, and $q = \sin\gamma = \cos\left(\dfrac{\pi}{2} - \gamma\right)$. If $\kappa = c \neq 0$, (1) gives $\psi = cs$ and then (2) gives $x = \dfrac{1}{c}\sin(cs + \gamma) - \dfrac{1}{c}\sin\gamma + c_1$, $y = -\dfrac{1}{c}\cos(cs + \gamma) + \dfrac{1}{c}\cos\gamma + c_2$. Let us set $\gamma = \pi/2$, so that the curve is initially perpendicular to the x-axis (vertical), and then choose c_1 and c_2 so as to eliminate the constant terms; then $x = \dfrac{1}{c}\sin\left(cs + \dfrac{\pi}{2}\right) = \dfrac{1}{c}\cos cs$, $y = -\dfrac{1}{c}\cos\left(cs + \dfrac{\pi}{2}\right) = \dfrac{1}{c}\sin cs$. This agrees with §7.4.5, Example 2 (second part), if $c > 0$ and $a = 1/c$.   **Q13.** At $A(a,0)$, $\psi = \alpha = 0$ and $\dot{x} > 0$, so we can put $\gamma = 0$. Also, $ds = a\psi\,d\psi$, so $x = a + a\int_0^\psi \psi\cos\psi\,d\psi$, $y = a\int_0^\psi \psi\sin\psi\,d\psi$. Integration by parts (§4.6.8.Q1(a)) yields $x = a(\cos\psi + \psi\sin\psi)$, $y = a(\sin\psi - \psi\cos\psi)$. These are parametric equations in the parameter ψ; equations in s are obtained by substituting $\psi = \sqrt{2s/a}$. Starting from these equations we can verify as follows that the curve they define is the involute. Given $P(x(\psi), y(\psi))$, let T be $(a\cos\psi, a\sin\psi)$; then OT and the curve at P have slope $\tan\psi$, while TP is calculated to have length $a\psi$ and slope $-1/\tan\psi$; thus it is tangent to the circle at T and equal in length to arc AT. It is also normal to the curve at P; and calculation of κ from the parametric equations gives $\rho = a\psi$, so T is the center of curvature at P, as hypothesized in Example 2.   **Q14.** $x = t$, $y = \frac{1}{2}t^2$ gives $\dot{s} = \sqrt{1 + t^2}$,

$\kappa = \dfrac{1}{(1 + t^2)^{3/2}}$. Then $\psi = \displaystyle\int_0^s \kappa\,ds = \int_0^t \kappa\dot{s}\,dt = \int_0^t \dfrac{1}{1 + t^2}\,dt = \arctan t$. Since $dy/dx = x$, $\alpha = \arctan x = \arctan t = \psi$.   **Q15.** From $x = r_0 e^{\alpha\theta}\cos\theta$, $y = r_0 e^{\alpha\theta}\sin\theta$, we calculate $\dot{s} = r_0\sqrt{\alpha^2 + 1}\,e^{\alpha\theta}$ and $\rho = 1/\kappa = r_0\sqrt{\alpha^2 + 1}\,e^{\alpha\theta}$. Then (a)

$s = \displaystyle\int_0^\theta \dot{s}\,d\theta = \dfrac{r_0\sqrt{\alpha^2 + 1}}{\alpha}(e^{\alpha\theta} - 1)$, so $\rho = \alpha s + r_0\sqrt{\alpha^2 + 1}$; or (b) $s = \displaystyle\int_{-\infty}^\theta \dot{s}\,d\theta =$

$\dfrac{r_0\sqrt{\alpha^2 + 1}}{\alpha}e^{\alpha\theta}$, so $\rho = \alpha s$. In terms of ζ (§3.3.13.Q9) equation (a) is $\rho = s\cot\zeta + r_0\csc\zeta$.   **Q16.** Let the positive direction along the curve be from $P(s)$ to $Q(s + \Delta s)$, let Δx be the corresponding increment of x, and let c be chord PQ. The x-

direction cosine of the secant arrow from P to Q is $\Delta x/c$, and as Δs and c approach 0 it approaches the x-direction cosine of the tangent arrow at P. But $\lim \dfrac{\Delta x}{c} =$

$$\lim \left(\frac{\Delta x}{\Delta s} \frac{\Delta s}{c} \right) = \lim \frac{\Delta x}{\Delta s} = \dot{x}(s). \text{ Similarly for } \dot{y}.$$

§7.5.8

Q1. Cf. §7.5.2, Example 3. $\displaystyle\int_1^2 x\sqrt{1+x^2}\,dx = \frac{1}{3}(1+x^2)^{3/2} \Big|_1^2 = \frac{1}{3}(5\sqrt{5}-2\sqrt{2})$.

Q2. Let the circle have radius b. (a) Parametrizing by θ, we have $ds = b\,d\theta$ and

the work is $\displaystyle\int_0^{2\pi} (\cos^2\theta)b\,d\theta = \frac{b}{2}(\theta + \sin\theta\cos\theta)\Big|_0^{2\pi}$ (§4.6.4, Example 8) $= \pi b$.

(b) Its kinetic energy is $\frac{1}{2}mv^2 = \pi b$, where m is its mass, so $v = \sqrt{2\pi b/m}$. (c) By §7.5.5(5), $a_N = \kappa v^2$; neither factor is 0. (d) The rail. **Q3.** As t increases from 0 to 2, (x,y) goes from (x_0, y_0) to $(x_0 + 2k, y_0 + 2l)$. Then the integral is

$$\int_0^2 (x_0 + kt)(y_0 + lt)\sqrt{k^2 + l^2}\,dt = \sqrt{k^2 + l^2}\int_0^2 [x_0 y_0 + (lx_0 + ky_0)t + klt^2]\,dt =$$

$2\sqrt{k^2 + l^2}\,(x_0 y_0 + lx_0 + ky_0 + \frac{4}{3}kl)$. **Q4.** (a) The conical surface is the surface of revolution of $y = \dfrac{r}{h}x$ from $x = 0$ to $x = h$, so its area is

$$\int_0^h 2\pi\left(\frac{r}{h}x\right)\sqrt{1 + \left(\frac{r}{h}\right)^2}\,dx = \pi r\sqrt{r^2 + h^2}.$$ (b) $l = \sqrt{r^2 + h^2}$, so the area is πrl.

After unrolling, the sector has radius l and curved boundary $2\pi r$; then comparing the whole circle of which it is a part, (area of sector)$/(\pi l^2) = (2\pi r)/(2\pi l)$, so area =

$\pi r l$. **Q5.** $\displaystyle\int_0^b 2\pi\sqrt{x}\sqrt{1 + \frac{1}{4x}}\,dx = 2\pi\int_0^b \sqrt{x + \frac{1}{4}}\,dx = \frac{4\pi}{3}\left(x + \frac{1}{4}\right)^{3/2}\Big|_0^b =$

$\frac{\pi}{6}[(4b+1)^{3/2} - 1]$. **Q6.** Since κ and v agree, so does $a_N = \kappa v^2$. But a_T is

0 for (ii), not necessarily 0 for (i). **Q7.** $a_T = \dot{v} = \dfrac{d}{dt}\sqrt{a^2\sin^2 t + b^2\cos^2 t} =$

$\dfrac{(a^2 - b^2)\sin t \cos t}{\sqrt{a^2\sin^2 t + b^2\cos^2 t}}$, $a_N = \kappa v^2 = \left(\dfrac{ab}{v^3}\right)v^2 = \dfrac{ab}{v} = \dfrac{ab}{\sqrt{a^2\sin^2 t + b^2\cos^2 t}}$. At A

and B $a_T = 0$, as symmetry implies: the particle speeds up from A to B, then slows down symmetrically, etc. As for a_N, it is a at A and b at B; it varies, inversely as v, between a and b. **Q8.** Let ρ be the radius and NS the x-interval $[-\rho, \rho]$. Then as in §7.5.4, Example 7, $d = y = \sqrt{\rho^2 - x^2}$ gives the integral

$$\int_{-\rho}^\rho 2\pi(1 + \sqrt{\rho^2 - x^2})\sqrt{\rho^2 - x^2}\sqrt{1 + \frac{x^2}{\rho^2 - x^2}}\,dx = 2\pi\rho\int_{-\rho}^\rho \left(1 + \sqrt{\rho^2 - x^2}\right)dx =$$

$$2\pi\rho\left[2\rho + \rho^2 \int_{-1}^{1} \sqrt{1-u^2}\,du\right] \quad \text{(by substituting } x = \rho u\text{). Using §4.6.4, Example 8,}$$

the last integral is $\frac{1}{2}\left(\arcsin u + u\sqrt{1-u^2}\,\right)\Big|_{-1}^{1} = \frac{\pi}{2}$, so the solution is $\pi\rho^2(4+\pi\rho)$.

Q9. Let A and B be opposite ends of a diameter of one of the circles in Fig. 7-41. Along one semicircle from A to B positive work is done; along the other, negative work. **Q10.** (a) By §5.1.1(3), $y(t) = 1 - \frac{1}{2}gt^2$. A similar argument gives $x(t) = t$ (replace g by 0 (§5.1.3.Q5(b))) and set $v_0 = 1$). Then $y = 1 - \frac{1}{2}gx^2$. (b) See Fig. A-57.

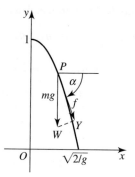

Fig. A-57

We have $\angle YWP = -\alpha$, so $f = mg\sin(-\alpha) = -mg\dfrac{\tan\alpha}{\sec\alpha} = -mg\dfrac{y'}{\sqrt{1+y'^2}}$

($§7.2.6.Q1(b)$) $= mg^2\dfrac{x}{\sqrt{1+g^2x^2}}$. (c) Since $ds = \sqrt{1+y'^2}\,dx$ the integral of $f\,ds$ is

$\int_0^{\sqrt{2/g}} mg^2 x\,dx = mg$. This is the same as $mg\cdot 1$, the work done on the particle falling from $y=1$ to $y=0$. **Q11.** The secant of the angle between the plane of dS and the xy-plane (cf. §7.2.6.Q1(b)). **Q12.** (a) The same as the first step of the cited remark. (b) As t_1 and $t_2 \to 0$, this ratio is "ultimately in the ratio of equality" ($§3.1.11.Q4$) with the ratio of the squares of the times, t_1^2/t_2^2; for $\dfrac{2x(t_1)}{t_1^2} \to \ddot{x}(0)$

and $\dfrac{2x(t_2)}{t_2^2} \to \ddot{x}(0)$, hence $\dfrac{2x(t_1)/t_1^2}{2x(t_2)/t_2^2} \to 1$, or $\dfrac{x(t_1)/x(t_2)}{t_1^2/t_2^2} \to 1$. Thus for small t the motion resembles motion at constant acceleration (§5.1.2). **Q13.** (i) If $a_N = 0$, $\kappa = a_N/v^2 = 0$ and the path is a straight line (§7.4.7, Example 1). (ii) If $a_N \neq 0$ but $a_T = 0$, $v = \int a_T\,dt$ is a non-zero constant and therefore $\kappa = a_N/v^2$ is one too: the path is a circle (same example). (iii) If $a_N \neq 0$ and $a_T \neq 0$, we have $v = \int a_T\,dt = a_T t + b$, where b is a constant, hence (measuring arc length from

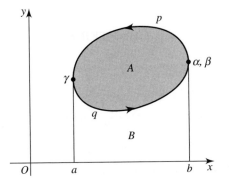

Fig. A-58

$t = 0$) $s = \int_0^t v\,dt = \frac{1}{2}a_T t^2 + bt = \dfrac{v^2 - b^2}{2a_T}$. Then $\kappa = \dfrac{a_N}{v^2} = \dfrac{a_N}{2a_T s + b^2}$, or

$\rho = \left(\dfrac{2a_T}{a_N}\right)s + \dfrac{b^2}{a_N}$, the equation of a logarithmic spiral (§7.4.8.Q15). **Q14.** (a) In

Fig. A-58 the point where $t = \alpha$ and β is located above the extreme x-value b; the general case is easily deduced from this one. Arc p is described as t increases from α to γ, arc q as t increases from γ to β. Considering only arc p, we have $A + B = \int_a^b y\,dx = \int_\gamma^\alpha y\dot{x}\,dt = -\int_\alpha^\gamma y\dot{x}\,dt$; considering only arc q, $B = \int_a^b y\,dx = \int_\gamma^\beta y\dot{x}\,dt$. Thus $A = -\int_\alpha^\gamma y\dot{x}\,dt - \int_\gamma^\beta y\dot{x}\,dt = -\int_\alpha^\beta y\dot{x}\,dt$. The other integral expression for A can be obtained similarly by dropping perpendiculars to the y-axis. These expressions remain valid if the curve is not all in the first quadrant, since neither A nor the integrals change if x is replaced by $x + h$ and y by $y + k$. E.g., in the first integral $y\dot{x}$ becomes $(y + k)\dot{x} = y\dot{x} + k\dot{x}$, so the integral is altered by the summand $-\int_\alpha^\beta k\dot{x}\,dt = k[x(\alpha) - x(\beta)] = 0$. (b) Average the two expressions for A. (c) Let $x = a\cos t$, $y = b\sin t$, then $A = -\frac{1}{2}\int_0^{2\pi}[(b\sin t)(-a\sin t) - (a\cos t)(b\cos t)]\,dt = \pi ab$. (d) Parametrize by θ, then $x = r(\theta)\cos\theta$, $y = r(\theta)\sin\theta$, so $A = -\frac{1}{2}\int_0^{2\pi}[(r\sin\theta)(\dot{r}\cos\theta - r\sin\theta) - (r\cos\theta)(\dot{r}\sin\theta + r\cos\theta)]\,d\theta = \frac{1}{2}\int_0^{2\pi} r^2\,d\theta$, which is the sum of $dA = \frac{1}{2}r^2\,d\theta$. (e) It is not the differential of an $f(x,y)$ having equal mixed partials (§4.5.5), because if $y = \dfrac{\partial f}{\partial x}$ and $-x = \dfrac{\partial f}{\partial y}$ then $\dfrac{\partial^2 f}{\partial y \partial x} = \dfrac{\partial}{\partial y}(y) = 1$ and $\dfrac{\partial^2 f}{\partial x \partial y} = \dfrac{\partial}{\partial x}(-x) = -1$. (f) As in (b) and (c) of the cited question. E.g., for rotation $dx = (\cos\eta)\,dx' - (\sin\eta)\,dy'$, etc.

CHAPTER 8

§8.2.10

Q1. The polynomials are 1, $1+x$, $1+x+\dfrac{x^2}{2}$, and $1+x+\dfrac{x^2}{2}+\dfrac{x^3}{6}$. The ones of orders 0 and 2 cross the curve. **Q2.** (a) Neither 1 nor $1-\dfrac{x^2}{2}$ crosses. (b) Both x and $x-\dfrac{x^3}{6}$ cross. (See Q19.) **Q3.** $x^3-x^2+x-1 = 2(x-1)+2(x-1)^2+(x-1)^3$. **Q4.** SIN $x =$ $\sin\dfrac{\pi}{180}\,x = \dfrac{\pi}{180}\,x - \left(\dfrac{\pi}{180}\right)^3\dfrac{x^3}{3!} + \left(\dfrac{\pi}{180}\right)^5\dfrac{x^5}{5!} - \cdots$. **Q5.** (a) These follow from $(f+g)^{(n)} = f^{(n)}+g^{(n)}$ and $(cf)^{(n)} = cf^{(n)}$. (b) Let p_n be the nth-order Taylor polynomial of f and q_n of g. Then the nth-order Taylor polynomial of $f + g$ is $p_n + q_n$, and of cf, cp_n. Since as $n \to \infty$ $p_n(x) \to f(x)$ and $q_n(x) \to g(x)$, $p_n(x) + q_n(x) \to f(x) + g(x)$ and $cp_n(x) \to cf(x)$. **Q6.**

$$1 - \frac{u^2}{2!} - \frac{v^2}{2!} + \frac{u^4}{4!} + \frac{u^2 v^2}{2!2!} + \frac{v^4}{4!} - \cdots - uv + \frac{u^3 v}{3!} + \frac{uv^3}{3!} - \cdots$$

$$= 1 - \frac{u^2 + 2uv + v^2}{2!} + \frac{u^4 + 4u^3 v + 6u^2 v^2 + 4uv^3 + v^4}{4!} - \cdots$$

$$= 1 - \frac{(u+v)^2}{2!} + \frac{(u+v)^4}{4!} - \cdots .$$

Q7. (a) By §8.2.7, Example 1, (ii), $\left|R_n\left(\dfrac{1}{2}\right)\right| < \dfrac{e^{1/2}}{(n + 1)!}\left(\dfrac{1}{2}\right)^{n+1}$. Since $e < 3 < 4$, $e^{1/2} < 2$, so $|R_n| < \dfrac{2}{(n + 1)!}\left(\dfrac{1}{2}\right)^{n+1} = \dfrac{1}{2^n(n + 1)!}$. (b) To get $\dfrac{1}{2^n(n + 1)!} <$ $0.0001 = \dfrac{1}{10,000}$, i.e. $2^n(n+1)! > 10,000$, we take $n = 5$. Then $\sqrt{e} \approx 1+\dfrac{1}{2}+\dfrac{1}{2!2^2}+$ $\dfrac{1}{3!2^3}+\dfrac{1}{4!2^4}+\dfrac{1}{5!2^5} = 1+\dfrac{1}{2}+\dfrac{1}{8}+\dfrac{1}{48}+\dfrac{1}{384}+\dfrac{1}{3840} = \dfrac{6331}{3840} \approx 1.6487$. **Q8.** $1-$ $\dfrac{1}{2}\left(\dfrac{\pi}{6}\right)^2 \approx 0.8629\ldots$; error $< \left(\dfrac{\pi}{6}\right)^4\dfrac{1}{4!} \approx 0.0031\ldots$, so possible error of 1 in the second place; $\cos(\pi/6) \approx 0.866$. **Q9.** (a) For the sine, the initial conditions are $y(0) = 0$, $y'(0) = 1$. Then $y = x + \dfrac{y''(0)}{2!}\,x^2 + \dfrac{y'''(0)}{3!}\,x^3 + \dfrac{y^{(4)}(0)}{4!}\,x^4 + \cdots$. Differentiating twice gives $y'' = y''(0) + \dfrac{y'''(0)}{1!}\,x + \dfrac{y^{(4)}(0)}{2!}\,x^2 + \cdots$. Then $y'' = -y$ means

$y''(0) = 0$, $y'''(0) = -1$, $y^{(4)}(0) = y''(0) = 0$, etc. The cosine series is similarly derived. (b) In the case of the sine series, for x near 0 and $\neq 0$, $\dfrac{y}{x} =$

$$\frac{1}{x}\left(x - \frac{x^3}{3!} + \frac{x^5}{5!} - \cdots\right) = 1 - \frac{x^2}{3!} + \frac{x^4}{5!} - \cdots;$$ the second step is valid because $p_n(x) \to$

y implies $p_n/x \to y/x$. The power series on the right has the value 1 at $x = 0$; assuming that it is continuous at $x = 0$, its limit as $x \to 0$ is then 1. The other limit goes similarly. (c) One could prove that as x increases from 0 the cosine series decreases and becomes negative, and then define $\pi/2$ as the first point at which it vanishes. **Q10.** (a) Substitute for x in $e^{ix} = \cos x + i \sin x$. (b) Let $z = e^{ix} = \cos x + i \sin x$. Then $z^2 = e^{2ix} = \cos 2x + i \sin 2x$. If this equals i, $\cos 2x = 0$ and $\sin 2x = 1$; then let $x = \pi/4$, so $z = \dfrac{1}{\sqrt{2}} + i \cdot \dfrac{1}{\sqrt{2}} = \dfrac{1}{\sqrt{2}}(1 + i)$. **Q11.** Let $g(x) = f(-x)$ and apply the first part to $g(x)$ on the interval $[-a, -b]$. Then since $g^{(n)}(x) = (-1)^n f^{(n)}(-x)$ we have

$$f(b) = g(-b) = g(-a) + \frac{g'(-a)}{1!}(-b-(-a)) + \frac{g''(-a)}{2!}(-b-(-a))^2 + \cdots$$

$$= f(a) + \frac{-f'(a)}{1!}(-(b-a)) + \frac{f''(a)}{2!}(-(b-a))^2 + \cdots$$

$$= f(a) + \frac{f'(a)}{1!}(b-a) + \frac{f''(a)}{2!}(b-a)^2 + \cdots.$$

Q12. Suppose $\sin 1 = p/q$. We have $\sin 1 = 1 - \dfrac{1}{3!} + \dfrac{1}{5!} - \cdots \pm \left(\dfrac{1}{(q-1)!} \text{ or } \dfrac{1}{q!}\right) + R_q$,

where $|R_q| \leq \dfrac{1}{(q+1)!}$ (Example 2) and $R_q \neq 0$ (by Taylor's theorem, because all derivatives of $\sin x$ are non-zero on the interval $0 < x < 1$). Then $q!R_q$ is an integer such that $0 < |q!R_q| \leq \dfrac{1}{q+1}$, which is impossible. **Q13.** We are given $\log u$ for $1 \leq u \leq 2$; if $0 < u < 1$, $1/u > 1$ and $\log u = -\log(1/u)$. **Q14.** (a) $1 - 2x + 3x^2 - 4x^3 + \cdots$ (b) $1 + \dfrac{1}{2}x - \dfrac{1}{2 \cdot 4}x^2 + \dfrac{1 \cdot 3}{2 \cdot 4 \cdot 6}x^3 - \dfrac{1 \cdot 3 \cdot 5}{2 \cdot 4 \cdot 6 \cdot 8}x^4 + \cdots = 1 + \dfrac{1}{2}x + \dfrac{1}{8}x^2 + \dfrac{1}{16}x^3 - \dfrac{5}{128}x^4 + \cdots$. **Q15.** $\sinh x = x + \dfrac{x^3}{3!} + \dfrac{x^5}{5!} + \cdots$, $\cosh x = 1 + \dfrac{x^2}{2!} + \dfrac{x^4}{4!} + \cdots$. **Q16.** Since the series has limit $-\infty$ for $x = -1$, it obviously has the same limit for $x < -1$. For $x > 1$ we have $|p_n(x)| = |\log(1+x) - R_n(x)|$; $\log(1+x)$ is fixed, so it suffices to show that $|R_n(x)| \to \infty$. We have $|R_n(x)| = \displaystyle\int_0^x \frac{t^n}{1+t}\,dt \geq \int_0^x \frac{t^n}{1+x}\,dt =$

$\dfrac{x^{n+1}}{(n+1)(1+x)} \to \infty$ as $n \to \infty$ (§3.3.13.Q8). **Q17.** (a)

$|(fg)^{(n)}(x)|$

$$= |f^{(n)}(x)g(x) + nf^{(n-1)}(x)g'(x)$$

$$+ \frac{n(n-1)}{2!} f^{(n-2)}(x)g''(x) + \cdots + f(x)g^{(n)}(x)|$$

$$\leq |f^{(n)}(x)||g(x)| + n|f^{(n-1)}(x)||g'(x)|$$

$$+ \frac{n(n-1)}{2!} |f^{(n-2)}(x)||g''(x)| + \cdots + |f(x)||g^{(n)}(x)|$$

$$\leq M^2 \left(1 + n + \frac{n(n-1)}{2!} + \cdots + 1\right) = M^2(1+1)^n = 2^n M^2.$$

(b) Writing f for $f(0)$, etc., we have

$$\left(f + \frac{f'}{1!} x + \frac{f''}{2!} x^2 + \frac{f'''}{3!} x^3 + \cdots\right)\left(g + \frac{g'}{1!} x + \frac{g''}{2!} x^2 + \frac{g'''}{3!} x^3 + \cdots\right)$$

$$= fg + \left(\frac{f'g + fg'}{1!}\right) x + \left(\frac{f''g + 2f'g' + fg''}{2!}\right) x^2$$

$$+ \left(\frac{f'''g + 3f''g' + 3f'g'' + fg'''}{3!}\right) x^3 + \cdots.$$

By Leibniz's rule this is $fg + \dfrac{(fg)'}{1!} x + \dfrac{(fg)''}{2!} x^2 + \dfrac{(fg)'''}{3!} x^3 + \cdots$. By (a), for this series $|R_n(x)| \leq \dfrac{2^{n+1} M^2}{(n+1)!} x^{n+1} = M^2 \dfrac{(2x)^{n+1}}{(n+1)!}$, which by the lemma in §8.2.6 approaches 0 as $n \to \infty$. **Q18.** (a) There are ξ and η between a and x such that $f(x) = \dfrac{f^{(n+1)}(\xi)}{(n+1)!} (x-a)^{n+1}$ and $g(x) = \dfrac{g^{(n+1)}(\eta)}{(n+1)!} (x-a)^{(n+1)}$. As $x \to a$, $g^{(n+1)}(\eta) \to g^{(n+1)}(a) \neq 0$, so $g(x) \neq 0$ for x near (but unequal to) a. Then as $x \to a$, $\dfrac{f(x)}{g(x)} = \dfrac{f^{(n+1)}(\xi)}{g^{n+1}(\eta)} \to \dfrac{f^{(n+1)}(a)}{g^{(n+1)}(a)}$. (b) Here $n = 1$, and the limit is $2/\cos 0 = 2$. **Q19.** We have $f(x) - p_n(x) = \dfrac{f^{(n+1)}(\xi)}{(n+1)!} (x-a)^{n+1}$, and since $f^{(n+1)}(\xi) \to f^{(n+1)}(a) \neq 0$ as $x \to a$, for x near a the sign of $f^{(n+1)}(\xi)$ is fixed. Hence near a the sign of $f(x) - p_n(x)$ is either everywhere the same as, or else everywhere opposite to, the sign of $(x-a)^{n+1}$, which changes as x passes a if n is even, but does not if n is odd.

Index

Princeton U-Store, NJ
Fri 17 Oct 2003
$38 list - $7.60 discount
= $30.40
+ 1.82 tax
$32.22 total